The Palgrave Macmillan Animal Ethics Series

Series Editors
Andrew Linzey
Oxford Centre for Animal Ethics
Oxford, UK

Clair Linzey
Oxford Centre for Animal Ethics
Oxford, UK

In recent years, there has been a growing interest in the ethics of our treatment of animals. Philosophers have led the way, and now a range of other scholars have followed from historians to social scientists. From being a marginal issue, animals have become an emerging issue in ethics and in multidisciplinary inquiry. This series will explore the challenges that Animal Ethics poses, both conceptually and practically, to traditional understandings of human-animal relations. Specifically, the Series will:

- provide a range of key introductory and advanced texts that map out ethical positions on animals;
- publish pioneering work written by new, as well as accomplished, scholars;
- produce texts from a variety of disciplines that are multidisciplinary in character or have multidisciplinary relevance.

For further information or to submit a proposal for consideration, please contact Amy Invernizzi, amy.invernizzi@palgrave-usa.com.

Michael J. Glover • Les Mitchell
Editors

Animals as Experiencing Entities

Theories and Historical Narratives

palgrave
macmillan

Editors
Michael J. Glover
University of the Free State
Bloemfontein, South Africa

Les Mitchell
University of the Free State
Bloemfontein, South Africa

ISSN 2634-6672 ISSN 2634-6680 (electronic)
The Palgrave Macmillan Animal Ethics Series
ISBN 978-3-031-46455-3 ISBN 978-3-031-46456-0 (eBook)
https://doi.org/10.1007/978-3-031-46456-0

This Palgrave Macmillan imprint is published by the registered company Springer Nature Switzerland AG.
The registered company address is: Gewerbestrasse 11, 6330 Cham, Switzerland

Paper in this product is recyclable.

To those beings who have always shared our worlds but who have seldom received our recognition of their deep, inner presences

Series Editors' Preface

This is a new book series for a new field of inquiry: Animal Ethics.

In recent years, there has been a growing interest in the ethics of our treatment of animals. Philosophers have led the way, and now a range of other scholars have followed from historians to social scientists. From being a marginal issue, animals have become an emerging issue in ethics and in multidisciplinary inquiry.

In addition, a rethink of the status of animals has been fuelled by a range of scientific investigations which have revealed the complexity of animal sentiency, cognition and awareness. The ethical implications of this new knowledge have yet to be properly evaluated, but it is becoming clear that the old view that animals are mere things, tools, machines or commodities cannot be sustained ethically.

But it is not only philosophy and science that are putting animals on the agenda. Increasingly, in Europe and the United States, animals are becoming a political issue as political parties vie for the "green" and "animal" vote. In turn, political scientists are beginning to look again at the history of political thought in relation to animals, and historians are beginning to revisit the political history of animal protection.

As animals grow as an issue of importance, so there have been more collaborative academic ventures leading to conference volumes, special journal issues, indeed new academic animal journals as well. Moreover, we have witnessed the growth of academic courses, as well as university posts, in Animal Ethics, Animal Welfare, Animal Rights, Animal Law, Animals and Philosophy, Human-Animal Studies, Critical Animal Studies, Animals and Society, Animals in Literature, Animals and Religion—tangible signs that a new academic discipline is emerging.

"Animal Ethics" is the new term for the academic exploration of the moral status of the non-human—an exploration that explicitly involves a focus on what we owe animals morally, and which also helps us to understand the influences—social, legal, cultural, religious and political—that legitimate animal abuse. This series explores the challenges that Animal Ethics poses, both conceptually and practically, to traditional understandings of human-animal relations.

The series is needed for three reasons: (i) to provide the texts that will service the new university courses on animals; (ii) to support the increasing number of students studying and academics researching in animal-related fields, and (iii) because there is currently no book series that is a focus for multidisciplinary research in the field.

Specifically, the series will

- provide a range of key introductory and advanced texts that map out ethical positions on animals,
- publish pioneering work written by new, as well as accomplished, scholars, and
- produce texts from a variety of disciplines that are multidisciplinary in character or have multidisciplinary relevance.

The new Palgrave Macmillan Series on Animal Ethics is the result of a unique partnership between Palgrave Macmillan and the Ferrater Mora Oxford Centre for Animal Ethics. The series is an integral part of the mission of the Centre to put animals on the intellectual agenda by facilitating academic research and publication. The series is also a natural complement to one of the Centre's other major projects, the *Journal of Animal Ethics*. The Centre is an independent "think tank" for the advancement of progressive thought about animals and is the first Centre of its kind in the world. It aims to demonstrate rigorous intellectual enquiry and the highest standards of scholarship. It strives to be a world-class centre of academic excellence in its field.

We invite academics to visit the Centre's website www.oxfordanimalethics.com and to contact us with new book proposals for the series.

Oxford, UK — Andrew Linzey
Oxford, UK — Clair Linzey

Praise for *Animals as Experiencing Entities*

"A groundbreaking collection of works that foreground the treatment, experiences, and standpoints of other animals in history. A brilliant and much needed advance in scholarship promoting a just and non-oppressive world for all."

—David Nibert, author of *Animal Oppression and Human Violence: Domesecration, Capitalism and Global Conflict*

Contents

Notes on Contributors

Sierra Apillanes is a graduate of the Sandra Day O'Connor College of Law at Arizona State University. She graduated in 2022.

Gwendolyn Bell is a graduate of the Sandra Day O'Connor College of Law at Arizona State University. She graduated in 2023.

Karen Bradshaw is Professor of Law at the Sandra Day O'Connor College of Law Arizona State University, where she teaches property, contracts, environmental law, natural resources and biodiversity. She is a senior sustainability scientist at the Global Institute of Sustainability and a Faculty Affiliate Scholar at the New York University School of Law Classical Liberal Institute. Bradshaw is the author of the internationally acclaimed book *Wildlife as Property Owners: A New Conception of Animal Rights. Wildlife as Property Owners* was featured in the Official 2022 GRAMMY gift bags and was "highly recommended" in a *Forbes* book review. Bradshaw earned a JD with honours from University of Chicago Law School, an MBA from California State University, Chico and a BS in Business Administration from University of California, Berkeley. She grew up near the Klamath and McCloud Rivers and now resides in the Sonoran Desert on rewilded land that is Certified Wildlife Habitat.

Samantha Brown graduated from Arizona State University's Sandra Day O'Connor College of Law in May 2022. She now works as a Family Law attorney in Scottsdale.

Heeral Chhabra is currently a post-doctoral research associate at Department of History, University of Liverpool, UK, as part of the ROH Indies Project. She was recently awarded PhD for her thesis *Animal 'Welfare', State Regulations and Questions of Cruelty c.1890-1940s* from University of Delhi, India. She was also a *Global History Fellow* at IISH, Amsterdam. Her research interests include Animal History, Global history, History of Education and Pedagogy of History.

Caitlin Doak is a judicial law clerk at the United States District Court for the District of Nevada. Previously, she served as a judicial law clerk at the Arizona Supreme Court. She graduated from Arizona State University's Sandra Day O'Connor College of Law in 2021. Doak's interest in natural resources law derives from her passion for the outdoors. She is especially drawn to issues involving public lands.

Challie Facemire is the Program Director of the Central Arizona Conservation Alliance at the Desert Botanical Garden where she oversees collaborative conservation working groups and projects. She graduated with a PhD from Arizona State University's Biology and Society program in Spring 2022. In her PhD program her areas of research were law and policy impacts on land and ecosystem management, the effects on ecosystems of differing management philosophies, and collaborative conservation. Prior to pursuing her PhD, Challie served as a writer, editor, and faculty research development specialist after obtaining a BA and MA in English.

Andrew Flack is Senior Lecturer in Modern and Environmental History at the University of Bristol. Dr Flack is an animal and environmental historian, working primarily on human engagements with the non-human animal world across the nineteenth and twentieth centuries. He is currently engaged in studies of nocturnal natures. He is the author of *The Wild Within: Histories of a Landmark British Zoo* (University of Virginia Press, 2018).

Michael J. Glover is an animal historian specialising in cattle histories in southern Africa during the colonial period. He is a postdoctoral research fellow at the International Studies Group, University of the Free State. He has published on cattle history, economic history, animal ethics, decoloniality theory and Activity Theory. He has a PhD on a cattle history of southern Africa, 1652–1980s from Leiden University (Netherlands) and University of the Free State (South Africa). Michael is an Associate Fellow of the Oxford Centre for Animal Ethics.

Terike Haapoja is a visual artist based in New York. She represented Finland in the 55. Venice Biennale with a solo show in the Nordic pavilion and exhibits her work internationally. Terike is co-editor of *The Helsinki Effect: A Public Alternative to the Guggenheim Model of Culture-Driven Development* (New York Terreform, 2016).

Kai Horsthemke, PhD, is Visiting Professor in Philosophy of Education at the University of the Witwatersrand, South Africa, and at KU Eichstätt-Ingolstadt, Germany, and a Fellow at the Oxford Centre for Animal Ethics, UK. He is the author of four monographs, *The Moral Status and Rights of Animals* (Porcupine Press, 2010), *Animals and African Ethics* (Palgrave Macmillan, 2015), *Animal Rights Education* (Palgrave Macmillan, 2018) and *Indigenous Knowledge: Philosophical and Educational Considerations* (Lexington Books, 2021). Together with Peggy Siyakwazi, Elizabeth Walton and Charl Wolhuter, he co-edited the first two editions of *Education Studies* (Oxford University Press Southern Africa, 2013 and 2016, respectively). His new book, *The Meaning of Death*, will be published by Lexington Books in 2024.

Clayton Kinsey is a student at Arizona State University's Sandra Day O'Connor College of Law, J.D. 2024. He is from Norman, Oklahoma, and received his BA at the University of California, Berkeley, Class of 2017. Clayton is interested in Environmental and Indian Law and is an enrolled member of the Choctaw Nation of Oklahoma.

Isaac Kort-Meade is a graduate of the Sandra Day O'Connor College of Law at Arizona State University. He graduated in 2022.

Pierce Libbey is a graduate of the Sandra Day O'Connor College of Law at Arizona State University. He graduated in Spring 2022.

Les Mitchell is a research fellow at the University of the Free State, International Studies Group and a Fellow of the Oxford Centre for Animal Ethics. He has lived and worked in the UK, Tanzania, Zambia, Zimbabwe, Malawi, and South Africa and has experience in pathology, biochemistry, community health, education, animal ethics, critical discourse analysis and is a facilitator for the Alternatives to Violence Programme. His research interests include ethics and non-human animals, discourses, power, genocide and moral disengagement. Les Mitchell is the author of *Reading the Animal Text in the Landscape of the Damned* (NISC, 2019).

Abraham Olivier is Professor of Philosophy at the University of Fort Hare, Visiting Professor at the University of Bayreuth and Co-Founder and Co-Chair of the *Centre for Phenomenology in South Africa*. He is the author of *Being in Pain* and editor/co-editor of the anthology *Phenomenology in an African Context* and several special journal issues, including, *The Southern Journal of Philosophy*, *International Journal of Philosophical Studies*, the *Journal of the British Society for Phenomenology* and *Angelaki*. He has published numerous peer-reviewed articles and book chapters on topics relating to phenomenology, philosophy of mind and African philosophy.

Ian S. Olivier is a doctor practicing at Tygerberg Hospital, Cape Town, and is completing his MSc in Neuroscience (Psychiatry) at the University of Cape Town (UCT) in the Department of Mental Health and Psychiatry. He graduated with a Bachelor of Medicine and Bachelor of Surgery (MBChB) from UCT, and an Honours in Physiology/Neuroscience researching structural brain changes in substance use disorders. His research work focuses on neural correlates of neuropsychiatric features such as impulsivity, compulsivity and others in various neuropsychiatric disorders, using functional MRI. He has been involved in both local and international research, including precision medicine and genomics in paediatric epilepsy, as well as work on functional MRI analyses for the global ENIGMA-OCD working group. He publishes clinical research in the fields of psychiatry and neurology. He has a keen interest in philosophy of mind, animal cognition and sentience, animal welfare and ethics, and the interplay between these topics and neuroscience.

Teya Brooks Pribac is an independent researcher. Research interests/activities include evolutionary thanatology, animal emotion, cognition and psyche, human-nonhuman animal relations, animal self-determination and liberation. She is the author of *Enter the Animal. Cross-species Perspectives on Grief and Spirituality* and numerous other publications.

Jonathan Saha is Associate Professor of History at the University of Durham. He is a historian of colonial Myanmar and his second book, *Colonizing Animals: Interspecies Empire in Myanmar*, was published with Cambridge University Press in 2021.

Sandra Swart is Professor of History at Stellenbosch University. She is a social and environmental historian of southern Africa, having received concomitantly both a doctorate in modern history at the University of Oxford and an MSc in Environmental Change & Management from Oxford. She has published on socio-environmental history and animals in history. She is the co-author, with Greg Bankoff, of *Breeds of Empire* (NIAS Press, 2007); co-editor, with Lance van Sittert, of *Canis Africanis: A Dog History of Southern* Africa (Brill, 2008) and sole author of *Riding High – Horses, Humans and History in South Africa* (Johannesburg: Witwatersrand University Press, 2010).

Marion Thomas is Senior Lecturer in History of Science at the University of Strasbourg, France. Her initial background is in biological sciences, and she then completed a PhD in the history of science at the University of Manchester in 2003. She specializes in the history of the modern life sciences, particularly animal behaviour and primatology. She also co-directed a three-year project on the emergence and reception of the cell theory in France and Germany, ca. 1800–1900. She recently completed a Habilitation Manuscript titled: "Of Chimpanzees and Men: A History of the Pasteur Institute in Kindia (Guinea), 1922-1965".

Samuel Wu is a graduate of Arizona State University's Sandra Day O'Connor College of Law, class of 2022. He currently is an attorney/officer with the U.S. Army Judge Advocate General's Corps. He is interested in military justice (criminal law) and military administrative law.

CHAPTER 1

Introduction: Animals 'caught with ourselves in the net of life and time'

Michael J. Glover and Les Mitchell

Introduction

The broader project we wish to contribute to in this volume is one where the *experiences* of those with little or no power, usually, although not exclusively, animals, are recognised and written about as sensate beings with inner worlds whose felt, subjective experiences are validated. The term animal is used here and elsewhere in the chapter to denote non-human animal. Being a form of critical research, we take cognisance of power, domination and oppression, and aim to highlight the perspectives of the dominated and oppressed, recognising them as possessing *umwelts*

Henry Beston. *The Outermost House: A Year of Life on the Great Beach of Cape Cod*. New York: The Viking Press, [1928] 1969, 25.

M. J. Glover (✉) • L. Mitchell
International Studies Group, University of the Free State,
Bloemfontein, South Africa
e-mail: hellomichaelspeaking@gmail.com; les@peacecaravan.co.za

M. J. Glover, L. Mitchell (eds.), *Animals as Experiencing Entities*,
The Palgrave Macmillan Animal Ethics Series,
https://doi.org/10.1007/978-3-031-46456-0_1

of their own.[1] Beyond recognition or validation of animals' experiences, some chapters in this volume in their own ways try to explore and in some cases recover, to the extents possible, the historical experiences of animals.[2]

The argument for animals' subjective historical experiences is straightforward. Animals are individual sentient entities, each with their own personally felt, subjective experiences. Further, animals were part of history. Therefore, animals had subjective historical experiences. This conclusion, coupled with the rarity of historians exploring animals' experiences of the past, forms the central interests of this volume. Our objective is to open new vistas into recognising and validating animals' experiences, via both theory and history-writing.

The theme of animals as experiencing entities is what links the chapters and characterises the volume. To date, pioneering animal historian Éric Baratay's work has been among the most explicit attempts to adopt animals' perspectives of historical events and processes. He argues that the lives of animals need not be a black hole in history.[3] He asserts that we "must also abandon the culturally constructed Western notion of animals as passive beings and see them instead as feeling, responding, adapting, and suffering".[4] Based on the idea that animals have historical subjective experiences that are to some degree recoverable, a number of authors in this volume explicitly seek to illuminate *what some animals' lived experiences might have been* in particular circumstances at specific times in the past.

[1] Timo Maran et al., *Animal Umwelten in a Changing World: Zoosemiotic Perspectives* (Tartu: University of Tartu Press, 2016); Morten Tønnessen, 'Making the Umwelt Bubble of the Modern Synthesis Burst', *Biosemiotics* 14, no. 1 (2021): 121–125; Kalevi Kull, 'Zoosemiotics Is the Study of Animal Forms of Knowing', *Semiotica* 2014, no. 198 (2014), 47–60; Ed Yong, *An Immense World: How Animal Senses Reveal the Hidden Realms Around Us* (London: The Bodley Head, 2022).

[2] For histories which endeavour to include animals' experiences, see discussion by Violette Pouillard. 'Animal Biographies: Beyond Archetypal Figures' *Journal of Animal Ethics* 12, 2 (2022): 182–178, and for histories which use animal biographical approaches, see John Simons, *Obaysch. A Hippopotamus in Victorian London* (Sydney: Sydney University Press, 2019), Éric Baratay. *Biographies animales: Des vies retrouvée* (France: Seuil, 2017) and Krebber, A., & Roscher, M. (Eds.) *Animal Biography. Re-framing Animal Lives* (Basingstoke: Palgrave Macmillan, 2018).

[3] Éric Baratay, *Le Point de Vue Animal: Une Autre Version de l'histoire* [*The Animal Perspective: Another Side of the Story*] (Paris: Seuil, 2012), 30.

[4] Éric Baratay translated by Stephanie Posthumus, 'Building an Animal History', in Louisa Mackenzie and Stephanie Posthumus, eds., *French Thinking About Animals*, The Animal Turn (Michigan: Michigan State University Press, 2015), 4.

Broadly, each author in this volume contributes in one of two ways. The first group, in Part I, theoretically explore animal subjectivity, animal experiences, and the ways in which these can be understood by humans (Brooks Pribac, Olivier & Olivier, Horsthemke, Flack & Swart, Haapoja). The second group of authors, in Part II, offer narrative accounts about specific animals or groups of animals and explore to some extent their subjective historical experiences (Saha, Thomas, Chhabra, Facemire et al., Mitchell). In summary, the first part diversely theorises about animal experiences and how they can be understood in various ways, while the second part's authors assume animals have subjective experiences and construct narratives that take into account how animals might have subjectively experienced the past.

The volume's principal contribution is that it endeavours to move the needle from abstractly recognising animals as subjects to theorising about animals' experiences and specifically exploring their historical experiences. It does so by undertaking enquiries which are daunting, even speculative at times, but using reason and evidence-based approaches. Animals' hidden and interior lives, intertwined with humans in myriad ways, comprise narratives which have been suppressed by dominant powers and ideologies at the time of the events and are often still marginalised today. We think decency demands that the animals' stories be told, and justice requires that we tell them as honestly as possible.

To gain understanding of the experiences and possible inner states of animals, our authors draw on a range of diverse disciplines. Animal historians, for example, converge on the idea that animal histories are multidisciplinary endeavours.[5] Before discussing the chapters in this volume, we first discuss some of the wide-ranging research into animals' experiences. After this, we consider how notions of human elitism have been a barrier to acknowledging and respecting animals as fellow sentient entities. We then briefly engage the common critique of anthropomorphism. Thereafter, we discuss how animal historians have begun to validate animals as sentient entities and how some have started to include animals' subjective experiences in their writing. We close by summarising the chapters in this volume.

[5] Ewa Domańska, 'Animal History.' *History & Theory* 56, 2 (2017): 267–287. Brett Walker, *The Lost Wolves of Japan* (Seattle: University of Washington Press, 2009), 26, 32, 176, 177. Pouillard, 2022, 178.

Scientific Research Into and Evidence of the Experiences of Animals

Charles Darwin was an early prominent exponent of the emotional and experiential commonalities between humans and animals. In 1873, he published *The Expression of Emotions in Man and Animals*, which blurred sharp distinctions between humans' and animals' emotional lives and experiences.[6] Especially in the past few decades, research into animals' emotions and experiences has increased quite dramatically.[7] Scholars from diverse disciplines including, but not limited to, biology, ethology, primatology, ichthyology, zoology, and ornithology, for example, have investigated feelings, emotions, and experiences in animals. Using various disciplines, Jeffrey Moussaieff Masson in 2004 wrote foresightfully about the emotional worlds of farm animals.[8] Jonathan Balcombe published *Pleasurable Kingdom: Animals and the Nature of Feeling Good* (2006), which investigated animals' capacities for pleasure, while Marc Bekoff explored animals' emotions including joy, sorrow, empathy, and why these emotional states matter.[9] Frans de Waal demonstrated that other animals too exhibit empathy, ornithologist Tim Birkhead explored the senses of birds in *Bird Sense: What It's Like to Be a Bird* (2012), and Teya Brooks Pribac investigated how animals have capacities for, and experiences of, grief.[10] Jonathan Balcombe wrote *The New York Times* best seller *What a Fish Knows* (2016), which demonstrated the remarkable social and

[6] Charles Darwin, *The Expression of Emotions in Man and Animals* (London: John Murray 1873).

[7] Marc Bekoff. *Animals Matter: A Biologist Explains Why We Should Treat Animals with Compassion and Respect.* New York: Random House, 2007; Carl Safina, *Beyond Words: What Animals Think and Feel.* New York: Henry Holt and Company, 2015; Frans de Waal. *Are We Smart Enough to Know How Smart Animals Are?* New York: W. W. Norton & Company, 2016; Peter Godfrey-Smith. *Other Minds: The Octopus, the Sea, and the Deep Origins of Consciousness.* New York: Farrar, Straus and Giroux, 2016.

[8] Jeffrey Moussaieff Masson, *The Pig Who Sang to the Moon: The Emotional World of Farm Animals.* (London: Jonathan Cape, 2004).

[9] Jonathan Balcombe, *Pleasurable Kingdom: Animals and the Nature of Feeling Good* (London: Macmillan, 2006), Marc Bekoff, *The Emotional Lives of Animals: A Leading Scientist Explores Animal Joy, Sorrow, and Empathy – And Why They Matter* (California: New World Library, 2007).

[10] Frans de Waal, *The Age of Empathy: Nature's Lessons for a Kinder Society* (New York: Three Rivers Press, 2009), Tim Birkhead, *Bird Sense: What It's Like to Be a Bird* (London: Bloomsbury Publishing, 2012), Teya Brooks Pribac, *Enter the Animal: Cross-species Perspectives on Grief and Spirituality* (Sydney: Sydney University Press, 2021).

emotional capabilities of fish.[11] Recently, Pulitzer Prize-winning author Ed Yong published *An Immense World: How Animal Senses Reveal the Hidden Realms Around Us* (2022). Positioned as a voyage through nature's *Umwelten*, the book articulates the diverse and wonderful sensory capacities of animals in terms of many different animals' perceptions.

Yong uses the notion of *Umwelt* as the book's unifying concept. *Umwelt* refers to each animal's "perceptual world".[12] This perceptual world is realised by a combination of the type of sensory system an animal has and the environment this system perceives. Thus, different animals' different sensoria in the same environment imply different experiences and thus different realities. He writes, "a multitude of creatures could be standing in the same physical space and have completely different *Umwelten*".[13] What Yong's book does so credibly is bring home to readers the enchantingly diverse sensory abilities of different animals. Butterflies can taste with their feet, catfish can taste with their skin, vampire bats can sense infrared radiation, turtles can track electromagnetic fields, while octopuses have nine partially autonomous nervous systems, and their arms likely taste and feel touch simultaneously.[14]

Crucially, by understanding animals' sensory systems, we are better able to begin to understand their experiences. In line with this understanding, Andrew Flack and Sandra Swart, in this volume, indeed argue for a "animal-oriented sensory turn" in animal history, to enable us to "draw closer to the lived realities of past animals".

Animals' internal lives and experiences are no longer considered an eternal black hole of unknowability. Today, we have diverse strategies for understanding, even if imperfectly, the ways in which animals experience their lives. In pedestrian ways, many people readily and intuitively recognise the difference between distress and joy in their companion animals. Marc Bekoff, for example, outlines some ways in which information about animals' emotional states might be readily obtained.

> The emotional states of many animals are easily recognizable. Their faces, their eyes and the ways in which they carry themselves can be used to make strong inferences about what they are feeling. Changes in muscle tone,

[11] Jonathan Balcombe, *What a Fish Knows: The Inner Lives of our Underwater Cousins* (London: Oneworld, 2016).

[12] Yong, *An Immense World*, 11.

[13] Yong, *An Immense World*, 5.

[14] Yong, *An Immense World*, 150–151, 330.

> posture, gait, facial expression, eye size and gaze, vocalizations and odors (pheromones) singly and together, indicate emotional responses to certain situations.[15]

In a number of disciplines where humans were hitherto the primary site of importance and value, animals are increasingly included. Multispecies ethnography, for example, has emerged from the conventionally human-centric discipline of anthropology. Multispecies ethnography theorists Eben Kirksey and Stefan Helmreich note that non-human living beings of all kinds, previously only on the edges of anthropology as food or symbols, are increasingly coming to the fore and that "[a]nimals, plants, fungi, and microbes ... have started to appear alongside humans in the realm of bios, with legibly biographical and political lives".[16]

Cognitive scientist Dominique Lestel and zoomusicologist Hollis Taylor write of probing the convergences and proximity of all life, and that what is central to this problem is "not that we must cohabit with others, but that we are the others and the others are us".[17] Indeed, at various moments in history, scholars have noticed that humans and animals are indeed fundamentally alike, or more alike than different. In a foresightful summation of what would later become cognitive ethology, in 1739, philosopher David Hume wrote:

> no truth appears to me more evident, than that beasts are endow'd with thought and reason as well as men. 'Tis from the resemblance of the external actions of animals to those we ourselves perform, that we judge their internal likewise to resemble ours; and the same principle of reasoning, carry'd one step farther, will make us conclude that since our internal actions resemble each other, the causes, from which they are deriv'd, must also be resembling.[18]

On the topic of animals' feelings and emotions, cognitive psychologist Mariska Kret and colleagues offer modern support for this view and claim

[15] Marc Bekoff, 'Animal Emotions: Exploring Passionate Natures', *BioScience* 50, no. 10 (2000): 863.

[16] Eben Kirksey and Stefan Helmreich, 'The Emergence of Multispecies Ethnography', *Cultural Anthropology* 25, no. 4 (2010): 545.

[17] Dominique Lestel and Hollis Taylor, 'Shared life: An introduction', *Social Science Information* 52, no. 2 (2013): 183.

[18] David Hume. *A Treatise of Human Nature*, ed. L. A. Selby-Bigge (Oxford: The Clarendon Press, [1739] 1951), 176.

it is important to apply ***evolutionary parsimony*** when accounting for the behaviour of evolutionarily related species in similar circumstances.[19] In other words, the hypothesis with the fewest assumptions is most reasonable to choose. Echoing Hume, they argue that until proven otherwise "we must assume that similar behavior in [various] species is paired with similar emotions and in some cases similar feelings".[20] Ecologist Carl Safina argues that the process of evolution does not suddenly generate totally unique and radically different creatures but that evolution is a process of slight variations to existing organisms including ourselves. "Before humans could be assembled", he argues, "evolution needed to have most of the parts in stock, and those parts were developed for earlier models. We inherited them".[21]

However, in *Other Minds* (2016), philosopher of science Peter Godfrey-Smith reminds us it is not only those animals who are relatively close to us in evolutionary terms who might share something analogous to what we describe as feelings, emotions, or experiences.[22] Convergent evolution can also produce similarities between organisms far apart in evolutionary terms. Mammals and cephalopods have a last common ancestor around 600 million years ago and yet their minds have evolved on both sides of this fork in evolution.[23] Zoologist Lars Chittka writing in *Scientific American* describes our growing understanding of insects and their lives:

> Bees, for example, can count, grasp concepts of sameness and difference, learn complex tasks by observing others, and know their own individual body dimensions, a capacity associated with consciousness in humans. They also appear to experience both pleasure and pain. In other words, it now

[19] Mariska E. Kret, Jorg J. M. Massen, and Frans B. M. de Waal, 'My Fear Is Not, and Never Will Be, Your Fear: On Emotions and Feelings in Animals', *Affective Science* 3, no. 1 (2022): 187.

[20] Kret, Massen, and de Waal, 187.

[21] Carl Safina, 'Animals Think and Feel: Précis of Beyond Words: What Animals Think and Feel (Safina 2015)', *Animal Sentience* 1, no. 2 (2016): 3.

[22] Peter Godfrey-Smith. *Other Minds: The Octopus, the Sea, and the Deep Origins of Consciousness*. New York: Farrar, Straus and Giroux, 2016.

[23] Anita Avramides, 'Other Minds', in *The Stanford Encyclopedia of Philosophy*, ed. Edward N. Zalta, Winter 2020 (Metaphysics Research Lab, Stanford University, 2020), https://plato.stanford.edu/archives/win2020/entries/other-minds/.

looks like at least some species of insects—and maybe all of them—are sentient.[24]

Over the last 20 years, intense research on the behaviour of a diverse range of animals has taken place in ethology, often employing ethograms to help catalogue in minute detail specific animal behaviours. For example, a paper by cognitive ecologist Lauren Stanton and colleagues about the construction of a standardised ethogram for the family of *Felidae* (cats) found 95 relevant publications relating to 30 species and 40 subspecies of felids.[25] The Facial Action Coding System adds a further dimension to such research. Originally used on humans it is now employed to understand the facial expressions of animals and their underlying emotions in cats, donkeys, dogs, horses, and marmosets, for example.[26]

Others have opted for creative and unconventional approaches to understanding animals' lives. In his book *Being a Beast* (2016), Charles Foster makes a heroic and very entertaining attempt to physically experience what it is like to live as certain animals.[27] He records, for example, his time wandering naked, moving on all fours in the grass of Wales, while living in a large hole in order to understand a little about the life of a badger. While this might appear bizarre, the book is deeply thoughtful, rigorous and offers important insights into this problem. Attempting to enter into the life world of another being requires much thought, introspection, empathy, courage, and humility as well as insights from different disciplines. In this volume, we accept that animals such as apes, dogs, horses, elephants, cats, and so on are in Tom Regan's phrase "subjects of a life"

[24] Lars Chittka, 'Insects Feel Joy and Pain', Scientific American, July 2023, accessed 14 July 2023. https://doi.org/10.1038/scientificamerican0723-26. See also David DeGrazia, "Sentience and Consciousness as Bases for Attributing Interests and Moral Status: Considering the Evidence and Speculating Slightly Beyond." In *Neuroethics and Nonhuman Animals*, eds. L. Johnson, A. Fenton, and A. Shriver (Cham: Springer 2020), 17–31.

[25] Lauren Stanton, Matthew Sullivan, and Jilian Fazio, 'A Standardized Ethogram for the Felidae: A Tool for Behavioral Researchers', *Applied Animal Behaviour Science* 173 (2015), 3–16.

[26] Frans de Waal. *Mama's Last Hug: Animal Emotions and What They Tell Us About Ourselves*. New York: W. W. Norton & Company, 2020; Catia Correia-Caeiro Anne Burrows, Duncan Wilson, Abdelhady Abdelrahman, and Tatako Miyabe-Nishiwaki. 'CalliFACS: The Common Marmoset Facial Action Coding System', *PLOS ONE* 17, no. 5 (2022): 1–44; Jen Wathan, Anne Burrows, Bridget Waller, and Karen McComb. 'EquiFACS: The Equine Facial Action Coding System', *PLOS ONE* 10, no. 8 (2015): 1–35.

[27] Charles Foster, *Being a Beast* (Suffolk: Profile Books, 2016).

and that they have experiential lives with all that this entails. To elucidate and hold a light to some of those experiences of our fellow beings is the inspiration for this project.

Ethologist Jonathan Balcombe, who extensively catalogues animals' experiences of pleasures, writes on those enjoyed in relation to play, food, sex, touch, love, and some chemicals.[28] We know that baby gorillas, dogs, rats, and other animals enjoy being tickled and rats let out high pitched squeaks when this happens; like dogs, cats, horses, primates, cows, and other animals, they also make "joy jumps".[29]

In terms of social relationships, a number of animals have been found to have triadic awareness, which is the ability to understand not only their own close relationships but those relationships which exist outside of their group. This suggests they possess an extended form of a practical theory of mind. Such animals include apes, monkeys, spotted hyenas, rooks, common ravens, and acorn woodpeckers.[30]

Research on the inner lives of animals is continually advancing. In a recent exchange Ronald Oldfield and Jonathan Balcombe examine which animals might experience betrayal by an entity close to them.[31] The authors consider degrees of awareness such as reflexivity, sentience, cognitive empathy, self-awareness, and theory of mind in animals from primates to fish. It is doubtful that such papers would have been published in an academic journal even as recently as 30 years ago. As Frans de Waal notes:

> *The Oxford Companion to Animal Behaviour published in 1987 maintained that there was absolutely no point studying animal emotions, because they tell us nothing new, and besides "animals are restricted to just a few basic emotions".*[32]

Contrary to this, as indicated above, there is a substantial and growing body of evidence suggesting that animals, other than humans, have

[28] Jonathan P. Balcombe, *Pleasurable Kingdom* (Basingstoke UK: Macmillan, 2007).

[29] de Waal, *Mama's Last Hug*.

[30] Michael A. Pardo et al., 'Wild Acorn Woodpeckers Recognize Associations Between Individuals in Other Groups', *Proceedings of the Royal Society B: Biological Sciences* 285, no. 1882 (2018): 1–7; de Waal, *Mama's Last Hug*.

[31] Ronald G. Oldfield, 'You Can't Betray a Fish: One Reason Eating Fish May Cause Less Harm Than Eating Cows.' *Journal of Animal Ethics* 12, no. 1 (2022): 51–58; Jonathan P. Balcombe, 'The Betrayed Fish: Reply to Oldfield', *Journal of Animal Ethics* 12, no. 1 (Spring 2022): 59–62.

[32] de Waal, *Mama's Last Hug*.

feelings, emotions, and experiences. Because they have diverse sensory systems, including some sensory faculties like echolocation that humans don't possess, it is reasonable to expect that animals have aspects of their inner lives for which we have no parallels. But, crucially, we can reasonably expect that what holds for animals' experiential capacities today would be similar in historic settings as well, thereby validating and supporting attempts to enquire into what their inner lives might have been like in the past. Because their senses are core to who they are and how they live and survive, it would be absurd to think that animals have only recently acquired sensory faculties. We acknowledge that studies of animals' experiences are a problematic and complex field for researchers but surely one worth pursuing. However, investigating animals' experiences is made even more challenging by three special features which require attention: (a) human elitism, (b) ideologies relating to animals and embedded in texts, and (c) the spectre of anthropomorphism.

Human Elitism, Language, and Anthropomorphism

The idea that humans have an elevated status placing them above all other living things is often taken as common sense. It is an ideology which is pervasive and runs deep. The idea of a hierarchical world composed of greater and lesser beings is one which stretches back at least to the time of Aristotle.[33] Tragically, in his scheme, those lower on the hierarchy are rightfully used by those above them, and so slaves are for the use of the master, women for the use of men, and animals and plants are to serve the needs and wishes of humans.[34] Despite evolutionary theory apparently being widely accepted, this self-serving hubris lives on with devastating consequences for animals who are confined and used in their billions for such things as food, power, transport, entertainment, and experimentation, while the same ideology helps to facilitate moral disengagement for those who support or carry out these oppressive actions.[35]

While it is common to accept other animals are living beings and products of an evolutionary journey as long as our own, considering them as

[33] Aristotle, 'Politics', accessed 20 March 2016, http://classics.mit.edu/Aristotle/politics.1.one.html.

[34] Chapter 5 Les Mitchell, *Reading the Animal Text in the Landscape of the Damned* (Grahamstown: NISC, 2019).

[35] Les Mitchell, 'Moral Disengagement and Support for Nonhuman Animal Farming', *Society & Animals* 19, no. 1 (2011): 38–58.

having meaningful lives and being important in themselves is often dismissed. Legal scholar Lisa Johnson claims that even though we might understand animals as *living beings*, this discourse is bifurcated and the conceptual delimitation of one part of it means that while we see animals as *living* we do not really see them as *beings* because this understanding of them is dimmed in our present world.[36] As philosopher Jonathan Birch noted in a recent review of Peter Singer's *Animal Liberation Now* (2023): "Our maltreatment of other animals stems not from speciesism as such, but from failing to respect other animals as sentient beings with lives of their own".[37]

Researchers encounter texts in many forms, and within these texts a form of reality is constructed by the producer of the text.[38] Language in use is a site of contestation about reality. Tools used in the portrayal of reality through text include ideological constructions, specific lexical selections, ideological common sense, highlighting and backgrounding, who speaks and who does not, what is included, what is excluded, and the silence of the text about particular things.[39]

References to animals are commonly used in derogatory descriptions such as *dirty pig* or calling a person a *donkey* or an *elephant*, and these reinforce negative animal stereotypes. Texts claim that animals *served* in combat or *assisted* in rural development, when in reality they were coerced into both of these actions to fulfil human desires. Euphemism is common as when animals in experiments are described as *models* which, never *who*, undergo *procedures* and are eventually *sacrificed*. Reports of experiments are written in the third person thereby obscuring any agency and seldom make mention of the struggles, pain, fear, cries, maiming, and panic of their victims. Everyday descriptions such as *pack horse*, *cart horse*, *pit pony*, *beef cattle*, *laboratory animals*, *cage bird*, and *race horse*, for example, imply the animals concerned have an inherent purpose for their lives and are naturally, rightfully, and even willingly being used for that purpose by humans.[40] In descriptions of 'animal farming', there is silence about the

[36] Lisa Johnson, *Power, Knowledge, Animals* (Basingstoke: Palgrave Macmillan, 2012), 100.

[37] Jonathan Birch, 'Animal Liberation – Then and Now', *Nature*, 618 (2023): 668. Peter Singer, *Animal Liberation Now* (London: Bodley Head, 2023).

[38] Norman Fairclough, *Language and Power*. (Harlow: Pearson Education, 2001).

[39] Fairclough, *Language and Power*.

[40] Joan Dunayer, *Animal Equality: Language and Liberation* (Derwood, Maryland: Ryce Publishing, 2001); Arran Stibbe, 'As Charming as a Pig: The Discursive Construction of the Relationship Between Pigs and Humans', *Society & Animals* 11, no. 4 (2003): 375–392; Les Mitchell, 'Nonhumans and the Ideology of Purpose', *Anthrozoös* 25, no. 4 (2012): 491–502.

horrific transportation and deaths of animals who are used for food; they are simply *sent to market* and *meat* is *produced*.

Everyday descriptions of exploitation, camouflaged in ideologically doctored language, go unnoticed as common sense entitlement which reinforces anthropocentric views of the world.[41] There is much more which might be said but reading and writing "against the grain" is for researchers in this arena an exacting but critically important process. It is also one which is often deeply revelatory.

Linked to human elitism and language use is the dire prohibition, especially in academic work, against what is described as "anthropomorphism". Originally meaning the ascription of human qualities to the gods, the meaning of the term changed to the ascribing of human characteristics to, amongst other entities, animals.[42] Kristin Andrews and Brian Huss point out that there is nothing at all wrong with anthropomorphism.

> *"anthropomorphism" can be defined as the attribution of human psychological, social, or normative properties to nonhuman animals. However, the term is often defined as an error—a misattribution of a human property to a nonhuman animal.*[43]

The word anthropomorphism has become code for *erroneous* anthropomorphism and carries the default position that it must be wrong no matter how compelling the evidence. Erroneous anthropomorphism takes two common forms. The most obvious is when animals have human characteristics assigned to them without any evidence to support such claims but Ed Yong also argues that one of the most prevalent and unnoticed forms of inappropriate anthropomorphism consists in *framing animals' experiences in terms of our sensory capabilities instead of theirs.*[44]

We assert that while it may be impossible, now or perhaps ever, to maximally and with complete accuracy understand the *precise contents* of

[41] Mitchell, 'Moral Disengagement and Support for Nonhuman Animal Farming'.

[42] 'Anthropomorphism | Etymology, Origin and Meaning of Anthropomorphism by Etymonline', accessed 16 June 2023, https://www.etymonline.com/word/anthropomorphism.

[43] Kristin Andrews and Brian Huss, 'Anthropomorphism, Anthropectomy, and the Null Hypothesis', *Biology & Philosophy* 29, no. 5 (2014): 711.

[44] Ed Yong, *An Immense World: How Animal Senses Reveal the Hidden Realms Around Us* (London: The Bodley Head, 2022), 12.

animals' experiences, it does not follow that no reasonable and evidence-based claims can be made about them. That we do not know everything about animals' experiences does not mean we can know nothing about them.

The dominance of a particularly rigid form of behaviourism, which understood animals' minds as black boxes and therefore impossible to understand, helped to fuel a fear of suggesting anything which might be construed as anthropomorphic.[45] Such an enduring attitude towards attempts to understand and describe animals' emotions, mental states, and experiences leads to the claim that any attempts to attribute similarity between human emotions and mental states and those of other animals is unacceptable, that attempts to understand animals' interiority and experiences are, in themselves, anthropomorphic. It prevents further investigation into animals' experiences and wrongfully abandons animals to states of non-experience.

Safina describes how competent research and even careers could be harmed by such ideological hegemony.

> *But as lesser intellects followed the Nobel Prize–winning pioneers, "anthropomorphism" became a pirate flag. If the word was hoisted, an attack was imminent. You wouldn't get your work published. And in the academic realm of publish or perish, jobs were at stake. Even the most informed, insightful, logical inferences about other animals' motivations, emotions, and awareness could wreck your professional prospects.*[46]

The portrayal of anthropomorphism as a transgression has led to (1) moving research activities away from a focus on the inner lives of animals, (2) skewing the interpretation of results in research away from any suggestion of anthropomorphic parallels, and (3) affecting researchers' career choices.

Overall, the threat of being found guilty of anthropomorphism, along with the self-censorship this invokes, has had a chilling effect on research into animal's lives from which we are only now slowly emerging. The increased academic interest in animals' experiences over the last several decades has slowly filtered into historical research.

[45] Raymond Nelson, "Behaviorism is false." *The Journal of Philosophy* 66, no. 14 (1969): 417–452.

[46] Carl Safina, 'Animals Think and Feel: Précis of Beyond Words: What Animals Think and Feel (Safina 2015)', *Animal Sentience* 1, no. 2 (2016): 3.

Animals' Experiences and Historical Research

Animal history, defined expansively as histories with animals as prominent themes or lenses, has grown from a subfield of historical studies over the past four decades and is today a firmly established field.[47] Since the publication of Frederick Brown's award-winning *The City Is More Than Human: An Animal History of Seattle* (2016), some historians and social scientists (including several in this volume) have increasingly conceptualised their investigations as more-than-human, including "multiple species and multiple voices".[48]

However, much of animal history has been interested in concerns other than the experiencing animals themselves.[49] In many ways, the sensing, sensate, experiencing animals have been largely absent from anglophone animal history. In the mid-1990s, historians such as Kathrine Grier, Erica Fudge, and Ann Norton Greene started to acknowledge animal sentience in their work.[50] But, as Andria Pooley-Ebert notes, these authors "hesitated to go further and to describe how horses and others experienced the past, because the problems they identified in asking about historical non-human actions and experience seemed overwhelming at the time".[51] Pooley-Ebert argues that the challenges of capturing animals' historical experiences are "significant" but that "we are beginning to overcome them", especially with the help of the sciences.[52]

Animal historians have indeed largely converged on the methodological claim that animal history requires support from numerous disciplines

[47] Harriet Ritvo, "Recent Work in Animal History (and How We Got Here)." *The Journal of Modern History* 94, 2 (2022): 404–419. Joshua Specht, 'Animal History after Its Triumph: Unexpected Animals, Evolutionary Approaches, and the Animal Lens', *History Compass* 14, 7 (2016), 326.

[48] Frederick Brown, *The City Is More Than Human: An Animal History of Seattle*. Seattle: University of Washington Press, 2016. See, for example, Emily O'Gorman and Andrea Gaynor, 'More-Than-Human-Histories', *Environmental History*, 25, 4 (2020): 711.

[49] Aaron Skabelund, 'Animals and Imperialism: Recent Historiographical Trends', *History Compass* 11, 10 (2013): 801–807. Specht 2016.

[50] Andria Pooley-Ebert, 'A Comparative Study of Horse–Human Relationships in Chicago and Rural Illinois' 'Introduction' in *The Historical Animal*, Susan Nance (ed). New York: Syracuse University Press, 2015, 152.

[51] Ibid.

[52] Pooley-Ebert in Susan Nance, ed., *The Historical Animal*, First edition (Syracuse, New York: Syracuse University Press, 2015), 152.

including sciences like zoology and veterinary science.[53] Historians who wish to validate and explore, however imperfectly, *the experiences of animals* require help from the sciences *a fortiori*. As Éric Baratay noted pithily: "To build an animal history, history needs the help of other sciences".[54] Or, as Susan Nance put it:

> interdisciplinary studies also help us to examine how particular contexts created particular kinds of animals and people, how humans and animals shared cultures, how we might approximate the outlines of an animal's experience, and how historical animals have always been individuals with unique lives.[55]

Historians like Brett Walker and Éric Baratay paved the way for historians to both recognise animals as sentient *and* draw on the sciences to explore what animals' historical experiences might have been.[56] One approach to writing about animals' subjective historical experiences is animal biographies in which often individual animals are focussed on to glean some sense of what the animal's felt life may have been like.[57] But, beyond rare exceptions, animal historians have frequently been hesitant to claim to know what animals' experiences of the past were. For example, in one prominent animal biography text, *Obaysch: A Hippopotamus in Victorian London* (2019), historian John Simons makes an oddly contradictory claim about accessing animals' past experiences. He argues that in historical scholarship, animals should be given "at least a symbolic form of agency".[58] But strikingly he continues to note that "We cannot reconstruct their [animals'] feelings or perceptions, of course, and nor should

[53] Erica Fudge, 'What Was It Like to Be a Cow? History and Animal Studies', in L. Kalof (ed.), *The Oxford Handbook of Animal Studies* (Oxford: Oxford University Press, 2017). Ewa Domańska, 'Animal History', *History & Theory* 56, 2 (2017), 282; Brett Walker, *The Lost Wolves of Japan* (Seattle: University of Washington Press, 2009), 26, 32, 176, 177.

[54] Éric Baratay translated by S. Posthumus 'Building an Animal History', in Louisa Mackenzie and Stephanie Posthumus (eds.), *French Thinking About Animals* (Michigan: Michigan State University Press, 2015), 5.

[55] Susan Nance, 'Introduction' in *The Historical Animal*, Susan Nance (ed). New York: Syracuse University Press, 2015, 7.

[56] Eric Baratay, *Le Point de Vue Animal: Une Autre Version de l'histoire* [*The Animal Perspective: Another Side of the Story*] (Paris: Seuil, 2012). Walker, 2009.

[57] John Simons. Obaysch. *A Hippopotamus in Victorian London*. (Sydney: Sydney University Press, 2019). Andre Krebber and Mieke Roscher (eds.) *Animal Biography. Re-framing Animal Lives* (Basingstoke: Palgrave Macmillan, 2018).

[58] Simons, *Animal Biography* , 9.

we think we can—except perhaps in the attempt to empathise with their sufferings". Two sentences later he writes that "In fact, the story of Obaysch is a story of misery and pain". Pain, misery, and sufferings *are* feelings, they are felt experiences. So, if we can know Obaysch endured suffering, pain, and misery, then we can know his past experiences at least to that extent. There is no reason that additional states such as boredom, panic, depression, and joy cannot be reasonably inferred if there is relevant evidence of events they endured or contexts in which they lived. Simons thus cannot say that we should not think we can reconstruct animals' feelings. The point is that animal historians have frequently been overly cautious, reticent, and timid about describing animals' subjective historical experiences.[59] A greater understanding of animals' sensoria, capabilities, and social lives can enable informed and more confident explorations of animals' experiences and subjectivity.

Indeed, sensory historians like Andrew Flack have focussed on the senses to investigate other animals' different experiences.[60] The subset of animal historians who include animal experiences in their narratives remains small but is growing.[61] Indeed, especially recently, a number of volumes and books have contributed to developing the field of animal history, and there are increasing signs of attempts to explicitly include

[59] For an interesting recent discussion, see Jacob Brandler, Do "Animals" Have Histor(ies)? Can/Should Humans Know Them? A Heuristic Reframing of Animal-Human Relationships. *Journal of Animal Ethics* 12 (2), 2022: 148–157.

[60] Andrew Flack, "Dark Trails: Animal Histories Beyond the Light of Day" *Environmental History* 27, no. 2 (2022): 215–241.

[61] These include but are not limited to Sandra Swart. "'It Is As Bad To Be a Black Man's Animal As It Is To Be a Black Man' – The Politics of Species in Sol Plaatje's Native Life in South Africa." *Journal of Southern African Studies* 40, 4 (2014): 689–705. Susan Nance, 'Introduction' in *The Historical Animal* (ed) Susan Nance. New York: Syracuse University Press, 2015, 7. Jonathan Saha, 'Colonizing Elephants: Animal Agency, Undead Capital and Imperial Science in British Burma', *BJHS Themes* 2 (2017), 169–189; Michael Glover. 'Cattle and colonialism: an animal-centred history of southern Africa 1652–1980s' (PhD dissertation) Leiden: Leiden University, 2021. Michael Glover. ""That other me, down and dreaming": an Animal Perspective Critique of Decoloniality Theory." *Social Dynamics*. Online First. (2023): 1–20.

animals' historical experiences.[62] Part II of our volume contributes to the daunting but increasing attempts of those writing about animals in history who aim to validate and explore animals' likely subjective historical experiences. This volume thus builds on the early endeavours of such scholars to affirm animals' historical experiences as also being meaningfully and relevantly part of history.

Looking Back to Look Forward

The chapters in our volume are diverse both in the disciples from which they arise and in the approaches of our authors. This is testimony to the breadth of the growing attention being given to the inner lives of animals, to the experiences of experiencing Others. And if the volume seems "uneven", not conforming easily to standard lines and boundaries, it is because of the varied landscape which is being traversed. For too long this has been a liminal space, an uncomfortable realm for many researchers, even a forbidden territory. Within this volume, we have embraced extraordinary but unorthodox texts such as Terike Haapoja's creative and imaginative essay 'History According to Cattle' (Chap. 6) and Challie Facemire and colleagues' 'Law Through the Eyes of Animals' (Chap. 10) as they significantly contribute to the project of affirming and incorporating diverse approaches to including animals' perspectives. So, if it is a somewhat nonconformist volume, we celebrate this as a strength and hopefully a call to bring about changes which are long overdue. In our view, failure to take into account the inner experiences of animals in past interactions with humans means our historical endeavours must necessarily remain

[62] Jennifer Bonnell and Sean Kheraj, (eds). *Traces of the Animal Past: Methodological Challenges in Animal History.* University of Calgary Press 2022. Roscher, Mieke, André Krebber, and Brett Mizelle, (eds). *Handbook of Historical Animal Studies.* Walter de Gruyter GmbH & Co KG, 2021. Hilda Kean and Philip Howell (eds.) *The Routledge Companion to Animal-human History* (Abingdon: Routledge, 2019), Clemens Wischermann and Aline Steinbrecher (eds). *Animal History in the Modern City: Exploring Liminality* (London: Bloomsbury Academic, 2018). For examples which include historical subjective experiences, see Violette Pouillard. "Visions of Concord: Wild animals and the Garden of the Revolution (Jardin des Plantes menagerie, 1793–c. 1820)." *Journal for the History of Environment and Society* 4 (2019): 11–40. J. Keri Cronin, '"And has not art promoted our work also?" Visual Culture in Animal-Human History' in *The Routledge Companion to Animal-Human History* (eds.) Hilda Kean and Phillip Howell. London: Routledge, 2019, 263. Sarah Cockram, 'History of Emotions' in *Handbook of Historical Animal Studies* (eds.) Mieke Roscher, André Krebber and Brett Mizelle (Berlin: DeGruyter, 2021), 409.

partial and anthropocentric. We cannot hope to understand our shared, mutually experienced multispecies history in order to forge a better way of being in the world, if we turn away from this moral, albeit often bleak, but necessary project.

Scholars who study processes of reconciliation after grave injustices against humans invoke the concept of truth-telling and argue that for reconciliation to occur, truth-telling is a requirement.[63] What happened to victims cannot be blanked out and denied. In some ways, interest in animals' past experiences can be regarded as a form of truth-telling about our shared histories with animals. This is not about Truth capital T but instead a qualified truth or truth to the extent possible, based on available evidence and methodologies. Nor is validating and documenting animals' historical experiences about providing another reason for despair or cataloguing horrors for their own sake. It is instead about better understanding who we as humans are and how we've related to the other sentient individual entities with whom we share conscious, experienced life. Recognising how we have affected animals, and how animals have experienced life, lays some groundwork for future, respectful, more harmonious, and just relations with other animals.

Overview of Chapters

This edited volume comprises two parts. Part I: 'Animals as Experiencing Entities, Theories, and Perspectives', and Part II: 'Animals' Experiences: Narratives and Histories.' The chapters are intentionally diverse and are not connected by periodisation, geographical location, or animal species. Instead, what connects all the chapters is their interest in animals' experiences and the extent to which these experiences can be recovered and grasped however imperfectly.

Part I: Animals as Experiencing Entities, Theories, and Perspectives

Chapter 2 opens Part I by tracking the 'animal developmental path' to indicate that animals possess human-comparable emotional, cognitive, social, cultural, and even creative capacities and needs. Emphasising animal grief, Teya Brooks Pribac conceptualises animal grief as an organismic

[63] Paulette Regan, *Unsettling the Settler Within: Indian Residential Schools, Truth Telling, and Reconciliation*. Toronto: UBC Press, 2010, 6.

response to loss that should be investigated at the individual animal-level rather than at the species-level. Significantly, she argues that *petits récits* or little narratives of animals' grief experiences enable relatable perceptions of human and other animals' emotional and experiential commonalities.

Via a phenomenological analysis, Chap. 3 by Dr Ian S. Olivier and Professor Abraham Olivier focusses on animal pain in context and how animal pain is indicated and expressed. They argue that animals are capable of indicating and expressing unique experiences of pain in relation to their particular species, evolution, and places of living, in human and natural contexts. One implication of this is that while general conditions for animal pain experiences exist, expressions of animal pain can vary according to individual animals' contexts.

In Chap. 4, Professor Kai Horsthemke argues for a form of critical animal histories as an off-shoot of critical animal studies. Arguing against epistemic relativism, he contends that animals have their own epistemic and experiential standpoint and that there is something that it is qualitatively like to be a particular animal.

Dr Andrew Flack and Professor Sandra Swart, in Chap. 5, close Part I by connecting it to the more narrative driven and historical investigations of Part II. Focusing on animal senses, they argue for an animal-orientated sensory turn, as one way of drawing closer to entangled humans' and animals' past sensory experiences. By emphasising sensory differences and capacities, they argue that being attuned to animals' sensory contexts and abilities can be generative of a richer understanding of animals' experiential pasts. Chapter 6, Terike Haapoja's creative and poetic work, closes Part I by exploring the limits and potentials of perceiving cattle history from a bovine perspective.[64] It argues that cattle lacking human language should not debar cattle from having their histories and experiences affirmed. The chapter imaginatively approximates a cattle perspective via adopting human language to describe cattle's contribution to history, cattle's conception of time, and to communicate that cattle ought not to be silenced in and debarred from history-writing. The chapter presents a cattle history periodisation from a cattle perspective including The Time

[64] The essay is originally from Laura Gustafsson and Terike Haapoja's book *History According to Cattle* (2015) and is an introduction to Gustafsson & Haapoja's project *Museum of the History of Cattle*, which is a travelling ethnographic museum aimed at acknowledging that alternative modes of experience exist. Laura Gustafsson and Terike Haapoja, 'History According to Cattle', in Laura Gustafsson and Terike Haapoja *History According to Cattle* (New York: Punctum Books, 2015), 2–3.

Before History, when cattle's ancestors—the Aurochs—roamed the earth, The Historical Time, from when cattle were domesticated, and the A historical Time, when cattle were subjected to industrial agriculture regimes. Although a creative writing essay, we regard this text as a seminal contribution to the development of animal history and in particular those histories which endeavour to include animals' perspectives.

Part II: Animals' Experiences: Narratives and Histories

Chapter 7 by A/Professor Jonathan Saha is a historiographical interrogation of the way the phrase 'great animal massacres' has featured in titles of historians' scholarly texts. By historicising 'great animal massacres', Saha argues that the animal victims' experiences can be recognised and validated in their own right, as opposed to being incidental to human experiences. Drawing critically on scientific knowledge of crows' experiential capacities and responses to conspecific deaths as well as historical documents, Jonathan Saha infers that the 1920 crow massacre of colonial Rangoon would have entailed distress, trauma, anxiety, fear, and anger for the crows of Rangoon.

In Chap. 8, Dr Marion Thomas traces the lives of two famous chimps, Rose and Kambi, who were captured in French Guinea. Rose was subjected to pioneering TB research in the country and Kambi was shipped to the United States for psychological and toxicological studies. Alongside themes of colonial science, empire, and power, Marion Thomas conceives of Rose and Kambi as sentient, experiencing beings and draws on archival sources to describe their likely felt experiences during their capture, captivity, and subjection to invasive biomedical research.

Chapter 9 by Dr Heeral Chhabra is a thoughtful investigation into how evidence of animals' subjectivities can be gleaned from colonial archival sources in India. Her focus on reading military transport manuals from the late nineteenth century against the grain enables her to make inferences about how colonial military officials conceptualised 'transport animals' like elephants, donkeys, camels, mules, and ponies, as well as the colonially prescribed treatment and 'training' of these animals for human purposes. In this way, Eurocentric and colonial conceptions of animals are brought to light while the animals' subjectivity and experiences are emphasised.

Chapter 10 by Challie Facemire and colleagues is a creative non-anthropocentric reimagining of canonical legal cases from the reasonably inferred perspectives of the animals involved in the cases. Combining insights from critical animal studies as well as animal and biodiversity law, and using a more-than-human perspective, the authors reconsider five major cases that include animals such as cattle, horses, burros, wolves, cetaceans, and monkeys. The authors offer a methodology for critiquing anthropocentric legal findings, reimagining legal cases to include the impacted animals' perspectives, and envisioning new ways to legally include the needs of non-human animals.

Finally, Chap. 11, authored by Dr Les Mitchell, is an investigation of pit ponies' subjective experiences of working in British coal mines between the early eighteenth and mid-twentieth centuries. Drawing on veterinary science and zoology as well as against the grain readings of archival material including official reports, newspapers, commissioned official enquiries, and a range of other sources, Mitchell conducts a wide-ranging investigation into pit ponies' experiences of forced mine labour, including their capture, the nature of their long, draining and dangerous shifts, injuries and abuse, and of vision impairments from lack of light and vitamin A.

Bibliography

Andrews, Kristin, and Brian Huss. 'Anthropomorphism, Anthropectomy, and the Null Hypothesis'. *Biology & Philosophy* 29, no. 5 (2014): 711–729.

'Anthropomorphism | Etymology, Origin and Meaning of Anthropomorphism by Etymonline'. Accessed 16 June 2023. https://www.etymonline.com/word/anthropomorphism.

Aristotle. 'Politics'. Accessed 20 March 2016. http://classics.mit.edu/Aristotle/politics.1.one.html.

Avramides, Anita. 'Other Minds'. In *The Stanford Encyclopedia of Philosophy*, edited by Edward N. Zalta, Winter 2020. Metaphysics Research Lab, Stanford University, 2020. https://plato.stanford.edu/archives/win2020/entries/other-minds/.

Balcombe, Jonathan P. 'The Betrayed Fish: Reply to Oldfield'. *Journal of Animal Ethics* 12, no. 1 (2022): 59–62.

Balcombe, Jonathan. *Pleasurable Kingdom: Animals and the Nature of Feeling Good*. London: Macmillan, 2006.

Balcombe, Jonathan. *What a Fish Knows: The Inner Lives of our Underwater Cousins*. London: Oneworld, 2016.

Baratay, Éric. *Le Point de Vue Animal: Une Autre Version de l'histoire* [*The Animal Perspective: Another Side of the Story.*] Paris: Seuil, 2012.
Baratay, Éric. "Building an Animal History." In *French Thinking about Animals*, edited by Louisa Mackenzie and Stephanie Posthumus, translated by Stephanie Posthumus, 3–14. Michigan: Michigan State University Press, 2015.
Baratay, Éric. *Biographies Animales: Des Vies Retrouvée*. France: Seuil, 2017.
Bekoff, Marc. *Animals Matter: A Biologist Explains Why We Should Treat Animals with Compassion and Respect.* New York: Random House, 2007a.
Bekoff, Marc. *The Emotional Lives of Animals: A Leading Scientist Explores Animal Joy, Sorrow, and Empathy – And Why They Matter.* California: New World Library, 2007b.
Bekoff, Marc. 'Animal Emotions: Exploring Passionate Natures'. *BioScience* 50, no. 10 (2000): 861–870.
Beston, Henry. *The Outermost House: A Year of Life on The Great Beach of Cape Cod.* New York: The Viking Press, [1928] 1969.
Birkhead, Tim. *Bird Sense: What It's Like to Be a Bird.* London: Bloomsbury Publishing, 2012.
Birch, Jonathan. 'Animal Liberation – Then and Now', *Nature*, 618 (2023): 668.
Bonnell, Jennifer and Sean Kheraj, (eds). *Traces of the Animal Past: Methodological Challenges in Animal History.* Alberta: University of Calgary Press, 2022.
Brooks Pribac, Teya. *Enter the Animal: Cross-species Perspectives on Grief and Spirituality* Sydney: Sydney University Press, 2021.
Brown, Frederick. *The City is More than Human: an Animal History of Seattle.* Seattle: University of Washington Press, 2016.
Chittka, Lars. 'Insects Feel Joy and Pain', *Scientific American*, July 2023, accessed 14 July 2023. https://doi.org/10.1038/scientificamerican0723-26.
Cockram, Sarah 'History of Emotions' in *Handbook of Historical Animal Studies,* edited by Mieke Roscher, André Krebber and Brett Mizelle. Berlin: DeGruyter, 2021.
Correia-Caeiro, Catia, Anne Burrows, Duncan Andrew Wilson, Abdelhady Abdelrahman, and Takako Miyabe-Nishiwaki. 'CalliFACS: The Common Marmoset Facial Action Coding System'. *PLOS ONE* 17, no. 5 (2022): 1–35.
Cronin, Keri. '"And has not art promoted our work also?" Visual Culture in Animal-Human History' in *The Routledge Companion to Animal-Human History* , edited by Hilda Kean and Phillip Howell. London: Routledge 2019.
Darwin, Charles. *The Expression of Emotions in Man and Animals.* London: John Murray 1873.
de Waal, Frans. *The Age of Empathy: Nature's Lessons for a Kinder Society.* New York: Three Rivers Press, 2009.
de Waal, Frans. *Are We Smart Enough to Know How Smart Animals Are?* New York: W. W. Norton & Company, 2016.

de Waal, Frans. *Mama's Last Hug: Animal Emotions and What They Tell Us About Ourselves.* New York: W. W. Norton & Company, 2020.

DeGrazia, David. "Sentience and Consciousness as Bases for Attributing Interests and Moral Status: Considering the Evidence and Speculating Slightly Beyond." *In Neuroethics and Nonhuman Animals*, edited by L Johnson, A Fenton, and A Shriver, 17–31. Cham: Springer, 2020.

Domańska, Ewa. 'Animal History.' *History & Theory* 56, no. 2 (2017): 267–287.

Dunayer, Joan. *Animal Equality: Language and Liberation.* Derwood: Ryce Publishing, 2001.

Fairclough, Norman. *Language and Power.* Harlow: Pearson Education, 2001.

Foster, Charles. *Being a Beast.* Suffolk: Profile Books, 2016.

Flack, Andrew. "Dark Trails: Animal Histories Beyond the Light of Day" *Environmental History* 27, no.2 (2022): 215–241.

Fudge, Erica. 'What Was It Like to Be a Cow? History and Animal Studies'. In *The Oxford Handbook of Animal Studies,* edited by L. Kalof. Oxford: Oxford University Press, 2017.

Glover, Michael. 'Cattle and colonialism: an animal-centred history of southern Africa 1652–1980s' (PhD dissertation) Leiden: Leiden University, 2021.

Glover, Michael. ""That other me, down and dreaming": an Animal Perspective Critique of Decoloniality Theory." *Social Dynamics.* 49, no.2 (2023): 260-279

Godfrey-Smith, Peter. *Other Minds: The Octopus, the Sea, and the Deep Origins of Consciousness.* New York: Farrar, Straus and Giroux, 2016.

Gustafsson, Laura and Terike Haapoja 'History According to Cattle'. In *History According to Cattle*, Laura Gustafsson and Terike Haapoja, 2–3. New York: Punctum Books, 2015.

Hume, David. *A Treatise of Human Nature*, edited by L. A. Selby-Bigge. Oxford: The Clarendon Press, [1739] 1951.

Johnson, Lisa. *Power, Knowledge, Animals.* Basingstoke: Palgrave Macmillan, 2012.

Kean, Hilda and Philip Howell (eds.) *The Routledge Companion to Animal-human History.* Abingdon: Routledge, 2019., Clemens Wischermann and Aline Steinbrecher (eds). *Animal History in the Modern City: Exploring Liminality.* London: Bloomsbury Academic, 2018.

Kirksey, Eben and Stefan Helmreich, 'The Emergence of Multispecies Ethnography', *Cultural Anthropology* 25, no. 4 (2010): 545–576.

Krebber, A., & Roscher, M. (Eds.) *Animal Biography. Re-framing Animal Lives.* Basingstoke: Palgrave Macmillan, 2018.

Kret, Mariska E., Jorg J. M. Massen, and Frans B. M. de Waal. 'My Fear Is Not, and Never Will Be, Your Fear: On Emotions and Feelings in Animals'. *Affective Science* 3, no. 1 (2022): 182–189.

Kull, Kalevi. 'Zoosemiotics Is the Study of Animal Forms of Knowing'. *Semiotica* 2014, no. 198 (2014): 47–60.

Lestel, Dominique and Hollis Taylor, 'Shared life: An introduction', *Social Science Information* 52, no. 2 (2013): 183–186.

Linzey, Andrew. *Creatures of the Same God. Explorations of Animal Theology.* Winchester: Winchester University Press, 2007.

Mackenzie, Louisa, and Stephanie Posthumus (eds.) *French Thinking about Animals.* Michigan: Michigan State University, 2015.

Mitchell, Les. 'Moral Disengagement and Support for Nonhuman Animal Farming'. *Society & Animals* 19, no. 1 (2011): 38–58.

Mitchell, Les. 'Nonhumans and the Ideology of Purpose'. *Anthrozoös* 25, no. 4 (2012): 491–502.

Mitchell, Les. *Reading the Animal Text in the Landscape of the Damned.* Grahamstown: NISC, 2019.

Moussaieff Masson, Jeffrey. *The Pig Who Sang to the Moon: The Emotional World of Farm Animals.* London: Jonathan Cape, 2004.

Nance, Susan. 'Introduction' in *The Historical Animal*, edited by Susan Nance. New York: Syracuse University Press, 2015a.

Nance, Susan, ed. *The Historical Animal.* First edition. Syracuse, New York: Syracuse University Press, 2015b.

O'Gorman, Emily and Andrea Gaynor, 'More-Than-Human-Histories', *Environmental History*, 25, no.4 (2020): 711-735.

Oldfield, Ronald. 'You Can't Betray a Fish: One Reason Eating Fish May Cause Less Harm than Eating Cows.', *Journal of Animal Ethics* 12, no. 1 (2022): 51–58.

Pardo, Michael A., Emilee A. Sparks, Tejal S. Kuray, Natasha D. Hagemeyer, Eric L. Walters, and Walter D. Koenig. 'Wild Acorn Woodpeckers Recognize Associations between Individuals in Other Groups'. *Proceedings of the Royal Society B: Biological Sciences* 285, no. 1882 (2018): 1–7.

Peter Singer, *Animal Liberation Now.* London: Bodley Head, 2023.

Pooley-Ebert, Andria. 'A Comparative Study of Horse–Human Relationships in Chicago and Rural Illinois' 'Introduction' in *The Historical Animal*, edited by Susan Nance. New York: Syracuse University Press, 2015.

Pouillard, Violette. 'Animal Biographies: Beyond Archetypal Figures' *Journal of Animal Ethics* 12, no.2 (2022): 182–178.

Pouillard, Violette. "Visions of Concord: Wild animals and the Garden of the Revolution (Jardin des Plantes menagerie, 1793–c. 1820)." *Journal for the History of Environment and Society* 4 (2019): 11–40.

Regan, Paulette. *Unsettling the Settler Within: Indian Residential Schools, Truth Telling, and Reconciliation.* Toronto: UBC Press, 2010.

Ritvo, Harriet. "Recent Work in Animal History (and How We Got Here)." *The Journal of Modern History* 94, no.2 (2022): 404–419.

Roscher, Mieke, André Krebber, and Brett Mizelle, (eds). *Handbook of Historical Animal Studies.* Berlin: Walter de Gruyter GmbH & Co KG, 2021.

Safina, Carl. Beyond Words: *What Animals Think and Feel.* New York: Henry Holt and Company, 2015.

Safina, Carl. 'Animals Think and Feel: Précis of Beyond Words: What Animals Think and Feel (Safina 2015)'. *Animal Sentience* 1, no. 2 (2016), 1–3.

Saha, Jonathan. 'Colonizing Elephants: Animal Agency, Undead Capital and Imperial Science in British Burma', *BJHS Themes* 2 (2017), 169–189.

Simons, John. *Obaysch. A Hippopotamus in Victorian London.* Sydney: Sydney University Press, 2019.

Skabelund, Aaron. 'Animals and Imperialism: Recent Historiographical Trends', *History Compass* 11, no.10 (2013): 801–807.

Specht, Joshua. 'Animal History after Its Triumph: Unexpected Animals, Evolutionary Approaches, and the Animal Lens', *History Compass* 14, no. 7 (2016), 326–336.

Swart, Sandra. "'It Is as Bad To Be a Black Man's Animal As It Is To Be a Black Man' – The Politics of Species in Sol Plaatje's Native Life in South Africa." *Journal of Southern African Studies* 40, no.4 (2014): 689–705.

Stanton, Lauren, Matthew Sullivan, and Jilian Fazio. 'A Standardized Ethogram for the Felidae: A Tool for Behavioral Researchers'. *Applied Animal Behaviour Science* 173 (2015): 3–16.

Stibbe, Arran. 'As Charming as a Pig: The Discursive Construction of the Relationship Between Pigs and Humans'. *Society & Animals* 11, no. 4 (2003): 375–392.

Tønnessen, Morten. 'Making the Umwelt Bubble of the Modern Synthesis Burst'. *Biosemiotics* 14, no. 1 (2021): 121–125.

Walker, Brett. *The Lost Wolves of Japan.* Seattle: University of Washington Press, 2009.

Wathan, Jen, Anne M. Burrows, Bridget M. Waller, and Karen McComb. 'EquiFACS: The Equine Facial Action Coding System'. *PLOS ONE* 10, no. 8 (2015): 1–35.

Yong, Ed. *An Immense World: How Animal Senses Reveal the Hidden Realms Around Us.* London: The Bodley Head, 2022.

PART I

Animals as Experiencing Entities, Theories, and Perspectives

CHAPTER 2

Je Suis, Je Suis—I am, I Follow: Formation of Animal Individual and Cultural Selves

Teya Brooks Pribac

Introduction: Route 66

Route 66, originally built to connect Chicago and L.A., runs through a beautiful park of petrified trees, the Petrified Forest National Park. The park attracts visitors from far and wide, many of whom take samples of the unusual stone wood formations home as souvenirs. To preserve the integrity of the park such destructive behaviour had to be discouraged. In order to do so, and given that people readily ignored the fact that such practice had been illegal since 1906, the park management displayed signs advising the public that their "heritage is being vandalized every day by theft losses of petrified wood of 14 tons a year, mostly a small piece at a time."[1] Like most other people, the park managerial team was most likely unaware—at least until social psychologist Robert Cialdini later instructed them—that

[1] Robert Cialdini, "Crafting Normative Messages to Protect the Environment," *Current Directions in Psychological Science* 12, no. 4 (2003): 107.

T. Brooks Pribac (✉)
Faculty of Arts and Social Sciences, School of Languages and Cultures, The University of Sydney, Camperdown, NSW, Australia

M. J. Glover, L. Mitchell (eds.), *Animals as Experiencing Entities*, The Palgrave Macmillan Animal Ethics Series,
https://doi.org/10.1007/978-3-031-46456-0_2

if you want people to abstain from a certain activity it may not be a good idea to tell them that it is a popular activity that most people engage in. You may find, in fact, that people will choose what is popular rather than what is ethically (or otherwise) desirable. Cialdini replaced the Park's signs accordingly, achieving a marked reduction in theft.

Humans are quick to condemn sheep as "stupid" for being "followers" without recognising the extent to which our own being and behaviour are impregnated with social learning, imitation, and other forms of dependency on the social environment, influencing our affective and decision-making dispositions and capacities. In the French language the first-person singular form of the verbs *to be* and *to follow* is identical: *je suis* = I am; *je suis* = I follow. The etymological roots of the forms may be unrelated to the powerful semantic outcome, which has been adapted by many, including Jacques Derrida[2] and more recently by the Charlie Hebdo movement with its moto *Je suis Charlie*. Like the French verb form, our development is also marked by a close affinity between *being* and *following*. To be-come is to follow first. Rewinding back to the time of our birth, we find our organisms substantially underdeveloped but ready to continue development in relationality with our socio-natural environment, which influences our maturing brain, our psycho-biological regulation, as well as the emergence of a sense of self. "We first know ourselves as reflected in the other."[3] We begin acquiring this "knowledge" before our reflective consciousness allows us to become aware of ourselves conceptually, and we continue to fill our internal databases with information on ourselves and the rest of the world following templates that emerge on the way and are heavily impacted by the social fabric.

In this chapter I track our (animal) developmental path to illustrate that the ensuing complexity of the process of becoming, as convergent evidence indicates, belongs to all animals, not only to humans. That is to say that other animals possess complex, human-comparable emotional, cognitive, social, cultural, and even creative capacities and needs. The first part of the chapter focuses on animals' cultural selves. All animals are embedded, from the beginning, in a social network of conspecifics (members of the same species) and others who will significantly influence who we are,

[2] Jacques Derrida, *The Animal that Therefore I Am* (New York: Fordham University Press, 2008).

[3] Daniel J. Siegel, *Mindsight: The New Science of Personal Transformation* (New York: Bantam Books Trade Paperbacks, 2011[2010]), 62.

how we feel, think, and behave. Animals shape culture and culture shapes us. Human prejudice against nonhuman animals and their perceived inferiority is an example of human cultural transmission. Culture informs what we look for and, as a consequence, what we find. Once humans began more systematic and integrative explorations of other animals' lives we uncovered human comparability on multiple levels, including in the intimate domain. The latter is the subject of the second part of the chapter, in which I discuss the experience of grief in more detail. Grief is often considered a complex emotion that other animals cannot experience as "fully" as humans, but a closer examination of the issue shows that grief is essentially an organismic response to loss, and its intensity will vary at an individual rather than species level.

Procedural and structural commonalities of animal organisms enable us to infer across species but also across temporal lines and historical periods. Technologically we are better equipped for the exploration of these commonalities than our ancestors may have been. However, the primary impediment to a more integrated view of nonhuman animals has been not technology but the master narrative that has kept the human "suspended between a celestial and a terrestrial nature (…) always less and more than himself."[4] This ontological quandary, which had for a long time fuelled misconceptions of, and violence against, other animals, is the second, parallel focus of the chapter. As we become more comfortable accepting our animal bodies and minds, fear of those *petits récits* or little stories[5] that challenge dominant frameworks recedes, and narratives change.

Lost in the Wilderness: Culture, Counterculture, and Being Caught in Between

Henry, one of the four rescued sheep living with us, has never been able to stand up on his hind legs. This left Henry deprived of one of the joys the others readily engaged in, namely eating the juicy sprigs, petals, and leaves from surrounding trees—the higher the plants the better. One day Henry came up with a creative solution to overcome this disadvantage: starting with the rose bush and later applying it to others, Henry realised

[4] Giorgio Agamben, *The Open: Man and Animal* (Stanford: Stanford University Press, 2004), 29.

[5] Jean-François Lyotard, *The Postmodern Condition: A Report on Knowledge*, trans. Geoff Bennington and Brian Massumi (Manchester: Manchester University Press, 1984).

that if he managed to grab any part of the branch with his mouth, he could, with patience and care, release it from his mouth and slip it under his chin; with the branch secured between the chin and the neck, he could now slowly move along the branch and eat the leaves and petals on the way.

I first wrote about Henry's achievement[6] within a context of problematising the emotional impact upon a resourceful mind of a setting that deprives such a mind of exercising its own potential as is the case of animal minds trapped in most forms of industrial agriculture. Can such a mind happily be put to sleep until the time comes when the potential can effectively be used or does potential equal *need*, so that the creative mind without adequate stimuli will cause pain like a broken limb does or a broken "heart"? The more we learn about other animals' minds and their entire organism in normative versus captive environments, the more evident the rhetorical nature of this question becomes.[7] I was reminded of Henry's spectacularly resourceful mind—he introduced other creative solutions to everyday problems such as knocking on the door when he desires cuddles or treats—while reading Carol Gigliotti's book on animal creativity,[8] in which she demonstrates that not only is creativity found in other-than-human animals but that it is at the very foundation of animal survival and existence. More recently, Henry and the rose bush came to mind again as I was reading about social learning and the rich cultural lives across animal species. Perhaps the most intriguing aspect of Henry's story is the speed at which the other three sheep adopted these behaviours (the rose bush trick, the knocking on the door)—not because they are sheep but because the behaviours are useful and produce results. Sheep do indeed follow but in most cases only when and whom they choose to, just like other animals.

This "following," also known as social learning, takes the form of horizontal (group level, e.g. from Henry to his friends) or vertical/oblique (cross-generational, e.g. from parents/teachers to children) transmission

[6] Teya Brooks Pribac, "Positive Sentience is Underrated," *Animal Sentience* 2019.220. https://doi.org/10.51291/2377-7478.1445.

[7] For example Bob Jacobs, Heather Rally, Catherine Doyle, Lester O'Brien, Mackenzie Tennison, and Lori Marino, "Putative Neural Consequences of Captivity for Elephants and Cetaceans," *Reviews in the Neurosciences* 33, no. 4 (2021):439–65. https://doi.org/10.1515/revneuro-2021-0100; Amanda C. Kentner, Kelly G. Lambert, Anthony J. Hannan, and S. Tiffany Donaldson, "Editorial: Environmental Enrichment: Enhancing Neural Plasticity, Resilience, and Repair," *Frontiers in Behavioral Neuroscience* 13, no. 75 (2019). https://doi.org/10.3389/fnbeh.2019.00075.

[8] Carol Gigliotti, *The creative lives of animals* (New York: New York University Press, 2022).

of information that can lead to the formation of cultures. Culture is characterised by the dissemination of "behavioural traditions through social learning"[9] and thus constitutes a second inheritance system with the primary inheritance system describing the inheritance of genes informing innate behaviours.[10] Sheep—bighorn sheep—were the first ungulates to equip human scientists with empirical evidence of the culturally based nature of the herds' seasonal migrations. Jesmer and colleagues[11] compared "historical populations" (those inhabiting an area for over 200 years), with populations translocated to novel landscapes and discovered a marked difference in migratory propensity between the two groups (65–100% for the former and under 9% for the latter). The research unveiled the value of cultural knowledge transmission underscoring these migrations, which are both relevant for migrants and mesmerising for viewers. It takes several generations and decades for a translocated society to acquire enough knowledge of the new territory's food growth cycles to bring migration back to significant levels. Other studies have highlighted the importance of elders as repositories of ecological (and other) knowledge. For instance, discussing elephants, Foley, Pettorelli, and Foley[12] observed that during a drought elders may remember sources of water/food that they had been taken to by other elders in past droughts and that are unknown to the young. This culturally transmitted knowledge may mark the difference between life and death for a family or herd.

Sheep, elephants, and humans are not the only animals who possess culture, others—vertebrates and invertebrates—do too. "Culture is everywhere," advised Andrew Whiten, author of a paper that evidences the presence of culture across an increasing number of animal species.[13] As with many other aspects of animal existence (tool use, emotional,

[9] Andrew Whiten, "The Burgeoning Reach of Animal Culture," *Science* 372, no. 46 (2021), 1.

[10] Brett Jesmer, Jerod A. Merkle, Jacob R. Goheen, Ellen O. Aikens, Jeffrey L. Beck, Alyson B. Courtemanch, Mark A. Hurley, Douglas E. McWhirter, Hollie M. Miyasaki, Kevin L. Monteith, and Matthew J. Kauffman, "Evidence of Social Learning from Translocated Animals," *Science* 361, no. 6406 (2018): 1023–25.

[11] Jesmer et al., "Evidence of Social Learning."

[12] Charles Foley, Nathalie Pettorelli, and Lara Foley, "Severe Drought and Calf Survival in Elephants," *Biology Letters* 4 (2008): 541–4.

[13] Whiten, "The Burgeoning Reach"; Andrew Whiten, "Culture Is So Common that Even Fish and Flies Have It," interviewed by Michael La Page in *New Scientist*, April 1, 2021. https://www.newscientist.com/article/2273450-animal-culture-is-so-common-that-even-fish-and-flies-have-it/.

cognitive, and social skills, etc.), as soon as we start looking for a particular trait traditionally thought to be exclusively human, we quickly find it in other animals as well. Culture in the rest of the animal realm is one of the newest areas of interest, with compelling discoveries reflecting cross-species comparability and adding further support to the proposition that, overall, similarities across animal species are indeed more significant than existing differences. For instance, the phenomenon of "mate choice copying," whereby the attractiveness of a potential mate is determined not directly but under the influence of how my conspecifics perceive that subject, has been observed in humans[14] but also in fruit flies, guppies, and others.[15] This phenomenon falls under the broader umbrella of public information use, which is a staple in social species, and describes the time- (and sometimes life-[16]) saving capacity to evaluate a resource (mate, food, etc.) not by directly interacting with it but by monitoring others doing so.[17] Advertising functions because of these principles. In marketing this is called "social proof," which refers to the increased proclivity of consumers to purchase a good or service if they know others are doing so. Our sociocultural environment also helps animals learn where and how to acquire comestibles, how to get from A to B, who is friend and who is foe, which behaviours to avoid and which to adopt, and many other aspects that determine who we are as humans, parrots, cuttlefishes, and others, and how well (or not) we do in this world. Human cultures may possess characteristics with no direct parallels in other animals' cultures, but as Mary Midgley aptly reminded: "[W]e can do justice to the miracle of the [elephant] trunk without pretending that nobody else has a nose."[18] The reverse could also be true: other animals' cultures may contain

[14] D. Waynforth, "Mate Choice Copying in Humans," *Human Nature* 18, no. 3 (2007): 264–71. https://doi.org/10.1007/s12110-007-9004-2.

[15] Sabine Nöbel, Mélanie Allain, Guillaume Isabel, and Etienne Danchin, "Mate Copying in Drosophila Melanogaster Males," *Animal Behaviour* 141 (2018): 9–15. https://doi.org/10.1016/j.anbehav.2018.04.019; Galef G. Bennett, and Kevin N. Laland, "Social Learning in Animals: Empirical Studies and Theoretical Models," *BioScience* 55, no. 6 (2005): 489–99.

[16] For example vicarious evaluation of new foraging sites to avoid unnecessary exposure to potential predators.

[17] K.N. Laland, N. Atton, and M.M. Webster, "From Fish to Fashion: Experimental and Theoretical Insights into the Evolution of Culture," *Philosophical Transactions Royal Society B* 366, no. 1567 (2011): 958–68. https://doi.org/10.1098/rstb.2010.0328.

[18] Mary Midgley, *Beast and Man: The Roots of Human Nature* (London: Routledge, 2002 [1978]), 198.

astonishing, unique elements yet to be discovered by human researchers. Culture, like other elements of our animal existence, has emerged in the process of evolution, built out of available shared material—traits and the underlying neurobiological machinery do not appear out of nowhere, that is why if we have something other animals are likely to have it too in one form or another—and adapted to the multitude of planetary micro-universes resulting in the wealth of behaviours, senses, and perceptions that we are slowly uncovering.[19]

Technological advancement has aided discoveries in important ways. For example, molecular analysis has enabled scientists to determine the existence of pair bonds and shared parental care in shoals of fishes—an environment in which observations of physical proximity cannot be considered a reliable indicator.[20] Albatross pairs, on the other hand, are comparatively easier to discern; due to the partners' longstanding commitment to each other some conservative political figures proclaimed them as good role models for humans, only to discover later, following DNA analysis, that a third of these couples consist of two females rather than a female and a male as previously imagined.[21] In the past, a significant proportion of the information found in Ed Yong's book on the spectacular richness of animal senses could perhaps have been imagined; however, its existence and validity have only been possible to confirm more recently with the help of technology.

> *The Achilles' heel of culture (…) can be that individuals end up blindly following the cultural norm when it's no longer relevant. But compared with genetically based evolution, cultural evolution can be very nimble and fast, as long as some individuals do innovate.*—Andrew Whiten[22]

Technology aside, numerous other factors determine knowledge and perception. Each investigation begins with an idea; behind this idea is a human, a human embedded in a culture, a socio-natural space that—along

[19] For example Ed Yong, *An Immense World: How Animal Senses Reveal the Hidden Realm Around Us* (New York: Random House, 2022).

[20] Tetsumi Takahashi, Haruki Ochi, Masanori Kohda, and Michio Hori, "Invisible Pair Bonds Detected by Molecular Analyses," *Biology letters* 8, no. 3 (2012): 355–7. https://doi.org/10.1098/rsbl.2011.1006.

[21] Anne Innis Dagg, *Animal Friendships* (Cambridge: Cambridge University Press, 2011), 68.

[22] Whiten, "Culture Is So Common," n.p.

with ontogenetic evolution, personal preferences, cognitive capacities, and even gender[23]—will affect what and how is investigated and how the results are interpreted. For instance, in designing experiments to test non-human animals' capacities of various kinds, the researchers will choose the animal species they wish to test. This means that some other species does not get tested for that specific capacity, and unless someone else does it, there will be no evidence for that capacity in other species. Sheep and farmed animals generally, along with fishes, have historically been neglected as research subjects for complex cognitions and behaviours in the belief (or perhaps out of the *desire to believe*, given the economic and ideological investment in the commodification of these species) that such complexity does not exist in them. When researchers began moving away from what Frans de Waal[24] termed "primate chauvinism" (the assumption that only primates are complex beings) and started to look for complexity outside primate groups, they found it—in sheep, pigs, chickens, fishes, and many others.[25]

Technical problems can also complicate research and results. For instance, the success (or lack thereof) of a research project depends on aspects such as the researchers' capacity to communicate the research question in a way that enables the tested subjects to understand the problem to be resolved since in most cases one cannot rely on human language. Many of these methodological problems can be overcome or at least minimised if researchers are immersed in the world of the animal individual/species/group of interest. If one does not know that world and how it

[23] Kevin Omland, Evangeline Rose, and Karan Odom, "Women Have Disrupted Research on Bird Song, and Their Findings Show How Diversity Can Improve All Fields of Science," *The Conversation,* September 11, 2020. https://theconversation.com/women-have-disrupted-research-on-bird-song-and-their-findings-show-how-diversity-can-improve-all-fields-of-science-142874.

[24] Quoted in Alison Abbott, "Animal Behaviour: Inside the Cunning, Caring and Greedy Minds of Fish," *Nature* 521, no. 7553 (2015): 412–4. https://doi.org/10.1038/521412a.

[25] For example Vinciane Despret, "Sheep Do Have Opinions," in *Making Things Public: Atmospheres of Democracy*, ed. Bruno Latour and Peter Weibel (Cambridge: MIT Press, 2005), 360–70; Jonathan Balcombe, *What a Fish Knows: The Inner Lives of Our Underwater Cousins* (London: Oneworld Publications, 2016); Lori Marino, and Christina M. Colvin, "Thinking Pigs: A Comparative Review of Cognition, Emotion, and Personality in *Sus Domesticus*," *International Journal of Comparative Psychology* 28, no. 1 (2015). https://doi.org/10.46867/ijcp.2015.28.00.04; Lori Marino, "Thinking Chickens: A Review of Cognition, Emotion, and Behaviour in the Domestic Chicken," *Animal Cognition* 20, no. 2 (2017): 127–47.

works, fish expert Culum Brown[26] noted, one is likely to miss things or misunderstand them. This idea finds strong support among students of other animal species. Over the past twenty years in particular, investigations into nonhuman animals' lives have become increasingly more holistic. It is now widely accepted that animals live intricate lives that are heavily influenced by the developmental and socio-natural environment more broadly. Old-fashioned labs and zoos do not offer the type of normativity necessary for the animals to express themselves and their potentials. As a consequence, human understanding of the nonhuman subjects' intimate and social lives, if relying on research from those environments, will also be compromised. When research is conducted in captivity such normativity should be approximated as much as possible.

Konrad Lorenz and other early ethologists at the beginning of the twentieth century were well aware of the value of normativity and they chose to study animals in their natural environments. Their purpose was to learn about other animals in order to dispel myths stemming from centuries of folklore and superstition as well as to establish ethology as a science. To this end, these early researchers limited themselves to record what they saw and abstained from interpreting intentions behind the observed behaviour. This was a useful technique for that particular purpose, as Safina notes,[27] but rapidly this methodology turned into an ideology; animal subjectivity became a taboo topic, banned from scientific and philosophical discourse. Implying that nonhuman animals may have mental states could cost people jobs and reputation; young researchers were warned to abstain from such deliberations.[28] Some took their chances, including Donald Griffin whose 1984 book *Animal Thinking* was labelled by a critic as "the satanic verses of animal behaviour."[29] Others conformed, openly or tacitly promoting a culture of violence against nonhuman animals and helping to consolidate the myth of animals as instinctual automata. If there is no scientific evidence that nonhuman animals have subjective lives (and how could there be evidence if no one is looking for it?), animal subjectivity does not exist. This has informed most human societies' *modus*

[26] Culum Brown, "Are Fish Far More Intelligent than We Realize?" interview by Joseph Stromber, *Vox*, August 4, 2014. https://bit.ly/3lrkIan.

[27] Carl Safina, *Beyond Words: What Animals Think and Feel* (New York: Henry Holt and Company, 2015).

[28] Sonja Yoerg, "Mentalist Imputations," *Science* 258, no. 5083 (1992): 830–1.

[29] Donald R. Griffin, *Animal Minds: Beyond Cognition to Consciousness* (Chicago: University of Chicago Press, 2001 [1992]), x.

operandi regardless of many humans' divergent intimate convictions of animal subjectivity.

While nonhuman animals' sentience is now a widely accepted reality and other aspects of human-nonhuman comparability are being progressively recognised, including culture, as discussed in the previous sections, in practice, in many areas, nonhuman animals continue to be treated as automatons. Clearly, one such area is animal agriculture, in which animals remain virtually without protection.[30] Another area is "wildlife" rehabilitation and orphan rearing. The fear of "humanising" free-living nonhuman animals, that is allowing them to become used to humans, which could potentially endanger both human and nonhuman life, has led to models of care based on detachment, whereby physical interaction between a rescued nonhuman animal and the human caregiver is largely limited to feeding, medical aid, and cage/enclosure cleaning. Furthermore, a substantial proportion of rehabilitation groups and organisations practice so-called hard release, a method that involves releasing an animal abruptly into an environment without allowing the animal the opportunity of prior familiarisation with the new place. The alternative to this method is known as "soft release," practiced commonly in kangaroo rehabilitation[31] and consisting in releasing the animals gradually into progressively larger pre-release enclosures bordering on the habitat that will eventually become their home. This enables animals to become acquainted with the new space, its sounds, smells, food sources, shelter opportunities, local social life, and other vital elements. The valuation of appropriateness of the hard release method is based on the assumption that a "wild animal" will know what to do in the "wild." However, cumulative evidence suggests that there is no such thing as "wild" animal or "wild" habitat: both animal and habitat are in their own ways cultural entities that need to learn or be learnt about. "Habitat quality," observes *Science Daily*,[32] summarising the previously mentioned research in the culture of bighorn sheep, "is best

[30] Elizabeth Ellis, *Australian Animal Law: Context and Critique* (Sydney: Sydney University Press, 2022).

[31] Steve Garlick, and Rosemary Austen, "Kangaroo Translocation: Program Efficiency and Welfare Goals" (Australian Wildlife Rehabilitation Conference, Adelaide, South Australia, 2010); Steve Garlick, and Rosemary Austen, "Learning about the Emotional Lives of Kangaroos: Cognitive Justice and Environmental Sustainability," *Relations* 2, no. 1 (2014). https://doi.org/10.7358/rela-2014-001-garl.

[32] University of Wyoming, "Big Game Animals Must Learn to Migrate and Pass Knowledge Across Generations," *Science Daily*, September 6, 2018, n.p. www.sciencedaily.com/releases/2018/09/180906141626.htm.

imagined as the physical landscape that animals occupy, in combination with the knowledge they have accumulated about how to use that landscape." Absence of such knowledge and gradual integration will inevitably negatively impact the released animals regardless of the wealth of potential food and other resources in that environment. It may also affect their cultures and ultimately conservation efforts more broadly, as recognised also by the United Nations Environment Programme, which has already made some relevant recommendations towards ethically and scientifically compliant approaches.[33]

Lost in the wilderness of nature or in the wilderness of farms, zoos, and labs, nonhuman animals may have always represented a powerful symbolic force and multileveled resource to the human animal, but they have rarely been considered for who they are or could be—other nations who "move finished and complete, gifted with the extension of the senses we have lost or never attained, living by voices we shall never hear," as Beston poeticised,[34] "caught with ourselves in the net of life and time, fellow prisoners of the splendour and travail of the earth." Nevertheless, throughout human history there have been innovators (from Pythagoras, within the Western context alone, to the nineteenth-century vegetarians to present-day intellectual and hands-on leaders) who have found these animals abandoned in the sometimes scary wild of the human mind and nurtured them, conceptually, back to a holistic self, challenging master narratives of oversimplification and oppression[35] and setting the ground for the current flourishing of research and knowledge about animals' minds, experiences, subjectivities, cultures, and learnings.

The Path Inward: Attachment Relationships and Grief

The detachment approach in conventional "wildlife" rehabilitation, discussed earlier, is particularly problematic when it is applied to orphan rearing. Infant development follows a species-specific normative process that

[33] Whiten "The Burgeoning Reach."

[34] Henry Beston, *The Outermost House* (New York: Henry Holt and Company, Inc., 1949 [1928]), 25.

[35] Norm Phelps, *The Longest Struggle: Animals Advocacy from Pythagoras to PETA* (New York: Lantern Books, 2007); Howard Williams, *The Ethics of Diet: An Anthology of Vegetarian Thought* (Guildford: White Crow Books, 2009).

human caregivers raising free-living (or any other animal) infants need to emulate as closely as possible for the in-care's optimal and long-term well-being. The fact that the infant is ultimately raised to be released into a community of conspecifics and that care should be taken to not "human-ise" them does not justify depriving them of the relational intimacy they need in this crucial formative process. Intimate transactions between the infant and caregiver (parents or surrogate) have equal biological value compared to alimentation and physical protection. In fact, when animals are born (or hatched) our brain and our organismic self-regulation are not fully formed yet, and they continue to develop in relation with our care-givers.[36] The quality of these relations determines the organisation of our maturing systems and ensuing psycho-physiological vulnerability or resilience.

While other animals' conceptual worlds—their thoughts, interpretations, etc.—will to a great extent always remain a mystery, their emotional life is more predictable because of shared underlying organismic structures and processes. A substantial amount of knowledge about and understanding of human attachment and loss (among other things) has been obtained through invasive experiments in other animals. Following the proclamation of the 1990s as "the decade of the brain" by then U.S. President George Bush, brain research flourished.[37] This research has revealed (and continues to reveal) brain comparability across species. Even when brains look substantially different from one another (most likely a result of adaptation to different living environments) evidence emerges of the existence of functional parallels. For example, avian brains lack the neocortex characteristic of mammalian brains. As a consequence, birds had been considered cognitively and otherwise unimpressive—the term "birdbrain" continues to be used by naïve individuals as an insult for someone with (presumed) diminished intellectual capacities. Cumulative research, along with anecdotal evidence, shows that birds are far from intellectually unsophisticated and their brains contain structures equivalent to the

[36] This has been demonstrated in mammals and birds, but emerging evidence indicates the presence and importance of attachment relations in other species, including reptiles, see Melissa Amarello, Jeffrey J. Smith, and J. Slone, "Family Values: Rattlesnake Parental Care Is More than Just Attendance" (Biology of the Rattlesnakes Symposium, Tucson, Arizona, 2011. https://www.snakes.ngo/ABSsm.pdf).

[37] Murray Goldstein, "Decade of the Brain—An Agenda for the Nineties," *Western Journal of Medicine* 161, no. 3 (1994): 239–41.

mammalian neocortex.[38] Furthermore, the shape of the neocortex itself differs from one mammalian species to another.[39] These differences have also fuelled human depreciation and oppression of other mammalian species in the belief that differences in neocortex shape automatically translate into differences in degrees of thinking as well as *feeling*. In fact, as neuroscientist Jaak Panksepp explained, traditionally the neocortex was believed to be not only

> the seat of conscious thought, but also of emotional feelings […] they [emotional feelings] were commonly deemed to be a form of thought, and affective and cognitive processes were envisioned to be completely interpenetrant in higher brain regions that generated certain higher cognitive processes such as frontal cortical regions.[40]

If one believes that a human-like neocortex is necessary not only to produce a literary or scientific masterpiece (which most humans, it is worth reminding, do not achieve) but also to be able to experience feelings, one can envision that certain mechanistically minded humans may have doubted that other animals' emotional lives could compare to those of humans or even that nonhuman animal emotional lives existed in the first place. Over the past decades the neocortex has lost its primacy, and the seat of phenomenological consciousness has been moved further down to the brain's evolutionary older parts, which are more similar across species. In other words, the weight of evidence indicates that affective states are already experienced in these ancient brain regions and they do not need to be "converted" into experience by evolutionary newer regions, such as the neocortex. The Cambridge Declaration on Consciousness[41] is a testimony to this shift.

[38] Martin Stacho, Christina Herold, Noemi Rook, Hermann Wagner, Markus Axer, Katrin Amunts, and Onur Güntürkün, "A Cortex-Like Canonical Circuit in the Avian Forebrain," *Science* 369, no. 6511 (2020). https://doi.org/10.1126/science.abc5534.

[39] For a comparative image, see brainmuseum.org.

[40] Jaak Panksepp, "Cross-Species Affective Neuroscience Decoding of the Primal Affective Experiences of Humans and Related Animals," *PLOS ONE* 6, no. 8(2011): e21236. https://doi.org/10.1371/journal.pone.0021236.

[41] Philip Low, Jaak Panksepp, Diana Reiss, David Edelman, Bruno Van Swinderen, and Christof Koch, "The Cambridge Declaration on Consciousness," Cambridge (UK), July 7, 2012. http://fcmconference.org/img/CambridgeDeclarationOnConsciousness.pdf.

While brain research has caused immense suffering to innumerous nonhuman animals, it has also revealed cross-species structural and functional comparability that enables us to appreciate the existence of phenomenological consciousness in other animals and in certain cases predict with moderate accuracy their subjective states. Grief is one of such states. Grief is often viewed as a complex emotion requiring substantial cognitive power for its generating and processing. In reality, as cumulative evidence suggests, grief is better viewed as an organismic response to loss. Cognitive elaboration can colour the experience in certain ways but the reason we can experience grief is *because* we are animals not *despite* the fact that we are animals—because we have an animal body and an animal brain, and they both develop in relation. We need relations the way we need food and water. To appreciate the potential depth of the experience of grief for humans and other animals, it helps to possess an adequate understanding of the significance and power of attachment relations.

Konrad Lorenz, mentioned previously, has over time inspired numerous individuals and their works, including John Bowlby who, with the help of the empirical undertakings of his colleague Mary Ainsworth, originated attachment theory.[42] In the 1950s when Bowlby began work in the area of infant-caregiver separation it was already recognised that absence of intimacy had a negative impact on the infants but the mechanisms behind this phenomenon were not known. Socially deprived human children (e.g. in orphanages) were not thriving emotionally and cognitively even when they received high-quality physical care and good nutrition. Harry Harlow[43] conducted invasive experiments on rhesus monkeys which provided further evidence of negative impacts of early social deprivation. These experiments also showed that deprivation of intimacy (and in some cases direct abuse forced upon nonhuman animals by the human experimenters) did not only affect the babies themselves—the direct recipients of this abuse or neglect—but it also transferred across generations, to their own offspring. For example, the mothers who were deprived of normative and loving care when they were infants, later became themselves negligent

[42] John Bowlby, *Attachment and Loss: Volume I: Attachment* (New York: Basic Books, 1982 [1969]); John Bowlby, *Attachment and Loss: Volume II: Separation Anxiety and Anger* (New York: Basic Books, 1973); John Bowlby, *Attachment and Loss: Volume III: Loss, Sadness and Depression* (New York: Basic Books, 1980).

[43] Harry F. Harlow, "The Nature of Love," *American Psychologist* 13(1958): 673–85.

or abusive mothers. However, if biological infants of abusive mothers are placed in foster care with a loving foster mother, these infants when they grow up will not be abusive to their children. Their own so-called mothering style will resemble that of the foster mother and not of the biological mother.[44] All this indicated to Bowlby that attachment relations and social bonds play a far more critical role than it had been imagined, and his discovery of Lorenz and the phenomenon of "imprinting" in precocial avian species lead to a conceptual breakthrough.

Precocial animals (e.g. ducks and sheep) are mobile at birth and are able to feed autonomously; they are not as dependent on the parents as altricial animals are (e.g. humans and rats). Nevertheless, even in precocial species—geese, in Lorenz's investigations—there was an obvious innate tendency to attach ("imprint"). This enabled Bowlby to envision the possibility that attachments may work in a similar fashion also in mammals (including humans): that both the infant and the caregiver are predisposed to form these attachments and that these attachments have serious biological value past nutrition and physical protection. Research over the following decades has confirmed Bowlby's supposition and it has also provided further evidence for the transgenerational effects mentioned earlier for humans and other animals.[45]

These investigations have unveiled, among other things, that the primary caregiver (usually the mother but it can be someone else, including members of different species) plays a crucial role in the development of the

[44] Harry F. Harlow, and Margaret Kuenne Harlow, "Social Deprivation in Monkeys," *Scientific American* 5 (1962): 136–46; Harry F. Harlow, Robert O. Dodsworth, and Margaret K. Harlow, "Total Social Isolation in Monkeys," *Proceedings of the National Academy of Sciences USA* 54, no. 1 (1965): 90–7.

[45] Dong Liu, Josie Diorio, Jamie C. Day, Darlene D. Francis, and Michael J. Meaney, "Maternal Care, Hippocampal Synaptogenesis and Cognitive Development in Rats," *Nature Neuroscience* 3, no. 8 (2000): 799–806; Michael J. Meaney, "Maternal Care, Gene Expression, and the Transmission of Individual Differences in Stress Reactivity across Generations," *Annual Review of Neuroscience* 24 (2001): 1161–92; Ian C.G. Weaver, Nadia Cervoni, Frances A. Champagne, Ana C. D'Alessio, Shakti Sharma, Jonathan R. Seckl, Sergiy Dymov, Moshe Szyf, and Michael J. Meaney, "Epigenetic Programming by Maternal Behaviour," *Nature Neuroscience* 7, no. 8 (2004): 847–54; Robert M. Sapolsky, "Mothering Style and Methylation," *Nature Neuroscience* 7, no. 8 (2004): 791–2; Stephen J. Suomi, "Attachment in Rhesus Monkeys," in *Handbook of Attachment: Theory, Research, and Clinical Applications*, ed. Jude Cassidy and Phillip R. Shaver, 3rd ed. (New York: The Guilford Press, 2016), 133–54.

infant's brain and psycho-biological regulation. Infant animals are not only dependent on the caregiver for basic physiological needs, such as food and physical protection; in infancy the brain and psycho-biological self-regulation are still developing. It is difficult to emphasise enough how important these early life relational transactions are for the wellbeing of an animal both in babyhood and throughout life, and that this holds true for all mammals and birds (and possibly wider), past and present, including those animals who routinely slip through the gaps of master narratives to suffer and die in the convenience of oblivion (farm animals, labour animals, experimental animals, animals persecuted as "pests," and others).

The brain and the patterns it forms are dependent on experience. What animals experience in these early stages of life will influence how our brain develops and gets "wired." The caregiver is a source of stimuli to which the infant responds; simultaneously the caregiver is also a superstructure[46] that guides the infant's responses. As the infant follows the gaze, the sound, the smell of the caregiver, the caregiver essentially functions as an external regulator of the infant's internal states.[47] Ideally, the caregiver will be able to soothe the baby animal if the baby is upset, and bolster the infants' positive states. If as an infant I experience a negative arousal state and my caregiver successfully regulates this state, it "teaches" me that negative states can be dealt with—they can be tolerated and regulated.[48] This "knowledge" does not get stored in my autobiographical memory as my brain is still developing this capacity; instead it goes directly into the rest of my system. Traumatic experiences can be particularly powerful and leave "emotional systems sensitized or desensitized, with permanent, epigenetically induced high-stress reactivity and excessive primary-process negativistic feeling."[49] In the long run, successful regulation leads to adap-

[46] Jonathan H. Polan, and Myron A. Hofer, "Psychobiological Origins of Infant Attachment and Its Role in Development," in *Handbook of Attachment: Theory, Research, and Clinical Applications*, ed. Jude Cassidy and Phillip R. Shaver, 3rd ed. (New York: The Guilford Press, 2016), 117–32.

[47] Allan N. Schore, "Attachment, Affect Regulation, and the Developing Right Brain: Linking Developmental Neuroscience to Pediatrics," *Pediatrics in Review* 26, no. 6 (2005): 204–17; Allan N. Schore, *The Science of the Art of Psychotherapy* (New York: W.W. Norton & Company, 2012).

[48] Schore, "Attachment."

[49] Jaak Panksepp, and Lucy Biven, *The Archaeology of Mind: Neuroevolutionary Origins of Human Emotions* (New York: W.W. Norton, 2012), 434.

tive physical and mental health, and vice versa: if the caregiver cannot adequately regulate and help the infant overcome the negative state, the infant learns that regulation is difficult or impossible. The success of such regulation, or lack thereof, will affect the recipient's *internal* system (e.g. levels of anxiety) but also how the individual is or becomes in relation to the *outside* world (e.g. social competencies).

The shaping of animals' autonomic nervous system represents a powerful example of attachment relations at work. The autonomic nervous system is the part of the nervous system that supplies internal organs (heart, lungs, bladder, kidneys, liver, etc.) and regulates certain body processes (blood pressure, breathing rates, digestion, etc.) automatically (autonomously), without our intentional help. It usefully helps us into a state of alert in a dangerous situation and prepares the organism to deal with such situation (e.g. heart rate increases, growth and other situational non-essentials are put on hold, etc.), and it helps us relax after the danger has passed. The developmental context affects individual differences in autonomic response to stressful events and situations. If we grow up in an adverse environment (e.g. if we are subjected to abuse or even neglect), our organism can become sensitised to stressors: it can become hyper-reactive and less efficient in dealing with challenging circumstances. The sympathetic branch (the fight-flight-freeze response) can become dominant, and that can lead to animals that are hypervigilant, anxious, aggressive, and have various physical and psychological issues; among other aspects it can affect animals' immunological resilience and make us more prone to inflammation later in life.[50] On the contrary, positive relational

[50] Katie A. McLaughlin, Margaret A. Sheridan, Sonia Alves, and Wendy Berry Mendes, "Child Maltreatment and Autonomic Nervous System Reactivity: Identifying Dysregulated Stress Reactivity Patterns Using the Biopsychosocial Model of Challenge and Threat," *Psychosomatic Medicine* 76, no. 7 (2014). https://doi.org/10.1097/PSY.0000000000000098; Manuela Gander, and Anna Buchheim, "Attachment Classification, Psychophysiology and Frontal EEG Asymmetry across the Lifespan: A Review," *Frontiers in Human Neuroscience* 9, no. 79 (2015). https://doi.org/10.3389/fnhum.2015.00079; Katherine B. Ehrlich, Gregory E. Miller, Jason D. Jones, and Jude Cassidy, "Attachment and Psychoneuroimmunology," in *Handbook of Attachment: Theory, Research, and Clinical Applications,* ed. Jude Cassidy and Phillip R. Shaver, 3rd ed. (New York: The Guilford Press, 2016), 180–201; Abbey W. Alkon, Thomas Boyce, Torsten B. Neilands, and Brenda Eskenazi, "Children's Autonomic Nervous System Reactivity Moderates the Relations Between Family Adversity and Sleep Problems in Latino 5-Year Old in the CHAMACOS Study," *Frontiers in Public Health* 5, no. 155 (2017). https://doi.org/10.3389/fpubh.2017.00155.

transmissions can reduce stress-related activity in this and related systems with beneficial outcomes.[51]

The infant's primary aim in the initial period of life is to establish emotional communication with the caregiver so that the infant can continue the development of these essentials. This innate need for connection is very strong, and in these early stages of animals' lives avoidance learning is inhibited and preference learning is dominant. This enables young animals to become attached to both loving and abusive caregivers. One of the problems with attachment to an abusive caregiver is that it makes animals vulnerable to abuse later in life. Polan and Hofer[52] discuss this based on research in rats but it is also applicable to other species, naturally including humans. They explain: "[E]arly maternal maltreatment is a double edge sword: one edge induces adult depression, whereas the other provides a safety signal that blunts fear learning and relieves the depressive behaviours."[53]

Based on our early experience in the world, animals form so-called *internal working models*, that is mental representations of the self, the socio-natural environment, and the self in relation to this environment. In this period we also develop attachment styles: we can develop a secure attachment style and three forms of insecure attachment styles: anxious, avoidant, and disorganised. The insecure anxious attachment style will be discussed in more detail later as it is more likely to lead to complicated grief.

Essentially, the caregiver's responses provide the infant with critical information about the self that the infant will adopt and internalise. Through this process animals learn, for example, whether we are worthy of love or not. Arguably, everyone is worthy of love, but that is not how the infant (and then the infant-turned adult) will feel if they get signals from the caregivers to the contrary, as Bowlby already observed.[54] This is what Siegel's observation about becoming to know ourselves as reflections in others' minds/behaviours, cited at the beginning of this chapter, means. The "self": in this context is the *experiential/relational self* not the *reflective self*;[55] it is not the self that *thinks* the self and watches itself in the mir-

[51] James A. Coan, "Toward a Neuroscience of Attachment," in *Handbook of Attachment: Theory, Research, and Clinical Applications*, ed. Jude Cassidy and Phillip R. Shaver, 3rd ed. (New York: The Guilford Press, 2016), 242–69.

[52] Polan and Hofer, "Psychobiological Origins."

[53] Polan and Hofer, "Psychobiological Origins," 123.

[54] Bowlby, *Attachment and Loss III*, 204–5.

[55] For example Cynthia Willett, "Can a Mirror Capture the Self?" *Animal Sentience* 25, no. 36 (2019). https://doi.org/10.51291/2377-7478.1467.

ror or reflected in a pool of water, it is the self that *feels* the self and acts according to these feelings and affective templates. It is a self that perhaps has different *kinds* but does not have *degrees*. It feels itself fully, in its entirety, whether it is a human self, a donkey self, or some other animal self.

Like infants, adult animals also form attachment relations, and these relations also have a regulatory function. An optimal dynamic will influence the psycho-biological regulation in a positive way, maximising positive arousal states and minimising negative ones, preventing under-stimulation or overstimulation, and, generally, keeping animals in an optimal arousal zone.[56] Attachment relations continue to influence and sustain animals on multiple psycho-physiological levels, which are affected when loss occurs and the animal is left without a co-regulator. Depending on the degree of intertwinement of these levels within an attachment dyad (i.e. how dependent I am, on the various levels, on my attachment figure such as a partner), following loss, a cumulative effect can occur bringing the organism to collapse and giving rise to complicated forms of grief, even death—indeed bereaved individuals are at high risk of mortality, including suicide.[57]

In front of the Shibuya Station in Tokyo, Japan, stands the statue of Hachiko, a dog who lived between 1923 and 1935 and has earned international fame for his presumed loyalty to his human companion, professor Ueno.[58] Ueno adopted Hachiko as a pup, the story goes, and the two were very close. Hachiko accompanied Ueno to the train station every morning; after a day of lectures and other university duties, Ueno returned to find Hachiko waiting for him, every day at 3pm in the same spot. One day Ueno died at work, and as a consequence never returned but Hachiko kept visiting the station every afternoon in the hope, it is believed, that Ueno eventually would return.

[56] Tiffany Field, "Romantic Breakups, Heartbreak and Bereavement," *Psychology* 2, no. 4 (2011): 382–7.

[57] J. Kaprio, M. Koskenvuo, and H. Rita, "Mortality after Bereavement: A Prospective Study of 95,647 Widowed Persons," *American Journal of Public Health* 77, no. 3 (1987): 283–7; Annette Erlangsen, Bernard Jeune, Unni Bille-Hrahe, and James W. Vaupel, "Loss of a Partner and Suicide Risks among Oldest Old: A Population-Based Register Study," *Age and Ageing* 33, no. 4 (2004): 378–83.

[58] Ignatius Koh, "Hachiko Statue in Shibuya," *Japan Travel*, June 23, 2019. https://en.japantravel.com/tokyo/hachiko-statue-in-shibuya/44644.

Technically Hachiko was not grieving because having not been given the opportunity to see or smell Ueno's body and determine his death, all along he expected his human companion to return. Nevertheless, from a grief studies perspective this story is relevant due to aspects paralleling complicated grief. Animals can experience "normal" grief, that is. grief that is acute for a while and then starts to recede, and we can experience complicated (also known as complex, prolonged, unresolved, or chronic grief), which is a form of grief that remains acute for prolonged periods of time, the bereaved animal literally becomes emotionally stuck. One can know very well that the beloved person is dead but this knowledge does not help overcome the powerful organismic response to the loss experienced. There is reason to believe that nonhuman animals, especially those in captivity, including companion animals, may be particularly vulnerable to developing complex grief. In fact, various situations or conditions make one more prone to complex grief.

One such condition is the experience of merged identities,[59] that is when the self is perceived and lived as part of a duo as opposed to independently or interdependently. Animals can develop merged identities with others—human or nonhuman. It is not uncommon in human couples who have been together for a long time and have failed to spread their emotional investment more broadly, outside the dyad. Dogs whose only companion and attachment figure is their human guardian are likely to develop merged identities, as probably occurred in Hachiko's case, and many other dogs that had been ascribed suicidal tendencies.[60] Conversely, if Hachiko had died before Ueno, Ueno himself could have developed complicated grief—human society can be unjustifiably dismissive of humans' love for their nonhuman companions.[61] Other captive animals may also be vulnerable to unresolved grief, including in sanctuaries. These

[59] Chris R. Fraley, and Phillip R. Shaver, "Attachment, Loss, and Grief: Bowlby's Views, New Developments, and Current Controversies," in *Handbook of Attachment: Theory, Research, and Clinical Applications,* ed. Jude Cassidy and Phillip R. Shaver, 3rd ed. (New York: The Guilford Press, 2016), 58.

[60] For example Antonio Preti, "Suicide among Animals: A Review of Evidence," *Psychological Reports* 101 (2007): 831–48.

[61] Kelly Oliver, "Pet Lovers, Pathologized," *New York Times,* October 30, 2011. https://archive.nytimes.com/opinionator.blogs.nytimes.com/2011/10/30/the-pathology-of-dependence-on-animals/; Teya Brooks Pribac, "Grieving at a Distance," in *Mourning Animals,* ed. Margo DeMello (East Lansing: Michigan State University Press, 2016), 193–9; Les Mitchell, *Reading the Animal Text in the Landscape of the Damned* (Grahamstown: NISC, 2019).

animals usually come from unfavourable environments where they would likely have been abused physically and psychologically; they may have difficulties trusting others and forming social bonds; if they eventually find security in someone (human or other) and this person dies, the loss can be traumatic.

Another condition that favours complex grief is attachment insecurity. Captive animals do not often experience the privilege of a normative species-specific upbringing that would promote security of attachment. If they are marked by insecure anxious attachment, as mentioned earlier, they are particularly prone to developing complex grief.[62] The insecure anxious style (also known as ambivalent) develops when the caregiver is anxious and unpredictable in their response: sometimes they are loving, other times harsh, sometimes too engaged, other times distant. The infant is not entirely sure what the caregiver's response will be; whether the caregiver is going to be available or not; the caregiver does not feel entirely reliable. As adults these individuals will likely be anxious and over-preoccupied with the availability of the attachment partner, they will fear abandonment and be chronically vigilant about separation, and they can become obsessed with the partner or attachment figure and show extreme jealousy.[63] Essentially, the brain becomes "wired" for clinging. The brain becomes organised in such a way that it chronically searches for cues that will signal the availability of the attachment figure and quickly detects cues of unavailability.[64] With such brain organisation, in the event of loss (through death or other modes of separation) the brain keeps clinging and the reflective mind has very little power over what the experiential self is doing or feeling.

Brain scans in human experimental subjects have shown that reminders of deceased loved ones activate brain regions linked to physical and social pain processing in both individuals suffering from non-complicated grief and those suffering from complicated grief, but in individuals with complicated grief along with pain regions *reward* centres also activate. These are structures and processes that we share with other animals. O'Connor and colleagues elaborate:

[62] Fraley and Shaver, "Attachment, Loss and Grief," 44–5.

[63] Linda Graham, "The Neuroscience of Attachment" (paper, Clinical Conversation, Community Institute for Psychotherapy, San Rafael, CA, 2008). https://bit.ly/36I7S1H.

[64] Fraley and Shaver, "Attachment, Loss and Grief."

> The addiction-relevant aspect of this neural response may help to explain why it is hard to resist engaging in pleasurable reveries about the deceased even though engaging in these reveries may prevent those with CG [complicated grief] from adjusting to the realities of the present. Many who suffer from addiction-like disorders experience them as afflictions; similarly we are not suggesting that reveries about the deceased are emotionally satisfying, but rather may serve as *craving responses*.[65] (*italics added*)

If the obsessive cycle is not broken, it can severely impact upon the individual's long-term psychological and physiological wellbeing.

This theoretical base facilitates cross-species discussions of grief as a personal experience and helps clarify how differences in the experience of grief manifest on an individual not species level, at least when it comes to mammals and birds. My grief will be different from the reader's grief or from a sheep's grief because we are individuals with different psychological constitutions, because the nature of the lost relationship was different, and because of other subjective and circumstantial factors, but not because I'm a human animal and someone else is another animal.

Other aspects of interest include various philosophical questions concerning perceptions of death and mortality as well as modes of mourning. Commentators often uncritically presuppose a human universality[66] in grief- and death-related matters that a closer examination proves non-existent.

To begin with, it is useful to make a distinction between grief as emotion and mourning as expression of such emotion. Mourning can occur without grief (e.g. ritualistic mourning in certain human societies) and vice versa; one can experience grief without external manifestation—a common occurrence in human societies with high mortality and little socio-economic support for the bereaved.[67] Most other animals also live in environments saturated with violence and persecution (increasingly of an anthropogenic nature) where showing weakness can increase danger. In

[65] Mary-Frances O'Connor, David K. Wellisch, Annette L. Stanton, Naomi I. Eisenberger, Michael R. Irwin, and Matthew D. Lieberman, "Craving Love? Enduring Grief Activates Brain's Reward Center," *NeuroImage* 42, no. 2 (2008): 972.

[66] For example Barbara J. King, *How Animals Grieve* (Chicago: Chicago University Press, 2013).

[67] For example Nancy Scheper-Hughes, *Death without Weeping: The Violence of Everyday Life in Brazil* (Berkeley: University of California Press, 1992); Dag Ø. Nordanger, "Coping With Loss and Bereavement in Post-War Tigray, Ethiopia," *Transcultural Psychiatry* 44, no. 4 (2007): 545–65.

some ways mourning can be perceived as a privilege. Furthermore, human expression of grief, when present, varies substantially across cultures—some weep, some laugh, some dance, others stand stoically seemingly emotionless.[68] Similarly, other animals may mourn in ways that humans do not recognise as an expression of grief. While death is all too common in the nonhuman world, humans do not often witness it firsthand and have the opportunity to observe other animals' reaction to the death of family members and friends (or the death happens in settings, such as farms and slaughterhouses, where humans do not prioritise these aspects). With the exception of companion animals, humans are not close enough to most animals—both physically and psychologically—to be able to detect nuances that may be indicative of grief. Unless the expression of grief is very emphatic and of a kind that is familiar to us, we can easily miss it, or attribute it to some other feeling.

In discussions of nonhuman animal grief the questions of understanding death and our own mortality also surface. These are certainly interesting questions and individuals' answers may colour the personal grief experience in certain ways. However, human-like understandings of these phenomena—and it is critical to use the term in the plural form because there is no universal understanding of these issues within the human species itself—are not a prerequisite for the experience of grief and they do not determine intensity of experience either. Nonhuman animals' experiences may indeed be more intense because unlike many humans they probably cannot find comfort in religious imagination and concepts of afterlives. Furthermore, concerning the question of one's mortality and the related fear of death some humans harbour, when a loved one dies, rather than worrying about their own mortality the bereaved oftentimes find it hard to carry on with their own lives, especially when complicated forms of grief develop, as noted earlier. Nevertheless, distinguishing between a living and a dead organism is an advantage so fundamental that it is hard to imagine other animals not being gifted with it. For instance, being able to tell a dead lion from a sleeping lion if I am his potential

[68] For example I. Neil Stevenson, "*Colerina:* Reactions to Emotional Stress in the Peruvian Andes," *Social Science and Medicine* 11, no. 5 (1977): 303–7; Carolyn Sargent, "Between Death and Shame: Dimensions of Pain in Bariba Culture," *Social Science and Medicine* 19, no. 12 (1984): 1299–304; Colin Murray Parkes, Pittu Laungani, and Bill Young (eds.), *Death and Bereavement across Cultures* (London: Routledge, 1997); Sjaak van Der Geest, "Dying Peacefully: Considering Good Death and Bad Death in Kwahu Tafo, Ghana," *Social Science and Medicine* 58, no. 5 (2004): 899–911.

dinner, or a dead child from a sleeping one, will inform behaviour and resource investment. Other animals may occasionally make a mistake in their judgement like humans do,[69] but most likely we all recognise death and understand it at least as cessation of agency[70] and what I called physical non-returnability (in the current form).[71] Ultimately all animals live in the real world, not only in human labs and imagination; they possess inherited wisdom and skill potentials for optimal life in a socio-environmental niche, some of which humans have uncovered, and the rest remains a mystery.

Conclusion: Pasts and Futures

The path from the house to the carport and other structures on the land we live on and share with other animals has been shaped by Henry and his comrades. It is unclear whether the routes had been chosen for their brevity or some other functional or even aesthetic purpose, but when sheep (and now humans) move between spaces, without failure we use the same routes. When other humans visit they quickly recognise the paths for what they are and follow them. It is fascinating to watch their brains' automatic categorisation system at work. It is this system that enables human and nonhuman animals to recognise a path as a path without prior close examination even if this is the first time they have encountered this particular path. This capacity for categorisation and generalisation emerges with the development of our brain and it is, like many other features, heavily dependent on experience. This is how it works: a neurotypical animal brain[72] processes sensory information through acquired concepts. "Concept" in this context describes a grouping of attributes and sensory details pertaining to individual familiar objects, whereas the term "object" encompasses actual objects/subjects such as paths, chairs, trees, or friends,

[69] For example Michael Inbar, "Mom's Hug Revives Baby that was Pronounced Dead," *Today*, September 4, 2010. http://bit.ly/3JPC3ZK.

[70] H. Clark Barrett, and Tanya Behne, "Children's Understanding of Death as the Cessation of Agency: A Test Using Sleep Versus Death," *Cognition* 96, no. 2 (2005): 93–108.

[71] Teya Brooks Pribac, "Animal Grief," *Animal Studies Journal* 2, no. 2 (2013): 67–90. http://ro.uow.edu.au/asj/vol2/iss2/5/; Teya Brooks Pribac, *Enter the Animal: Cross-Species Perspectives on Grief and Spirituality* (Sydney: Sydney University Press, 2021).

[72] Autistic brains, for example, divert from this trend.

but also complex phenomena, situations, and events.[73] Once a concept has formed, details of the concept elude our conscious awareness and we tend to focus on the bigger picture. In sum, if we acquire the concept of a path, in most cases we will quickly recognise a path whether we are at someone's house, in the forest, a city, or elsewhere.

This is an incredibly useful capacity because it enables animals a moderately smooth navigation through the socio-ecological environment; life would be impossible if we had to stop and examine every single thing we encounter on our daily routes. However, this automatisation also makes us vulnerable to illusion and prejudice since the function of this system is to offer the best possible hypothesis based on experience. For example, if we have been brought up in a carnist[74] society that conceptualises human animals as superior to other animals, we are likely to adopt such conceptualisation without serious scrutiny. This is what enables a human to perceive two completely comparable beings, such as a cow and a dog, in two completely different lights, one as food, the other as companion. In other human cultures categories may be different, for instance a cow could be sacred and venerated, while a dog could be unclean and avoided. This is also how other types of oppressive narratives consolidate and thrive—the narratives of racism, genderism, ableism, and others. Understanding this process, understanding that there is little volition, little comprehensive analysis, and ethical evaluation behind certain ideas that we may be attached to but are damaging to others may help us open up to other narratives, other possible worlds, and ways of being.

Historically, nonhuman animals have suffered immensely at human hands. "The look, even of the dying animal, tells its 'rational' caretaker or killer," wrote Gotthilf Heinrich von Schubert, a nineteenth-century physician and naturalist, "that it may not understand the spooky depth of the human being, but that it can sense it."[75] Science and common sense allow us to infer knowledge across species and time lines. What we know about

[73] Allan Snyder, Terry Bossomaier, and D. John Mitchell, "Concept Formation: 'Object' Attributes Dynamically Inhibited From Conscious Awareness," *Journal of Integrative Neuroscience* 3, no. 1 (2004): 31–46; Giorgio Vallortigara, Allan Snyder, Gisela Kaplan, Patrick Bateson, Nicola S. Clayton, and Lesley J. Rogers, "Are Animals Autistic Savants?" *Plos Biology* 6, no. 2 (2008): 0208-14. https://doi.org/10.1371/journal.pbio.0060042.

[74] Melanie Joy, *Why We Love Dogs, Eat Pigs and Wear Cows* (San Francisco: Conari Press, 2009).

[75] Reported in Kocku von Stuckrad, *A Cultural History of the Soul: Europe and North America from 1870 to the Present* (New York: Columbia University Press, 2022), 11.

nonhuman animals now was equally valid in Ancient Greece and beyond since neither humans nor other animals have changed significantly in this relatively short timeframe of our planetary existence. While we cannot know exactly what someone else's experience is like—human or nonhuman, past or present—extant commonalities enable us a moderately accurate guess of how, for example, the horse felt while she/he was being beaten in the middle of Turin that day in early January 1889, and possibly also how Nietzsche felt witnessing it.[76] We cannot change history, but we can shape the routes of the future and consolidate kinder narratives for others to follow.

Bibliography

Abbott, Alison. "Animal Behaviour: Inside the Cunning, Caring and Greedy Minds of Fish." *Nature* 521, no. 7553 (2015): 412–4. https://doi.org/10.1038/521412a.

Agamben, Giorgio. *The Open: Man and Animal*. Stanford: Stanford University Press, 2004.

Alkon, Abbey W., Thomas Boyce, Torsten B. Neilands, and Brenda Eskenazi. "Children's Autonomic Nervous System Reactivity Moderates the Relations between Family Adversity and Sleep Problems in Latino 5-Year Old in the CHAMACOS Study." *Frontiers in Public Health* 5, no. 155 (2017). https://doi.org/10.3389/fpubh.2017.00155.

Amarello, Melissa, Jeffrey J. Smith, and J. Slone. "Family Values: Rattlesnake Parental Care Is More than Just Attendance." Biology of the Rattlesnakes Symposium, Tucson, Arizona, 2011. https://www.snakes.ngo/ABSsm.pdf.

Balcombe, Jonathan. *What a Fish Knows: The Inner Lives of Our Underwater Cousins*. London: Oneworld Publications, 2016.

Barrett, H. Clark, and Tanya Behne. "Children's Understanding of Death as the Cessation of Agency: A Test Using Sleep Versus Death." *Cognition* 96, no. 2 (2005): 93–108.

Bennett, Galef G., and Kevin N. Laland. "Social Learning in Animals: Empirical Studies and Theoretical Models." *BioScience* 55, no. 6 (2005): 489–99.

Beston, Henry. *The Outermost House*. New York: Henry Holt and Company, Inc., 1949 [1928].

Bowlby, John. *Attachment and Loss: Volume I: Attachment*. New York: Basic Books, 1982 [1969].

[76] David Brooks, *Turin: Approaching Animals* (Blackheath: Brandl & Schlesinger, 2021), 9.

Bowlby, John. *Attachment and Loss: Volume II: Separation Anxiety and Anger.* New York: Basic Books, 1973.
Bowlby, John. *Attachment and Loss: Volume III: Loss, Sadness and Depression.* New York: Basic Books, 1980.
Brooks Pribac, Teya. "Animal Grief." *Animal Studies Journal* 2, no. 2 (2013): 67–90. http://ro.uow.edu.au/asj/vol2/iss2/5/.
Brooks Pribac, Teya. "Grieving at a Distance." In *Mourning Animals*, edited by Margo DeMello, 193–9. East Lansing: Michigan State University Press, 2016.
Brooks Pribac, Teya. "Positive Sentience is Underrated." *Animal Sentience* 2019.220. https://doi.org/10.51291/2377-7478.1445.
Brooks Pribac, Teya. *Enter the Animal: Cross-Species Perspectives on Grief and Spirituality.* Sydney: Sydney University Press, 2021.
Brooks, David. *Turin: Approaching Animals.* Blackheath: Brandl & Schlesinger, 2021.
Brown, Culum. "Are Fish Far More Intelligent than We Realize?" Interview by Joseph Stromber. *Vox*, August 4, 2014. https://bit.ly/3lrkIan.
Cialdini, Robert. "Crafting Normative Messages to Protect the Environment." *Current Directions in Psychological Science* 12, no. 4 (2003): 105–9.
Coan, James A. "Toward a Neuroscience of Attachment." In *Handbook of Attachment: Theory, Research, and Clinical Applications*, edited by Jude Cassidy and Phillip R. Shaver, 3rd ed., 242–69. New York: The Guilford Press, 2016.
Derrida, Jacques. *The Animal that Therefore I Am.* New York: Fordham University Press, 2008.
Despret, Vinciane. "Sheep Do Have Opinions." In *Making Things Public: Atmospheres of Democracy*, edited by Bruno Latour and Peter Weibel, 360–70. Cambridge: MIT Press, 2005.
Ehrlich, Katherine B., Gregory E. Miller, Jason D. Jones, and Jude Cassidy. "Attachment and Psychoneuroimmunology." In *Handbook of Attachment: Theory, Research, and Clinical Applications,* edited by Jude Cassidy and Phillip R. Shaver, 3rd ed., 180–201. New York: The Guilford Press, 2016.
Ellis, Elizabeth. *Australian Animal Law: Context and Critique.* Sydney: Sydney University Press, 2022.
Erlangsen, Annette, Bernard Jeune, Unni Bille-Hrahe, and James W. Vaupel. "Loss of a Partner and Suicide Risks among Oldest Old: A Population-Based Register Study." *Age and Ageing* 33, no. 4 (2004): 378–83.
Field, Tiffany. "Romantic Breakups, Heartbreak and Bereavement." *Psychology* 2, no. 4 (2011): 382–7.
Foley, Charles, Nathalie Pettorelli, and Lara Foley. "Severe Drought and Calf Survival in Elephants." *Biology Letters* 4 (2008): 541–4. https://doi.org/10.1098/rsbl.2008.0370.
Fraley, Chris R., and Phillip R. Shaver. "Attachment, Loss, and Grief: Bowlby's Views, New Developments, and Current Controversies." In *Handbook of*

Attachment: Theory, Research, and Clinical Applications, edited by Jude Cassidy and Phillip R. Shaver, 3rd ed., 40–62. New York: The Guilford Press, 2016.
Gander, Manuela, and Anna Buchheim. "Attachment Classification, Psychophysiology and Frontal EEG Asymmetry across the Lifespan: A Review." *Frontiers in Human Neuroscience* 9, no. 79 (2015). https://doi.org/10.3389/fnhum.2015.00079.
Garlick, Steve, and Rosemary Austen. "Kangaroo Translocation: Program Efficiency and Welfare Goals. Australian Wildlife Rehabilitation Conference, Adelaide, South Australia, 2010.
Garlick, Steve, and Rosemary Austen. "Learning about the Emotional Lives of Kangaroos: Cognitive Justice and Environmental Sustainability." *Relations* 2, no. 1 (2014). https://doi.org/10.7358/rela-2014-001-garl.
Gigliotti, Carol. *The creative lives of animals.* New York: New York University Press, 2022.
Goldstein, Murray. "Decade of the Brain—An Agenda for the Nineties." *Western Journal of Medicine* 161, no. 3 (1994): 239–41.
Graham, Linda. "The Neuroscience of Attachment." Paper, Clinical Conversation, Community Institute for Psychotherapy, San Rafael, CA, 2008. https://bit.ly/36I7S1H.
Griffin, Donald R. *Animal Minds: Beyond Cognition to Consciousness.* Chicago: University of Chicago Press, 2001 [1992].
Harlow, Harry F. "The Nature of Love." *American Psychologist* 13 (1958): 673–85.
Harlow, Harry F., and Margaret Kuenne Harlow. "Social Deprivation in Monkeys." *Scientific American* 5 (1962): 136–46.
Harlow, Harry F., Robert O. Dodsworth, and Margaret K. Harlow. "Total Social Isolation in Monkeys." *Proceedings of the National Academy of Sciences USA* 54, no. 1 (1965): 90–7.
Inbar, Michael. "Mom's Hug Revives Baby that was Pronounced Dead." *Today,* September 4, 2010. http://bit.ly/3JPC3ZK.
Innis Dagg, Anne. *Animal Friendships.* Cambridge: Cambridge University Press, 2011.
Jacobs, Bob, Heather Rally, Catherine Doyle, Lester O'Brien, Mackenzie Tennison, and Lori Marino. "Putative Neural Consequences of Captivity for Elephants and Cetaceans." *Reviews in the Neurosciences* 33, no. 4 (2021): 439–65. https://doi.org/10.1515/revneuro-2021-0100.
Jesmer, Brett, Jerod A. Merkle, Jacob R. Goheen, Ellen O. Aikens, Jeffrey L. Beck, Alyson B. Courtemanch, Mark A. Hurley, Douglas E. McWhirter, Hollie M. Miyasaki, Kevin L. Monteith, and Matthew J. Kauffman. "Evidence of Social Learning from Translocated Animals." *Science* 361, no. 6406 (2018): 1023–25. https://doi.org/10.1126/science.aat098.
Joy, Melanie. *Why We Love Dogs, Eat Pigs and Wear Cows.* San Francisco: Conari Press, 2009.

Kaprio, J., M. Koskenvuo, and H. Rita. "Mortality after Bereavement: A Prospective Study of 95,647 Widowed Persons." *American Journal of Public Health* 77, no. 3 (1987): 283–7.

Kentner, Amanda C., Kelly G. Lambert, Anthony J. Hannan, and S. Tiffany Donaldson. "Editorial: Environmental Enrichment: Enhancing Neural Plasticity, Resilience, and Repair." *Frontiers in Behavioral Neuroscience* 13, no. 75 (2019). https://doi.org/10.3389/fnbeh.2019.00075.

King, Barbara J. *How Animals Grieve*. Chicago: Chicago University Press, 2013.

Koh, Ignatius. "Hachiko Statue in Shibuya." *Japan Travel*, June 23, 2019. https://en.japantravel.com/tokyo/hachiko-statue-in-shibuya/44644.

Laland, K.N., N. Atton, and M.M. Webster. "From Fish to Fashion: Experimental and Theoretical Insights into the Evolution of Culture." *Philosophical Transactions Royal Society B* 366, no. 1567 (2011): 958–68. https://doi.org/10.1098/rstb.2010.0328.

Liu, Dong, Josie Diorio, Jamie C. Day, Darlene D. Francis, and Michael J. Meaney. "Maternal Care, Hippocampal Synaptogenesis and Cognitive Development in Rats." *Nature Neuroscience* 3, no. 8 (2000): 799–806.

Low, Philip, Jaak Panksepp, Diana Reiss, David Edelman, Bruno Van Swinderen, and Christof Koch. "The Cambridge Declaration on Consciousness." Cambridge (UK), July 7, 2012. http://fcmconference.org/img/CambridgeDeclarationOnConsciousness.pdf.

Lyotard, Jean-François. *The Postmodern Condition: A Report on Knowledge*. Translated by Geoff Bennington and Brian Massumi. Manchester: Manchester University Press, 1984.

Marino, Lori, and Christina M. Colvin. "Thinking Pigs: A Comparative Review of Cognition, Emotion, and Personality in *Sus Domesticus*." *International Journal of Comparative Psychology* 28, no. 1 (2015). 10.46867/ijcp.2015.28.00.04.

Marino, Lori. "Thinking Chickens: A Review of Cognition, Emotion, and Behaviour in the Domestic Chicken." *Animal Cognition* 20, no. 2 (2017): 127–47.

McLaughlin, Katie A., Margaret A. Sheridan, Sonia Alves, and Wendy Berry Mendes. "Child Maltreatment and Autonomic Nervous System Reactivity: Identifying Dysregulated Stress Reactivity Patterns Using the Biopsychosocial Model of Challenge and Threat." *Psychosomatic Medicine* 76, no. 7 (2014). https://doi.org/10.1097/PSY.0000000000000098.

Meaney, Michael J. "Maternal Care, Gene Expression, and the Transmission of Individual Differences in Stress Reactivity across Generations." *Annual Review of Neuroscience* 24 (2001): 1161–92.

Midgley, Mary. *Beast and Man: The Roots of Human Nature*. London: Routledge, 2002 [1978].

Mitchell, Les. *Reading the Animal Text in the Landscape of the Damned*. Grahamstown: NISC, 2019.

Nöbel, Sabine, Mélanie Allain, Guillaume Isabel, and Etienne Danchin. "Mate Copying in Drosophila Melanogaster Males." *Animal Behaviour* 141 (2018): 9–15. https://doi.org/10.1016/j.anbehav.2018.04.019.

Nordanger, Dag Ø. "Coping With Loss and Bereavement in Post-War Tigray, Ethiopia." *Transcultural Psychiatry* 44, no. 4 (2007): 545–65.

O'Connor, Mary-Frances, David K. Wellisch, Annette L. Stanton, Naomi I. Eisenberger, Michael R. Irwin, and Matthew D. Lieberman. "Craving Love? Enduring Grief Activates Brain's Reward Center." *NeuroImage* 42, no. 2 (2008): 969–72.

Oliver, Kelly. "Pet Lovers, Pathologized." *New York Times*, October 30, 2011. https://archive.nytimes.com/opinionator.blogs.nytimes.com/2011/10/30/the-pathology-of-dependence-on-animals/

Omland, Kevin, Evangeline Rose, and Karan Odom. "Women Have Disrupted Research on Bird Song, and Their Findings Show How Diversity can Improve All Fields of Science." *The Conversation,* September 11, 2020. https://theconversation.com/women-have-disrupted-research-on-bird-song-and-their-findings-show-how-diversity-can-improve-all-fields-of-science-142874.

Panksepp, Jaak, and Lucy Biven. *The Archaeology of Mind: Neuroevolutionary Origins of Human Emotions.* New York: W.W. Norton, 2012.

Panksepp, Jaak. "Cross-Species Affective Neuroscience Decoding of the Primal Affective Experiences of Humans and Related Animals." *PLOS ONE* 6, no. 8 (2011): e21236. https://doi.org/10.1371/journal.pone.0021236.

Parkes, Colin Murray, Pittu Laungani, and Bill Young (Eds.). *Death and Bereavement across Cultures.* London: Routledge, 1997.

Phelps, Norm. *The Longest Struggle: Animals Advocacy from Pythagoras to PETA*. New York: Lantern Books, 2007.

Polan, Jonathan H., and Myron A. Hofer. "Psychobiological Origins of Infant Attachment and Its Role in Development." In *Handbook of Attachment: Theory, Research, and Clinical Applications*, edited by Jude Cassidy and Phillip R. Shaver, 3rd ed., 117–32. New York: The Guilford Press, 2016.

Preti, Antonio. "Suicide among Animals: A Review of Evidence." *Psychological Reports* 101 (2007): 831–48.

Safina, Carl. *Beyond Words: What Animals Think and Feel.* New York: Henry Holt and Company, 2015.

Sapolsky, Robert M. "Mothering Style and Methylation." *Nature Neuroscience* 7, no. 8 (2004): 791–2.

Sargent, Carolyn. "Between Death and Shame: Dimensions of Pain in Bariba Culture." *Social Science and Medicine* 19, no. 12 (1984): 1299–1304.

Scheper-Hughes, Nancy. *Death without Weeping: The Violence of Everyday Life in Brazil.* Berkeley: University of California Press, 1992.

Schore, Allan N. "Attachment, Affect Regulation, and the Developing Right Brain: Linking Developmental Neuroscience to Pediatrics." *Pediatrics in Review* 26, no. 6 (2005): 204–17.

Schore, Allan N. *The Science of the Art of Psychotherapy.* New York: W.W. Norton & Company, 2012.

Siegel, Daniel J. *Mindsight: The New Science of Personal Transformation.* New York: Bantam Books Trade Paperbacks, 2011 [2010].

Sjaak van Der Geest. "Dying Peacefully: Considering Good Death and Bad Death in Kwahu Tafo, Ghana." *Social Science and Medicine* 58, no. 5 (2004): 899–911.

Snyder, Allan, Terry Bossomaier, and D. John Mitchell. "Concept Formation: 'Object' Attributes Dynamically Inhibited From Conscious Awareness." *Journal of Integrative Neuroscience* 3, no. 1 (2004): 31–46.

Stacho, Martin, Christina Herold, Noemi Rook, Hermann Wagner, Markus Axer, Katrin Amunts, and Onur Güntürkün. "A Cortex-Like Canonical Circuit in the Avian Forebrain." *Science* 369, no. 6511 (2020). https://doi.org/10.1126/science.abc5534.

Stevenson, I. Neil. "*Colerina:* Reactions to Emotional Stress in the Peruvian Andes." *Social Science and Medicine* 11, no. 5 (1977): 303–7.

Suomi, Stephen J. "Attachment in Rhesus Monkeys." In *Handbook of Attachment: Theory, Research, and Clinical Applications*, edited by Jude Cassidy and Phillip R. Shaver, 3rd ed., 133–54. New York: The Guilford Press, 2016.

Takahashi, Tetsumi, Haruki Ochi, Masanori Kohda, and Michio Hori. "Invisible Pair Bonds Detected by Molecular Analyses." *Biology letters* 8, no. 3 (2012): 355–57. https://doi.org/10.1098/rsbl.2011.1006.

University of Wyoming. "Big Game Animals Must Learn to Migrate and Pass Knowledge across Generations." *Science Daily*, September 6, 2018. www.sciencedaily.com/releases/2018/09/180906141626.htm.

Vallortigara, Giorgio, Allan Snyder, Gisela Kaplan, Patrick Bateson, Nicola S. Clayton, and Lesley J. Rogers. "Are Animals Autistic Savants?" *Plos Biology* 6, no. 2 (2008): 0208–14. https://doi.org/10.1371/journal.pbio.0060042.

von Stuckrad, Kocku. *A Cultural History of the Soul: Europe and North America from 1870 to the Present.* New York: Columbia University Press, 2022.

Waynforth, D. "Mate Choice Copying in Humans." *Human Nature* 18, no. 3 (2007): 264–71. https://doi.org/10.1007/s12110-007-9004-2.

Weaver, Ian C.G., Nadia Cervoni, Frances A. Champagne, Ana C. D'Alessio, Shakti Sharma, Jonathan R. Seckl, Sergiy Dymov, Moshe Szyf, and Michael J. Meaney. "Epigenetic Programming by Maternal Behaviour." *Nature Neuroscience* 7, no. 8 (2004): 847–54.

Whiten, Andrew. "Culture Is So Common that Even Fish and Flies Have It." Interview by Michael La Page. *New Scientist*, April 1, 2021a. https://www.newscientist.com/article/2273450-animal-culture-is-so-common-that-even-fish-and-flies-have-it/.

Whiten, Andrew. "The Burgeoning Reach of Animal Culture." *Science* 372, no. 46 (2021b). https://doi.org/10.1126/science.abe6514.
Willett, Cynthia. "Can a Mirror Capture the Self?" *Animal Sentience* 25, no. 36 (2019). https://doi.org/10.51291/2377-7478.1467.
Williams, Howard. *The Ethics of Diet: An Anthology of Vegetarian Thought.* Guildford: White Crow Books, 2009.
Yoerg, Sonja. "Mentalist Imputations." *Science* 258, no. 5083 (1992): 830–1.
Yong, Ed. *An Immense World: How Animal Senses Reveal the Hidden Realm Around Us.* New York: Random House, 2022.

CHAPTER 3

Pain in Context: Indicators and Expressions of Animal Pain

Ian S. Olivier and Abraham Olivier

Introduction

This chapter aims to contribute to the endeavour of investigating nonhuman animals[1] as experiencing subjects in their own right with their own species-specific expressions. We focus on contemporary empirical studies and try to offer a phenomenological analysis of the indications and expressions of pain experience in animals. To do this, we utilise findings from experimental studies in the fields of neurobiology, evolutionary-developmental biology, comparative psychology, and cognitive ethology.

[1] Hereafter referred to as animals for the sake of brevity.

I. S. Olivier
Department of Psychiatry and Mental Health, Neuroscience Institute,
University of Cape Town, Cape Town, South Africa

A. Olivier (✉)
Department of Philosophy, University of Fort Hare, East London, South Africa
e-mail: aolivier@ufh.ac.za

M. J. Glover, L. Mitchell (eds.), *Animals as Experiencing Entities*,
The Palgrave Macmillan Animal Ethics Series,
https://doi.org/10.1007/978-3-031-46456-0_3

We attempt to show how different animals are capable of indicating and expressing specific and unique experiences of pain bound to their particular species, evolution, and places of living in human and natural contexts. We think this is a topic that is of great import in furthering our understanding of pain in animals, both historically and presently. To the best of our knowledge, it is a topic not much considered in the standing literature.

Until recently, there was still a widely held belief both within and outside academic circles that pain was a uniquely human experience.[2] While it is now generally accepted that the capacity to feel physical pain is shared widely across animal species,[3] debate amongst scientists and philosophers continues on whether some animals, like fish, feel pain too.[4] Consequently, academic debates commonly focus on comparisons between human and animal pain.[5] Such comparisons typically comprise two basic common aspects. The first is that animals and humans have anatomical, physical, and behavioural similarities, which support the view that animals suffer pain in a similar way to humans.[6] Second, many argue that the physical comparisons made between humans and animals give no reason to believe that animals suffer pain in any way similar to humans, and for each similarity, one can find dissimilarity that weakens the analogy.[7] Notably, some maintain that despite significant physical similarities, animals lack the capacity of language which is arguably essential to human consciousness (or "higher order" thinking and the possession of a theory of mind), thus denying any consciousness or pain experience comparable to humans.[8]

A seminal aspect of such comparisons concerns which taxonomic groups experience nociception (the neural processing of a stimulus that causes or has the potential to inflict damage) accompanied by conscious pain experience.[9] It is generally accepted by pain researchers that nociception and pain are different phenomena.[10] Compared to nociception, con-

[2] Rollin, "Animal Pain", 425; Breed, *Ethology and Animal Behaviour*, 197.

[3] Gregory, *Physiology and Behaviour of Animal Suffering*, 1.

[4] Key, "Why Fish Do Not Feel Pain".

[5] Allen, "Animal Pain", 621–637.

[6] Allen, "Animal Pain", 621–637; Bekoff, "Animal Emotions", 18; Rollin, "Animal Pain", 425; Breed, *Ethology and Animal Behaviour*, 197.

[7] Allen, "Discovery of Animal Consciousness", 217; Allen and Bekoff, "Animal Consciousness", 66.

[8] Carruthers, *Phenomenal Consciousness*.

[9] Barrot, Salvat, and Yalcin, "Animal-Based Translational Research", 375.

[10] Raja et al., "Definition of Pain".

scious or subjective pain experience remains what philosophers like to call a hard problem. Indeed, assessing pain, or negative internal mental states, in animals is notoriously difficult as is amply demonstrated by Thomas Nagel's famous essay "What Is It Like to Be a Bat?" Pain counts commonly as an example of the hard problem of knowing what it is like for others to feel pain, particularly what pain is like for animal others. Animals have adapted to different environments both evolutionarily and within their lifetime, which shape their capacities and experiences.[11] So how can we compare their pain experience to ours?

While there are likely significant physical and experiential similarities between human and animal pain, this comparative approach for inferential purposes regarding which species experience pain is insufficient.[12] We argue that further understanding of animal pain can be gained by closer analysis of the comprehensive context, including social dynamics, non-linguistic indicators, and linguistic and non-linguistic expressions of pain in animals. More specifically, we examine the relationship between cognitive ethology, environmental factors, and the forms of indicators and expressions related to physical pain that apply to animals.

We take a phenomenological approach by concentrating on the seminal features of experience of pain in animals and indicators and expressions thereof in species-specific social contexts. Instead of developing theoretical arguments, our method will be to focus on empirical studies to make our case.[13] Consequently, we attempt to conduct an empirical phenomenological investigation that draws on experimental approaches and cognitive ethology. We try to show how different animals are capable of expressing specific and unique experiences of pain bound to their particular species, evolution, and places of living.

[11] Gregory, *Physiology and Behaviour of Animal Suffering*, 8; Sneddon et al., "Defining and Assessing Animal Pain", 204–209.

[12] Allen, "Animal Pain", 623.

[13] While we cite important research that has been conducted into animal pain, this does not mean we generally condone research that inflicts pain or other forms of suffering on animals (e.g., through social deprivation, lack of environmental enrichment, stress, fear). Instead, we believe the great majority of animal research is not sufficiently justified and we do not condone such research, especially where harm is involved. Nonetheless, we want to use the insights gained in these past experiments to learn more about animal pain, hopefully without further supporting unjustifiable harm caused to animals in the name of the very scientific ideology that has harmed so many animals in the first place (Mitchell 2019).

Again, our focus on the ways in which animals indicate, express, and communicate their pain in lived contexts addresses an issue not much considered in the standing literature on animal pain. From this perspective, we do not so much attempt to counter existing research on the similarities between human and animal pain but rather hope to explore experiential similarities and differences between different animals themselves. In phenomenological terms, this is an attempt to "return to the animals themselves".

We start by discussing the main definitions of and approaches to pain in animals in the first section. Sections "Defining Pain in Animals", "Distinguishing Indications and Expressions of Animal Pain", "Assessing Indicators of Animal Pain", and "Animal Responses Indicative of Conscious Pain: Vertebrates and Invertebrates" examine indicators and expressions of pain. Section "The Social Context of Pain Indicators and Expressions in Animals"[14] considers how indicators and expressions relate to the social context of animal pain.

Defining Pain in Animals

For the sake of clarity, we need to specify exactly what is understood as pain. We focus on physical pain rather than psychological or social pain as is mostly done in mainstream pain research.[15]

The most widely used definition of pain is offered by the International Association for the Study of Pain (IASP). The IASP has recently, in 2020, updated its previous definition formulated in 1979 and it is supposed to be applicable to humans and nonhuman animals.[16] The definition states

[14] We do not here discuss the impact of language, discourses, or human expressions of pain in animals, despite its profound historical importance and impact of how animal pain has been perceived (or denied and ignored). While we argue for species-specific expression of pain in animals, human use of language and associated meaning has been used in the history of animal pain and animal experimentation to promote the power of ideology and deny the capability of animals to experience pain in many ways (e.g., in animal experimentation descriptions of emotional states or painful states in animals were generally avoided, using reductive terms instead to only indicate physiological or behavioural responses, avoiding the idea of the animal as a subject or feeling being, in a similar way that animals or parts of animals are referred to as "goods", or "commodities", or "products" in animal agriculture). For more detailed and historical descriptions of the way human discourse has shaped our understanding and views of animal pain and suffering, see Mitchell (2019) and Rollin (1989, 2011).

[15] Olivier, *The Social Dimension of Pain*, 2022.

[16] Raja et al., "The Revised International Association for the Study of Pain Definition of Pain", 3.

that pain is "an unpleasant sensory and emotional experience typically caused by, or resembling that caused by, actual or potential tissue injury".[17]

This definition includes several notes, some of which are relevant to our chapter. The first note states that "pain is always a personal experience that is influenced to varying degrees by biological, psychological, and social factors".[18] This illustrates that the IASP generally agrees on pain being a subjective phenomenon, but also that the social environment is significant.

The second note highlights the difference between pain and nociception, and states that pain is not identical to activity in nociceptors (sensory neurons responsible for nociception). This is a recurrent issue in studies of animal pain, as the presence of nociceptors is considered a necessary albeit insufficient condition for subjective pain experience.[19]

In the third note, the IASP mentions that "through their life experiences, individuals learn the concept of pain and its applications".[20] This points to the effect of prior pain experience on pain as well as the functional role thereof, which is a frequent topic of discussion[21]— we explore this in sections "Distinguishing Indications and Expressions of Animal Pain", "Assessing Indicators of Animal Pain", and "Animal Responses Indicative of Conscious Pain: Vertebrates and Invertebrates".

Importantly, for all "non-verbal" animals the IASP states in an additional note that "verbal description is only one of several behaviours to express pain; inability to communicate does not negate the possibility that a human or a non-human animal experiences pain."[22] This addresses criticism of the previous IASP definition including a misleading clause that gives the impression that a necessary condition for pain is its verbal expression, thus not including non-verbal infants or animals who have their own species-specific modes of expression.[23]

[17] Raja et al., "The Revised International Association for the Study of Pain Definition of Pain", 2.

[18] Raja et al., "The Revised International Association for the Study of Pain Definition of Pain", 2.

[19] Walters and Williams, "Evolution of mechanisms", 2019, 3.

[20] Raja et al., "The Revised International Association for the Study of Pain Definition of Pain", 2.

[21] Allen et al., "Deciphering Animal Pain", 357–361.

[22] Raja et al., "Definition of Pain", 1977.

[23] Aydede, "IASP definition of Pain", 2.

Notably, the main emphasis of the IASP's definition of pain is the notion that pain is always a subjective experience as personally reported by individuals.[24] While this definition is claimed to be "applicable" to animals, it is still (despite its revision) unhelpful for defining pain in a large subset of species.[25] Furthermore, it still omits the motivational function of pain that is essential to the evolutionary understanding of pain.[26]

The *definition for animal pain*, which is most commonly used,[27] refers to an early functional definition described by Manfred Zimmermann. Zimmermann defines pain as:

> [A]n aversive sensory experience caused by actual or potential injury that elicits protective and vegetative reactions, resulting in learned avoidance, and may modify species-specific behaviour, including social behaviour.[28]

It is postulated that many species that possess central neural states associated with nociceptive plasticity exhibit most of these features,[29] including fish[30]and crustaceans,[31] as well as insects and molluscs.[32] Several important components of pain are thus implicated in diverse species. One of these components is a complex aversive state. The aversive state can be observed in, for example, avoidance learning and conditioned place avoidance or preference. This again indicates an affective-motivational component of pain.[33] Another important component of the definition is modified "social behaviour" which arguably excludes many asocial vertebrate and invertebrate animals.[34] Furthermore, there is also a functional component of motivating avoidance or protective behaviour. Importantly, as Walters

[24] Raja et al., "The Revised International Association for the Study of Pain Definition of Pain", 1978.

[25] Sneddon et al., "Defining and Assessing Animal Pain", 202.

[26] Walters and Williams, "Evolution of mechanisms", 2.

[27] Sneddon et al., "Defining and Assessing Animal Pain", 202.

[28] Zimmermann, "Physiological mechanisms", 2; as quoted by Walters and Williams, "Evolution of mechanisms", 2; and quoted by Sneddon et al., "Defining and Assessing Animal Pain", 202.

[29] Walters and Williams, "Evolution of mechanisms", 2.

[30] Sneddon, "Evolution of nociception", 2.

[31] Elwood, "Pain in Crustaceans", 1.

[32] Walters, "Nociceptive Biology of Molluscs", 1.

[33] Walters and Williams, "Evolution of mechanisms", 2; Sneddon et al., "Defining and Assessing Animal Pain", 203.

[34] Walters, "Evolution of mechanisms", 2.

points out, the purely functional definitions of pain do not sufficiently include a core feature of pain, for both humans and animals: the distressing subjective experience, the negative valence, or suffering associated with it. Thus, the two differing, complementary functional and subjective approaches to defining pain can be difficult to combine[35] and it continues to be a problem to assess which species feel pain or not.[36]

The aversive motivational state associated with nociception and its similarity to humans in some animals has been the main argument for some researchers to show that animals experience pain.[37] The key function of this aversive motivational state, compared to nociception, is the formation of a powerful and durable motivation to avoid situations that led to the pain,[38] thereby providing long-term protection and increasing fitness.[39] Nociception usually functions to immediately alleviate tissue damage through reflex actions, with pain enabling protection in the long term. However, nociception can also be long lasting in the form of long-term nociceptive sensitisation without a corresponding subjective experience, meaning that long-term behavioural changes on their own are not sufficient to prove the experience of a pain in an animal.[40] Thus, it is widely agreed that there is not one single criterion or factor that provides evidence or definite proof for a subjective sense of pain in a being.[41] Instead, it has been suggested to follow a framework of criteria that can help to identify and assess animal pain. These we discuss in the following sections on assessing indicators of animal pain.

Distinguishing Indications and Expressions of Animal Pain

Drawing on phenomenological literature, specifically the work of Husserl and Heidegger, we would like to distinguish between expressions and indicators of pain in animals in the following sections.

[35] Sneddon et al., "Defining and Assessing Animal Pain", 202; Bateson, "Assessment of Pain", 827.

[36] Key, "Why Fish Do Not Feel Pain".

[37] Sneddon et al., "Defining and Assessing Animal Pain", 202.

[38] Sneddon et al., "Defining and Assessing Animal Pain", 202.

[39] Bateson, "Assessment of Pain", 827.

[40] Sneddon et al., "Defining and Assessing Animal Pain", 202.

[41] Bateson, "Assessment of Pain", 834.

To explain this distinction, consider the following comparison that Husserl makes between indications and expressions as two different kinds of signs in the First Investigation of his *Logical Investigations*. Indications pertain to something that refers to something else, for instance, smoke indicating fire, or a fossil as a sign of a mammal, or a flag referring to a nation,[42] or so one can add, withdrawal, or nursing indicating pain experience. According to Husserl, indications do not express meaning. Expressions, on the contrary, carry meaning, that is, they show how things appear to the subject. Expressions, according to Husserl, primarily manifest in terms of acts of speech, excluding gestures and facial expressions. The function of expressions is more particularly, according to Husserl, to signify and communicate what sense objects make as they appear to subjects. Thus, for Husserl, expressions consist of signifiers such as written or spoken words or sentences.[43] On their own, signifiers say nothing. Only through an act of signification, when a signifier is used to confer meaning to an object that can be communicated, does it become an expression.

To frontload some of our main arguments in the next three sections, note that we make use of but also differ from Husserl's distinction between indications and expressions as follows.

(a) We make use of the term indications, but contrary to Husserl, we attribute meaning also to indications. To make this point strong, and following the animal pain literature, we speak of "indicators". As we hope to show in sections "Distinguishing Indications and Expressions of Animal Pain" and "Assessing Indicators of Animal Pain", animals of different species indicate clearly with responses to nociception signs that can be considered to mean pain to them. Also in humans some gestures such as nursing clearly signify what we can take to mean pain.

(b) We deploy the term "expressions", but unlike Husserl, we take non-linguistic indications such as gestures or facial grimaces and other forms of body language also as expressions as far as they signify and *communicate* meaning. We show in detail in section "Animal Responses Indicative of Conscious Pain: Vertebrates and Invertebrates" how in human animals such non-linguistic expressions can be considered to communicate meaning.

[42] See Husserl, *Logical Investigations I* § 2 (2001).

[43] See Husserl, *Logical Investigations I* § 7 (2001).

(c) We also speak of "linguistic expressions"; however, as we show in section "Animal Responses Indicative of Conscious Pain: Vertebrates and Invertebrates", we go beyond Husserl and drawing on current field research on "animal linguistics" (e.g., Suzuki 2021, 221)[44] and on Heidegger's notion of language, we take expressions to be linguistic signifiers in a broad sense. As such they include signs and acts of communication such as non-verbal vocalisations (utterances or tone of voice). Accordingly, we speak of "animal language" which in part manifests in terms of non-verbal linguistic expressions.

Animals seem to signify or have their very own modes of signification, both through indicators and through expressions of pain. As we show in detail in sections "Distinguishing Indications and Expressions of Animal Pain" and "Assessing Indicators of Animal Pain", indicators of animal pain refer to the various non-linguistic physiological and behavioural signs that animals are in pain. As above, expressions of animals can be considered species-specific linguistic and non-linguistic signs that animals use to communicate that they are in pain. Again, linguistic signs refer to what we call, in line with some field research on animal linguistics, "animal language".[45] We take animal language to include in a broad sense non-verbal acts of communication in animals. However, the distinction between indicators and expressions is much more overlapping. It is hard to understand the specific signs animals use to signify pain, and how to distinguish non-verbal linguistic expressions from non-linguistic expressions and indicators. Conceding this lack, in the following sections we try to use empirical (experimental) studies to explore the workings of both indicators and expressions.

Assessing Indicators of Animal Pain

Due to our inability to understand the language animals use to communicate, we humans mostly rely on the behavioural, physiological, and anatomical indicators of pain exhibited by animals in order to study it.[46]

[44] Suzuki, "Animal linguistics", 221.

[45] Following Suzuki (2021) and others we take animal linguistics to mean theory of animal language.

[46] Bateson, "Assessment of Pain", 827; Sneddon et al., "Defining and Assessing Animal Pain", 203.

Importantly, we utilise the study of the capacity to learn, which is a key function associated with pain. A combination of all of these aspects with the animals' evolutionary and ecological contexts taken into consideration allows for the assessment of indicators of animal pain.[47]

Bateson devised a popular framework that offers physical and behavioural criteria to identify pain in animals based on similarities to its identification in humans.[48] One has to keep in mind that this framework uses a human view of pain as a point of departure and may not sufficiently take into account how animals may behave and respond completely differently to pain due to their evolution and individual functional adaptation to different environmental and social contexts.[49] Nonetheless, these criteria suggest possible *indicators* of pain and are helpful in further studying pain in the absence of verbal communication.

The criteria for pain originally devised by Bateson include presence of nociceptors, connections of these to the brain, brain structures similar to the human cerebral cortex processing pain, opioid receptors in the neural system, analgesia changing behaviour and physiology associated with pain, learning avoidance of potentially painful stimuli in a rapid and in-elastic manner.[50] Interestingly, criteria including social indicators of pain are omitted despite mention of the importance later in Bateson's work.[51]

These criteria have since been updated as more research has become available, and this should continue. Additional criteria were thus developed and are separated into two key principles with overlapping features. The *first set* of criteria includes so-called whole animal responses, which are indicative of pain. These include central processing of nociception in brain areas regulating motivated behaviour, nociceptive processes that are sensitive to endogenous modulation, nociception activating physiological responses, responses more complex than reflexes, longer-term changes in behaviour that decrease the chance of the stimulus, protective behaviour (e.g., guarding wound, limp, rubbing or licking), and reduction of all

[47] Allen et al., "Deciphering Animal Pain"; Allen and Bekoff, "Animal Consciousness", 66; Walters, "Evolution of Mechanisms", 5; Sneddon et al., "Defining and Assessing Animal Pain"; Elwood, "Pain in Crustaceans", 1.

[48] Bateson, "Assessment of Pain", 827.

[49] Bateson, "Assessment of Pain", 827; Bekoff, "Animal Emotions"; Sneddon et al., "Defining and Assessing Animal Pain", 202; Walters and Williams, "Evolution of Mechanisms", 2–6.

[50] Bateson, "Assessment of Pain", 834.

[51] Bateson, "Assessment of Pain", 837.

these by analgesia.[52] It is worth emphasising that nociceptors are consistently found in all animal taxa tested experimentally across the animal kingdom, from vertebrates to invertebrates (e.g., molluscs, annelids, arthropods such as crustaceans and insects), as reviewed by Sneddon et al. in 2014 and 2018.[53] However, some consider animals not in possession of certain regions found in the human cortex to be unable to experience pain.[54] There is no logical reason that painful sentience must be localised to a central brain, and in animals different structures can fulfil similar functions.[55] For example, the octopus has the largest nervous system of all invertebrates and is most often implied to have conscious pain, but most of its neurons are found in multiple complex ganglia in each arm, and not in a central brain, raising the possibility of each arm feeling pain independently.[56] Furthermore, subcortical brain regions which are phylogenetically older structures are likely to be the primary subtract of pain experience.[57] While central processing of nociceptive information does not guarantee conscious pain, it does indicate integration of information and is found from vertebrate to invertebrate species.

The *second set* of criteria refers to change in motivation. This includes self-administration of analgesia, paying cost to access analgesia, selective attention to noxious stimuli over other stimuli, altered behaviour such as conditioned place avoidance and avoidance learning, relief learning, long-lasting changes in numerous responses, paying costs to avoid noxious stimulation, and modification of the avoidance of noxious stimulus by trade-offs with other motivational costs.[58] Notably, there is still no mention of pain-related social behaviour or communication changes, highlighting the need for its inclusion in future adaptations of these criteria.

As Sneddon et al. point out, these multimodal criteria must be considered in *combination* when assessing animal pain.[59] A high number of criteria being met indicate a higher likelihood of pain experience. The accurate

[52] Sneddon et al., "Defining and Assessing Animal Pain", 203.

[53] Sneddon et al., "Defining and Assessing Animal Pain", 203–204; Sneddon, "Comparative Physiology of Nociception and Pain", 63–74.

[54] Rose et al., "Can Fish Really Feel Pain?", 97; Key, "Why Fish Do Not Feel Pain".

[55] Walters, "Defining Pain and Painful Sentience in Animals", 3–4.

[56] Walters, "Defining pain and painful sentience in animals", 3.

[57] Devor, "Where Is Pain in the Brain?", 5; Damasio and Damasio, "Pain and Other Feelings", 3; Panksepp, "Cross-Species Affective Neuroscience", 7.

[58] Sneddon et al., "Defining and Assessing Animal Pain", 203.

[59] Sneddon et al., "Defining and Assessing Animal Pain".

utilisation of these criteria is limited by a lack of data on the behavioural and social responses of certain species to pain. The absence of such data is likely reflective of lack of research rather than a lack of pain experience in these animals.[60] Thus, a multimodal approach is required, with the principle of triangulation, which entails taking all indices together to provide evidence for the internal affective state.[61]

Another way to categorise pain measures in animals has been proposed more recently by Mogil (2019).[62] Categories include biomarkers, pain-stimulated behaviour (including grimacing), pain depressed behaviour, measures of pain-related functional disability, and methods involving conditioning and motivational conflicts.[63]

Multiple measures have gained popularity in the laboratory setting. One of these includes measuring ultrasonic vocalisation by rats in response to aversive contexts such as pain,[64] where they emit 22kHz calls in place of the 50kHz calls emitted when playing or being tickled.[65] Also grimacing and facial expressions have been used in multiple mammalian species to recognise and measure pain.[66] We return to this in the section on pain expressions in animals.

Lastly, when considering capacity for pain, the "precautionary principle" is often invoked, where an animal should be given the benefit of the doubt when there is uncertainty about whether or not that animal can feel pain.[67] Arguably, "absolute certainty" is currently unattainable, which is no reason, however, to ignore the significance of the clear criteria of indicators of pain in animals.

Animal Responses Indicative of Conscious Pain: Vertebrates and Invertebrates

As above, the hard problem of human pain is its conscious individual (first-personal) experience. Again, another term for this used by the IASP is subjective experience. This is a hard problem because of the limited

[60] Sneddon et al., "Defining and Assessing Animal Pain", 203; Walters, "Defining Pain and Painful Sentience in Animals", 3–4.

[61] Sneddon et al., "Defining and Assessing Animal Pain", 209.

[62] Mogil, "Measurement of Pain in the Laboratory Rodent".

[63] Mogil, "Measurement of Pain in the Laboratory Rodent", 29–59.

[64] Kurejova, "Ultrasound Vocalizations".

[65] Panksepp and Burgdorf, "Unconditioned Tickle-Reward in Rats".

[66] Mogil et al., "Facial Grimace Scales for Pain Measurement", 489.

[67] Birch, "Animal Sentience and the Precautionary Principle", 2; Sneddon et al., "Assessing Animal Pain", 209; Elwood, "Evidence for Animal Pain", 2.

access to others' subjective experience. As the IASP concedes, we access subjective experience through personal reports of pain. To state the obvious, animals do not report pain as humans do. Hence the question is, what can we take to be indicators of conscious (subjective) animal pain?

The Institute for Laboratory Animal Research (ILAR) formed the "Committee on the Recognition and Alleviation of Pain in Laboratory Animals", which suggested which animal responses may indicate nociception and/or pain.[68] Some responses do not indicate subjective experience as in humans: autonomic responses, simple avoidance responses, some behavioural responses (e.g., movements), and other measures such as cerebral blood flow increases. On the other hand, the ILAR committee identified which animal responses are likely to indicate what the IASP would consider subjective pain experience. These include discrimination of painful states (e.g., pain being a discriminative stimulus, which likely requires awareness) and making decisions based on this discrimination, motivations to avoid pain or noxious stimuli, affective states of fear or anxiety when threatened with noxious stimuli, and behavioural changes (e.g., vocalisations, apathy, sleep disruption).

Several argue for the importance of studying some kinds of learning from pain in order to elucidate the functions of conscious pain experience, which includes states of conscious awareness in which attention is paid to the pain, its causes, and what it causes.[69] This includes complex learning behaviour like trace conditioning (classical conditioning involving a time period between the conditioned stimulus and the unconditioned stimulus), sophisticated kinds of operant learning, as well as conditioned place preference (animals choosing to spend time in places previously associated with relief from pain) and conditioned place aversion (animals choosing to avoid places associated with noxious stimuli).[70]

It is widely agreed that mammals and birds show evidence of a conscious experience of pain, and most support that all other vertebrates are capable of experiencing pain, although more research is needed.[71] Notably, the ILAR committee regards all vertebrates as capable of experiencing conscious pain but has not yet taken this stance on invertebrates. The

[68] National Research Council (US) Committee on Recognition and Alleviation of Pain in Laboratory Animals, "Recognition and Alleviation of Pain in Laboratory Animals", 19–21.

[69] Allen, "Animal Pain", 636; Allen et al., "Deciphering Animal Pain", 356.

[70] Walters, "Defining pain and painful sentience in animals", 4; Allen, "Animal Pain", 633.

[71] National Research Council (US) Committee on Recognition and Alleviation of Pain in Laboratory Animals, "Recognition and Alleviation of Pain in Laboratory Animals", 21.

discussion below makes a point to address both vertebrates and invertebrates.

There is ample evidence that mammals and birds have complex learning behaviour associated with conscious pain.[72] Birds, amphibians, and reptiles show pain responses that are reduced by analgesia, such as guarding, reduced activity, social withdrawal, and behavioural decisions regarding hunger and exploring new habitats, as well as cognitive bias in birds where environmental conditions impact affective states.[73] However, studies on pain in amphibians and reptiles are still lacking, particularly regarding long-term motivational and behavioural change.

Importantly, one needs to consider the vastly different environments that these taxa inhabit. Even mammals exhibit profound differences in pain nociception according to their environment. For example, while many mammals have nociceptive responses to high carbon dioxide (CO_2) levels causing peripheral tissue acidosis, this function is lost in the naked mole rat which lives in burrows rich in CO_2, and thus does not possess nociceptors that respond to acids.[74]

Fish are the most diverse vertebrate group and show significant species-specific differences in pain behaviour. However, they typically show similar physiological responses, withdrawal, conditioned place aversion and preference (with motivational trade-off for food), altered activity and feeding behaviour both current and in the future, and these are also affected by analgesia.[75]

The fact that pain indicators are species-specific is particularly apparent in the case of invertebrates, to which we turn now.

Studying pain in invertebrates has been more challenging. Molluscs have been understudied, but cephalopods especially show large complex nervous systems and advanced learning abilities, while squids show long-term changes to localised injury, with evidence of associative learning, as well as enhanced sensitivity with associated changes in defensive responses following injury, which was reduced by analgesics.[76] Some molluscs display motivational states and cognitive abilities that could be consistent with states that are functionally similar to pain, although divergence in

[72] Sneddon et al., "Defining and Assessing Animal Pain", 205.

[73] Sneddon et al., "Defining and Assessing Animal Pain", 205–206.

[74] Park, "Pain Insensitivity in the African Naked Mole-Rat".

[75] Sneddon et al., "Defining and Assessing Animal Pain", 206–207.

[76] Sneddon et al., "Defining and Assessing Animal Pain", 207; Crook and Walters, "Nociceptive Behavior and Physiology of Molluscs", 185.

evolution and differences in neuroanatom, physiology and lifestyle indicate that if pain experiences are present in molluscs, they would likely be vastly different from those in mammals.[77] This makes the assessment of pain indicators in molluscs rather challenging.

Data in arthropods is also somewhat lacking, although a few species in the subphylum of crustaceans and the class of insects have been studied to some degree.[78] Crustaceans have largely been regarded as incapable of experiencing pain and only able to respond through reflexes. However, several behavioural and physiological criteria show that they have capabilities to experience pain that go beyond mere nociception (as reviewed by Elwood, 2019).[79] Crustaceans show motivational trade-off (and prolonged motivational change) rather than consistent reflex responses to pain, with long-term memory of the motivational change. For example, hermit crabs usually quickly abandon their shells (a highly valuable resource) when they receive an electric shock, but when the odour of a predator is present, they are less likely to evacuate their shells. Evacuation depends on what one can deem their assessment of the quality of the shells, which affected future likelihood of changing shells, even after a prolonged period of time.[80] Furthermore, they show complex, prolonged grooming, rubbing, and wound-guarding of the specific site locating the noxious stimulus, as well as fitness-enhancing, anxiety-like states, such as increased weariness or risk averse behaviour (found in crayfish), which is associated with serotonin levels in the CNS.[81] In addition, crabs show rapid avoidance learning, switching shelters, or emerging into light after only two trials when exposed to shock in an alternative shelter beforehand, which is associated with high risk of predation.[82] While there is no way of telling whether these cases indicate the negative feelings implied in human pain, these experiments show at least consistency with the discussed criteria for pain as well as indicators of conscious pain discussed in this section. Thus, pain experience is clearly indicated although this indication should not be confused with claiming proof of what may be deemed a conscious, subjective pain experience that is the same as in humans.[83]

[77] Crook and Walters, "Nociceptive Behavior and Physiology of Molluscs", 185.

[78] Elwood, "Pain in Crustaceans", 1; Sneddon et al., "Defining and Assessing Animal Pain", 207–208.

[79] Elwood, "Pain in Crustaceans", 1.

[80] Elwood, "Pain in Crustaceans", 2.

[81] Elwood, "Pain in Crustaceans", 2.

[82] Elwood, "Pain in Crustaceans", 1.

[83] Elwood, "Pain in Crustaceans", 5.

Lastly, other arthropods such as insects show clear responses to noxious stimuli that can be suppressed (e.g., while feeding) or intensified, and show similar peripheral nociceptors to mammals, as in other invertebrates. However, how this nociceptive information is processed within the CNS in insects is still mostly unknown, despite some parallels between insect brains and vertebrate brains that could support the claim that insects have subjective pain experience.[84] Insects are also able to avoid places linked to noxious stimuli, can learn from electric shocks, and show pain relief learning (learning to associate normally to neutral stimuli with pain cessation and to prefer these stimuli). This indicates the capability of assessing motivational variables where nociception is a strong motivator. However, the degree to which insects are capable of processing affective information is still unclear. Notably, there is some indication of possible anxiety-like states in bees (where agitated honeybees may exhibit pessimistic cognitive biases in decision-making) and motivational components of behaviour, which may or may not require only advanced sensory processing.[85] A recent study also showed that bees display motivational trade-off of nociceptive responses, using learned colour clues and thus central processing to modulate nociceptive behaviour based on contextual information.[86] On the other hand, insects often show behaviour that seems incompatible with pain experience, for example, inconsistent protective behaviour and feeding despite severe injuries. Insects comprise a vast variety of species, and only few studies on pain in a handful of species exist thus far.

Finally, an important methodological point to be made is that researchers need to avoid adhering blindly to the principle of parsimony in their methodology of studying animal behaviour. Referred to as Morgan's canon, this principle is to explain behaviour in the simplest way until other reasons can be found has been misunderstood to encourage studying animal behaviour in terms of mindless stimulus-response machines.[87] Thus, a convincing case has been made for considering sensation, emotion, and intention when studying animals.[88] On the other hand, studies of animal pain have often been limited by comparison to human pain experience,

[84] Sneddon, "Defining and Assessing Animal Pain", 208.

[85] Sneddon, "Defining and Assessing Animal Pain", 208.

[86] Gibbons et al., "Motivational trade-offs and modulation of nociception in bumblebees".

[87] Bateson, "Assessment of Pain", 836.

[88] Bateson, "Assessment of Pain", 836; Burghardt, "Animal Awareness"; Lockwood, "Anthropomorphism"; Panksepp, "Cross-Species Affective Neuroscience", 13; de Waal, "What is an Animal Emotion?", 191–199.

and should move to a more impartial, species-appropriate approach that considers features observed in other species that might not be found in humans.[89] This, together with the importance of studying animals in their natural context, is widely utilised in cognitive ethology, which will be discussed in section "The Social Context of Pain Indicators and Expressions in Animals[90]".

Expressions of Pain in Animals[91]

We have seen that pain in animals requires studying multiple physiological and behavioural indicators, often requiring experimental paradigms. What about direct ways of expressing and communicating pain in animals?

As outlined in the section "Defining Pain in Animals", we distinguish between linguistic expressions and non-linguistic expressions in animals. Again, linguistic expressions include signs and acts of communication such as non-verbal vocalisations (utterances or tone of voice). Non-linguistic expressions include in particular the use of body language to communicate meaning. This means, various kinds of physical behaviour, as opposed to

[89] Elwood, "Pain in Crustaceans", 5.

[90] We do not here discuss the impact of language, discourses, or human expressions of pain in animals, despite its profound historical importance and impact of how animal pain has been perceived (or denied and ignored). While we argue for species-specific expression of pain in animals, human use of language and associated meaning has been used in the history of animal pain and animal experimentation to promote power of ideology and deny the capability of animals to experience pain in many ways (e.g., in animal experimentation descriptions of emotional states or painful states in animals were generally avoided, using reductive terms instead to only indicate physiological or behavioural responses, avoiding the idea of the animal as a subject or feeling being, in a similar way that animals or parts of animals are referred to as "goods", or "commodities", "products" in animal agriculture). For more detailed and historical descriptions of the way human discourse has shaped our understanding and views of animal pain and suffering, see Mitchell (2019) and Rollin (1989, 2011).

[91] We do not here discuss the impact of language, discourses, or human expressions of pain in animals, despite its profound historical importance and impact of how animal pain has been perceived (or denied and ignored). While we argue for species-specific expression of pain in animals, human use of language and associated meaning has been used in the history of animal pain and animal experimentation to promote power of ideology and deny the capability of animals to experience pain in many ways (e.g., in animal experimentation descriptions of emotional states or painful states in animals were generally avoided, using reductive terms instead to only indicate physiological or behavioural responses, avoiding the idea of the animal as a subject or feeling being, in a similar way that animals or parts of animals are referred to as "goods", or "commodities", "products" in animal agriculture). For more detailed and historical descriptions of the way human discourse has shaped our understanding and views of animal pain and suffering, see Mitchell (2019) and Rollin (1989, 2011).

words, are used to communicate meaning. This includes facial expressions, gestures, body posture, grimacing, eye movement, touch, and the use of space. While animals cannot express and communicate pain in human language, absence of more human signs of pain is no reason to deny animals pain, especially considering the diverse indicators of animal pain as discussed in the previous section. What is not much considered in the standing literature on animal pain, however, are the species-specific forms of linguistic expression that animals use to communicate. We attempt to address this lacuna particularly in the following section.

As indicated in section "Defining Pain in Animals", in our reference to linguistic expressions, we make use of Heidegger's broad phenomenological understanding of language in terms of the Greek term *logos* in *Being and Time*.[92] Heidegger renders logos in its basic meaning of language as speech or discourse (*Rede*). Speech here denotes demonstration, yet not in the sense of reasoning or proving or judging, but of pointing to, of letting something be seen from itself as far as it allows that. As such speech has a revelatory character. So, things can "say" something in terms of the meaning their sounds might have. We do not just hear meaningless sounds, which we put into meaningful words. Rather sounds always have meaning and tell us something prior to having this described in terms of sentences or statements. Heidegger writes that, for instance, we hear a creaking waggon, the north wind singing, the woodpecker tapping, the fire crackling, and in each case hear meaningful sounds such that we feel encountered, addressed, spoken to, by these things, albeit not necessarily in the form of words.[93]

Consider the application of Heidegger's broad phenomenological notion of language to the expression of pain in animals. This requires as venture-point the phenomenological attitude of letting animals show their pain from themselves. In Heidegger's view this would be openness to what they may say in their very own species-specific way. In this sense, we argued that the fact that we lack knowledge of animal language is no reason for denial but rather all the more reason to be open to explore the possibility of animal-specific non-verbal linguistic expressions of pain.

There is nothing obscure in attributing species-specific linguistic expressions to animals. An example is offered by Suzuki's award paper on field

[92] Heidegger, *Being and Time*, 2001:§7. For a detailed discussion, see Olivier, "On the Nature of Language", (2008: 311ff.).

[93] Heidegger, *Being and Time*, 207.

research of what he calls "animal linguistics" and his exploration of "referentiality and compositionality in bird calls".[94] Suzuki writes,

> Field experiments have revealed that Japanese tits produce unique alarm calls when encountering predatory snakes, which serve to enhance the visual attention of call receivers with respect to snake-like objects. Moreover, tits often combine discrete types of meaningful calls into fixed-ordered sequences according to an ordering rule, conveying a compositional message to receivers. These findings indicate that two core capabilities of language, namely, referentiality and compositionality, have independently evolved in the avian lineage.[95]

Thus, in an animal sense, with the focus on animals themselves, one can identify signs, their meaningful reference to something (semiotics), and their compositionality in systems of signification and communication (syntax), from which one can infer what may be called animal language. Suzuki seems to focus particularly on vocal communication, for instance, acoustically discrete alarm calls in Japanese tits.[96] Such acoustic expressions may be combined in messages as is demonstrated in the case of birds which produce alarm calls in "fixed ordered sequences according to an ordering rule" to warn other birds of a predator. Notably, referring to a long list of references to support his field research, Suzuki points out that such combination of sound elements to communicate does not only apply to Japanese tits but are also widely detected in bats, mice, mongooses, cetaceans, gibbons, and gorillas.

Arguably, one will do more justice to animals if one concedes their very own use of language. As we humans lack knowledge of animal linguistics, we tend to speak of non-verbal animal language in human terms. The importance here is to avoid the anthropocentric approach of modulating animal languages on human language as the gold standard of language. As Suzuki indicates, we rather need to explore new ways to talk about semiotics and syntax, thus core aspects of language.[97] In any case, keep in mind that also in humans the pain of others is often known by observing non-verbal linguistic expressions in the form of vocal communication such as groans, moans, grunts, calls, and cries. This goes also for non-linguistic

[94] Suzuki, "Animal Linguistics", 221.
[95] Suzuki, "Animal Linguistics", 221.
[96] Suzuki, "Animal Linguistics", 223.
[97] Suzuki, "Animal Linguistics", 225.

expressions such as grimaces and gestures, for instance, tears, coming before the words that are so integral to evaluating others' emotions.[98]

Due to our lack of knowledge of animal language, it remains hard to distinguish clearly between signs that indicate or express pain. We do not claim that a consistent distinction can be made between animal-specific non-verbal linguistic expressions and non-linguistic expressions and indications. We simply draw this distinction by taking a phenomenological approach that seems to find support in empirical research. Thus, ultimately, by making this distinction, we want to advocate the phenomenological attitude of openness and acknowledge the capability of animals to use their species-specific means to express their pain.

In this line of thinking, one may thus observe that animals have rich forms of non-linguistic expressions and non-verbal linguistic expressions to communicate meaning between each other, including bird song, displays of peacocks, the dance of bees, elephant calls, acoustically discrete alarm calls of monkeys for different predators, and so on.[99] Much of this is carefully studied in zoo-semiotics. However, to our knowledge, there is no literature on zoo-semiotics and pain expression.

Animals related more closely to humans show easily recognisable expressive signs of pain. Some of these that are readily observable include what we may call non-linguistic expressions of pain: "escape reactions; abnormal posture, gait or speed, guarding behaviour"; "withdrawal and recoil responses; licking, biting, chewing or scratching; frequent changes in body position—restlessness, rolling, writhing, kicking, tailflicking", "impaired breathing pattern, shallow breathing", "increased rate of breathing; muscle tension, tremor, twitching, spasm, straining; depression, sluggishness, hiding, withdrawal, lying motionless, seeking cover, sleeplessness; avoidance behaviour and aversion to the scene of the trauma".[100]

We can take some non-verbal linguistic expressions of pain to refer to vocal forms of communication; this may include, for instance, "vocalising—groaning, whimpering, crying, squealing, screaming, growling, hissing, barking"; "vocalising or aggression during movement or manipulation"; "groaning during breathing".[101]

[98] Moscoso, *Pain: A Cultural History*, 9.
[99] Mitchell, *Reading the Animal Text*, 152; Suzuki, "Animal Linguistics", 224.
[100] Gregory, *Physiology and Behaviour of Animal Suffering*, 99.
[101] Gregory, *Physiology and Behaviour of Animal Suffering*, 99.

Due to our lack of understanding of animal language, we may tend to take these expressive signs mostly to be non-linguistic indicators or expressions. However, a significant part of them is what one can call an animal-specific linguistic repertoire that serves to express and communicate specific meanings. Animals will know this better themselves and ultimately we need to resort to making inferences.

Prima facie, pain and other related sensory, affective, and motivational states can be easily recognised in some animals. Strong inferences about what they are feeling can be made from their faces, eyes, posture, the way they carry themselves, muscle tone, gait, facial expressions, gaze, odours, pheromones, or else vocalisations.[102] People who have very little experience observing animals are mostly in agreement with what an animal is most likely feeling, and training can drastically improve both accuracy and intersubjective agreement, inter- and intra-reliability.[103] These intuitions about animals' (linguistic and non-linguistic) expression of their emotional states are quite reliable and often accurately predict future behaviour. Bekoff (2005) states that "this predictive value offers one of the strongest arguments for the existence of emotions and feelings in nonhuman species".[104]

Importantly, some species are more expressive than others about their pain. For example, sheep and deer have limited or only subtle vocal responses, which can lead to the belief that they have only limited capabilities of experiencing pain. Even nonhuman primates show relatively little expressive reaction to injury or pain, particularly with human presence, and look unwell only with grave illness or severe injury, then, for example, they may look miserable, dejected, huddled with a crouched posture, grimace and moan or scream, clench teeth, shake, and even self-mutilate, stop grooming, and avoid companions. Birds with acute pain may vocalise it through distress calls, but with chronic pain may show similar passive immobility with crouched posturing, closed eyes and head drawn inwards, wings flat against their body, and the neck retracted, resulting in a drooping, miserable appearance. Dogs may be more expressive and thus evoke more care and sympathy from caregivers.[105] Still yet, dogs may suppress signs of injury or pain to avoid looking vulnerable in front of other dogs,

[102] Bekoff, "Animal Emotions".

[103] Mogil et al., "Facial Grimace Scales for Pain Measurement", 482–490.

[104] Bekoff, "Animal Emotions", 18.

[105] Gregory, *Physiology and Behaviour of Animal Suffering*, 101.

and thus avoid flinching, limping, or whimpering, but this does not mean that they are not in pain. Animal stoicism is widespread in the animal kingdom and may have evolved as part of maintaining social rank, as weakness perceived by others can be detrimental to that animal in some societies.[106] Dogs then may show pain expression when alone; then they may limp, lick incessantly, shrink away, shiver, whimper or even howl, or use facial expressions as a communicative function with humans.[107]

In addition, pain expression in new-borns can be quite different to adult animals. While adult dogs withdraw the specific limb when receiving a mild shock on the leg, puppies under five days show generalised flexion of all limbs, whining and attempts to crawl away, and human and rat infants also only show generalised whole-body writhing movement with pain.[108]

Recognising pain thus can be difficult even for experienced researchers. A major difficulty is differentiating pain from distress or general lack of well-being. This is particularly challenging in the case of reptiles, where chronic or persistent pain is associated with inappetence, lethargy, weight loss, and acute pain by flinching, muscle contractions, attempts to bite, and aversive movements. However, it is difficult to associate these signs specifically with pain, as with most other animals.

A particularly active development in recent research over the past decade is the assessment of facial expressions as signs of pain in multiple species. As above, we take facial expressions in animals to belong to their species-specific non-linguistic expressive repertoire—arguably, this overlaps with what we call non-linguistic pain indicators. Facial grimace scales, which are based on facial expressions that indicate pain, have been developed for at least ten different species of mammals (reviewed by Mogil et al., 2020),[109] including mice and rats, based on orbital tightening, nose/cheek flattening, ear and whisker changes.[110] Other species added over the last few years include horses, rabbits, cats, sheep, ferrets, seals, cows, and pigs.[111]

Interestingly, these facial expressions are closely related to facial expressions of pain in humans and infants (the characteristic universal pain

[106] Breed, *Ethology and Animal Behavior*, 197–198.

[107] Mota-Rojas, "Dog's Emotions, Facial Expressions", 4.

[108] Gregory, *Physiology and Behaviour of Animal Suffering*, 101.

[109] Mogil et al., "Facial Grimace Scales for Pain Measurement", 480.

[110] Klune et al., "Comparing the Rat Grimace Scale", 2.

[111] Mogil et al., "Facial Grimace Scales for Pain Measurement", 486.

expression) that occur with potentially painful stimuli, and thus may be conserved across mammalian taxa,[112] which was already proposed by Darwin in 1872. Indeed, observing facial expressions for pain has been used as a method to measure pain in humans unable to express pain verbally, such as neonates and young children.[113]

It has been suggested that these facial expressions even measure the affective component of pain, and emotional states could be clearly distinguished in mice using facial expressions.[114] Notably, the development of the first of these facial expression scales for pain, the *Mouse Grimace Scale*,[115] was first introduced during the discovery that mice were capable of the emotional contagion of pain, which is related to empathy.[116] Langford et al. found that emotional contagion occurred when mice observed other mice being in pain. It is still uncertain, however, to what degree facial expression in mice or rats serve a purposeful communicative function for pain.[117]

It is still under discussion which pain behaviours in animals are what may be considered instinctive responses that may not indicate or express conscious pain, and which pain behaviours have evolved for the purpose of consciously communicating danger signals and threat of pain to their conspecifics. Caution is once again vital here. While the presence of recognisable behaviour may be convincing, the absence of human-like or other pain indicators and expressions is not yet evidence of the absence of pain.

The Social Context of Pain Indicators and Expressions in Animals

We have tried to show how animals display a wide variety of indicators and expressions of pain with many features common to humans but also species-specific aspects. We argue in this closing section that these indicators and expressions are particular to the environment, evolutionary development, and, importantly, social factors that shape the experience of animals.

[112] McMillan, *Mental Health and Well-being in Animals*, 334.
[113] Mogil et al., "Facial Grimace Scales for Pain Measurement", 482.
[114] Dolensek et al., "Facial expressions of emotion states", 89.
[115] Langford et al., "Facial Expressions of Pain", 447.
[116] Langford et al., "Social Modulation of Pain as Evidence for Empathy in Mice", 1967.
[117] Mogil et al., "Facial Grimace Scales for Pain Measurement", 483.

The indication and expression of animal pain, and thus its study and assessment, is very much dependent on the evolutionary, ecological, and social contexts that the individual animal is situated in. One area of study that supports this observation is cognitive ethology, coined by Donald Griffin. Griffin insisted that studies of animal consciousness should be conducted together with its evolutionary basis, thereby revolutionising the science of animal behaviour.[118] Cognitive ethology can be defined as "the comparative, evolutionary, and ecological study of nonhuman animal minds, including thought processes, beliefs, rationality, information processing, intentionality, and consciousness".[119] Cognitive ethology is not to be confused with classical ethology in early Europe, which, like behaviourism, did not consider the study of internal states of minds in animals, even though early proponents like Konrad Lorenz suggested rich mental lives in animals.[120] Unlike cognitive ethology, classical ethology considered only instinct and innateness, mostly disregarding learning or behavioural flexibility, which is of particular importance in cognitive ethology, especially in Griffin's arguments for conscious awareness.[121] Thus, cognitive ethology aims to know how brains and mental capabilities in animals evolved in natural conditions, the selective pressures involved, and how these resulted in the large diversity of brains, mental abilities, and behaviours seen in so many species.[122] Essentially, the goal may be to understand the animal's point of view, even what it would be like to be another animal.[123]

Cognitive ethology also aims to counter the belief that experimental equipment and designs are sufficient to study animal minds, and advocate niche-related specialisation, highly tuned to animal subjects in their natural environment.[124] On the other hand, cognitive ethology is also criticised for being insufficiently experimental, requiring more adequate experimental control measures.[125] To address this critique, cognitive ethology is challenged to ensure careful observation, which may be resource- and

[118] Allen and Bekoff, "Animal Minds, Cognitive Ethology", 299; Griffin, "Prospects for a Cognitive Ethology", 527–538.

[119] Allen and Bekoff, "Animal Minds, Cognitive Ethology", 299.

[120] Allen and Bekoff, "Animal Minds, Cognitive Ethology".

[121] Allen and Bekoff, "Animal Minds, Cognitive Ethology".

[122] Bekoff, "Animal Emotions", 24; de Waal, "What is an Animal Emotion?", 191.

[123] Bekoff, "Animal Emotions", 24.

[124] Allen and Bekoff, "Animal Minds, Cognitive Ethology", 308.

[125] Allen and Bekoff, "Animal Minds, Cognitive Ethology", 305.

time-intensive, as it is ideally located in the animals' natural environment rather than in artificial laboratory settings. This challenge shows in fact an advantage of cognitive ethology, which is to demand recognition of the diversity that arises from species adapting to their specific environments.[126]

The emphasis of studying behaviour of animals in the environments that they evolved or reside in is specifically important in order to account for the complexities and nuanced ways in which animals indicate and express their experiences.[127] If one takes seriously the subjective nature of experience, it is important to understand the individual perspectives of animals and how they live in their own world. Thus, the specific and unique situations that they evolved and live in have to be studied; the researcher must attempt to avoid understanding them only from their own human perspective.[128] Each animal has a specific way of interacting with its environment, their "self-world", which is the starting point to understanding the life of an animal, and is captured by the notion of *Umwelt* by the twentieth-century biologist Jakob von Uexküll.

One of the main components in cognitive ethology is this "private experience", as suggested by Burghardt (1997),[129] which includes the deliberate effort to understand mental states such as perceptions and emotions of other animals.[130] Instead of avoiding the study of mental states such as emotions for fear of being non-scientific, it becomes a centre piece, together with the environment that shapes these states.[131] Emotions, just as other behavioural phenotypes, have evolved in specific environments, which is why field research is so essential for their study.[132]

In the same line of thought, we argue that, like emotions and other mental states, pain experience has evolved in specific contexts, with species-specific differences. This means that understanding animal pain and its indicators and expressions requires studying the specific contexts,

[126] Allen and Bekoff, "Animal Minds, Cognitive Ethology", 308; de Waal, "What is an Animal Emotion?", 191.

[127] Bekoff, "Animal Emotions", 24.

[128] Bekoff, "Animal Emotions", 25; Allen and Bekoff, *Species of Mind*; de Waal, "What is an Animal Emotion?", 191.

[129] Burghardt, "A Fifth Aim for Ethology", 276.

[130] Bekoff, "Animal Emotions", 24.

[131] Bekoff, "Animal Emotions", 25.

[132] Panksepp, "Cross-Species Neuroscientific Understanding of the Affective Mind", 548; de Waal, "What is an Animal Emotion?", 191; Bekoff, "Animal Emotions", 25.

ecological and social, in which animals live. We focus specifically here on the *social* context.

Strikingly, there is little to no literature on animal pain in cognitive ethology, even though cognitive ethology has been widely applied more broadly in research on animal welfare and well-being.[133] This kind of study requires an intense but broad approach with coordination from the fields of ethology, neurobiology, psychology, and philosophy.[134]

In the closing subsection we try to address this lacuna by linking the approaches in ethology with pain indication and expression in animals.

Pain has social significance both through its effect on the animal's social environment and vice versa through the impact of the social environment on their pain experience. On the one hand, pain affects the animal's social environment. In some social contexts, pain can act as a warning signal that helps animals to meet their social roles and expectations; in others it can be an impediment that hinders them to do that. Pain that is extreme or disabling can prevent the individual from being socially useful, show vulnerability, and can result in social marginalisation or abandonment if prolonged, which is partly why so many animals are able to hide their pain so well. On the other hand, social contexts affect how animals experience, indicate, express, and communicate their pain,[135] for instance, through social contagion and empathy.[136]

Using some examples, we specifically try to demonstrate in the following how through social contagion and empathy social contexts influence pain and vice versa how pain shapes these same contexts in return. Ironically, the studies available are mostly not from natural social contexts as is advocated by cognitive ethology, but from experimental sources. Thus, our argument still needs more support from such studies.

Church (1995) was the first to demonstrate through experiments that rats behaved in ways that terminated the distress of a conspecific.[137] Similar behaviour has been observed in pigeons.[138] The first experimental discovery that mice are capable of empathy in the sense of "emotional conta-

[133] Dawkins, "Motivation, Fitness, and Animal Welfare", 1; Desire, Bossy, and Veissier, "Emotions in Farm Animals", 165.

[134] Bekoff, "Animal Emotions", 23.

[135] For a further going discussion of the shaping effect of the social context and language on pain experience in human animals, see Olivier, *The Social Dimension of Pain*, 2022.

[136] De Waal and Preston, "Mammalian Empathy", 498.

[137] Church, "Emotional Reactions of Rats to the Pains of Others", 132.

[138] Watanabe and Oho, "Experimental Analysis of 'Empathic' Response", 269.

gion" was only in 2006 by Langford and colleagues.[139] This study showed that the pain sensitivity of mice was modulated by exposure to their cage mates being in pain, with no effect observed with strangers. When given an equally noxious stimulus, these mice showed increased pain behaviour, and this was dependent on the visual observation of their cage mate being in pain. There was also a bidirectional influence on pain behaviour by their cage mate's status when given noxious stimuli of different intensities (i.e., if the neighbour received a stronger noxious stimulus, the mouse with the lighter noxious stimulus displayed more severe pain than they usually would, and vice versa). The authors also found that general pain mechanisms would have been sensitised due to pain sensitivity in different modalities also being altered solely by observing a cage mate in pain. They ruled out that this was just a generalised stress response.[140]

Other studies with similar findings showed that these effects are not explained by general distress responses, and hyperalgesia (increased sensitivity to feeling pain and an extreme response to pain) does not co-occur with anxiety, changes in corticosterone (stress) or stress-induced hyperalgesia.[141] Indeed, multiple studies have shown this social transfer and modulation of pain experience in mice,[142] even on a persistent basis.[143] Depending on the social context and multiple factors, distress in a conspecific can both cause pain hypersensitivity as described and cause antinociception or act as analgesia in rodents. The proposed mechanism here is interestingly olfactory[144] and oxytocin (a neuropeptide hormone involved in social bonding) in the cage mate.[145]

These findings are not just limited to empathic sharing and pain modulation, but also involve prosocial behaviours towards their conspecifics in distress, grooming and consoling each other,[146] helping to avoid further pain/distress,[147] and freeing their cage mate when in a restrainer. In this

[139] Langford et al., "Social Modulation of Pain as Evidence for Empathy in Mice", 1967.

[140] Langford et al., "Social Modulation of Pain as Evidence for Empathy in Mice", 1967.

[141] Smith et al., "Social Transfer of Pain in Mice".

[142] Kim et al., "Social Transmission of Fear in Rats"; Smith et al., "Social Transfer of Pain in Mice", Smith et al., "Social Transfer of Pain and Analgesia", 153; Lu et al., "Social Interaction with a Cagemate in Pain", 385.

[143] Baptista-de-Souza, "Mice Undergoing Neuropathic Pain", 664.

[144] Fanselow, "Stressed Rats Produce Opioid Analgesia", 589.

[145] Burkett, "Oxytocin-Dependent Consolation Behavior", 375.

[146] Lu et al., "Social Interaction with a Cagemate in Pain" 385.

[147] Church, "Emotional Reactions of Rats", 132.

latter study, when there was a second restrainer with chocolate, rats typically freed their cage mate first and then shared the chocolate, and rodents are known to sacrifice personal gain to relieve pain or distress in fellow rodents. This indicates the importance of pain-eliciting behaviours in animals that serve as social cues and that can be transferred socially through interaction, with empathic sharing of painful experiences being a motivating factor leading to altruistic behaviour.[148] Thus, through social interaction, important information about environmental threats are learned or transferred to conspecifics, enhancing survival, while also reducing pain experience.[149] Empathic pain can even "include sensory information when the stimuli and task instructions emphasize the specific nature of the feeling at the location of the injury".[150]

Multiple studies have investigated the neural correlates of the social transfer of pain.[151] Smith et al. (2017) have indicated the importance of the anterior cingulate cortex (ACC), involved in the affective component of pain, with another study (Smith et al. 2021: 1) showing the importance of ACC circuitry to the nucleus accumbens (important for motivation) in the social transfer of pain and analgesia, but not in the social transfer of fear. Furthermore, the medial prefrontal cortex is critical for the processing of enhanced pain perception in the spine as a result of empathy of pain.[152]

The seminal question is how animals may *communicate* these pain experiences resulting in these empathic responses, given the vital role of animal pain indication and expression to assess and recognise animal pain.

In the social transfer of pain, it seems that multiple sensory modalities are involved. The facial grimacing discussed in section "Animal Responses Indicative of Conscious Pain: Vertebrates and Invertebrates", as a mode of the expression of pain, may be involved as a key facilitator in the social communication of pain in various animals, although there is still not enough definite evidence that animals use grimacing specifically for communication.[153]

[148] Lu et al., "Social Interaction with a Cagemate in Pain", 387.

[149] Lu et al., "Social Interaction with a Cagemate in Pain", 387.

[150] De Waal and Preston, "Mammalian Empathy".

[151] Smith et al., "Social Transfer of Pain in Mice", Smith et al., "Social Transfer of Pain and Analgesia", 153; Li et al., "Cagemate in Pain Facilitates Subsequent Spinal Nociception", 1253.

[152] Li et al., "Cagemate in Pain Facilitates Subsequent Spinal Nociception", 1253.

[153] Mogil et al., "Facial Grimace Scales for Pain Measurement".

Notably, mice clearly responded to visual images of conspecifics,[154] discriminate between and show preferences for videos of different social behaviours in conspecifics.[155] Similarly, rats avoided a space that showed photographs of rats experiencing foot shock compared to photographs of rats with neutral expressions, and this avoidance disappeared when either the face or the body of the rats in the photographs were airbrushed.[156]

Another study has shown that olfactory cues can be sufficient for the social transfer of pain,[157] but are not necessary.[158] Importantly, emotions that impact pain experience include fear, and the social transmission of fear is mainly dependent on ultrasonic vocalisation.[159] Such vocalisation can be considered a significant form of species-specific non-verbal linguistic expression used to communicate pain.

Ultimately, the way that animals communicate their pain and how this results in the social transfer of pain is likely to vary between different species and their ecological, including social, environments. While we used these specific examples from studies on mice and rats, pain and social environment are similarly intertwined in other animals. This is particularly evident in social animals like dogs, and the impact of shared sensory, emotional, and motivational features of physical pain on their social bonds.[160] Notably, de Waal brings to the fore indirect but important research on the social impact of affective components of pain, such as anxiety-like states and distress, and its motivation for altruistic behaviour, among non-mammalian animals as well.[161] Thus, while there are few if any studies on the social transfer and relationships of pain in animals other than rodents, there seems to be some evidence at least of the social communication of pain and its effect also in other species.

These observations reveal the complex, multifaceted relationship between pain experience and its indicators and expressions in the social environment of animals, directly shaping their pain experience. We suggest that combined approaches and considerations are important to

[154] Yakura et al., "Visual Recognition".

[155] Watanabe et al., "Conspecific Social Behaviour in Mice", 523.

[156] Nakashima et al., "Emotional Signal of Pain from Conspecifics".

[157] Smith et al., "Social Transfer of Pain in Mice".

[158] Langford et al., "Social Modulation of Pain as Evidence for Empathy in Mice", 1967.

[159] Kim et al., "Social Transmission of Fear in Rats".

[160] Eisenberger, "The Pain of Social Disconnection", 421; McMillan, "The Psychobiology of Social Pain", 154.

[161] De Waal and Preston, "Mammalian Empathy".

further our understanding of animal pain, and that observations of phenomena should further take place in the animals' natural ecological environments. This has important implications for understanding and studying pain experience in animals, its indicators and expressions, as well as their treatment, highlighting the need to pay close attention to the social environment and dynamics.

Conclusion

Animal pain is still a contentious issue of discussion, despite multiple lines of evidence from anatomical, physiological, behavioural, and psychological data. We show that social context and dynamics, widely recognised to play a major role in the formation and experience of pain in humans, also constitute a crucial component in the recognition and assessment of animal pain, as the individual animal experiences pain within their species-specific social environment. This can be observed by studying the multiple variable indicators and expressions of pain in animals. We tried to show the importance of taking seriously the socio-ecological environment and species-specific and context-specific indicators and expressions in the study of animal minds. However, there is a lack of research on indicators and expressions of animal pain and social context, with very few studies investigating the involvement of the social environment and pain in animals, apart from a handful of very telling studies involving exclusively rodents. Thus, there is a dire need to observe how pain affects the social environment and how the social environment in turn affects pain in animals. This requires special awareness of its specificity according to species, their particular environment, and form of social life, expression, and communication. Ultimately, our chapter is a call for further studies of pain in the animal's natural socio-ecological environment to assist in the recognition of which animals likely experience pain, how it is indicated and expressed, the social context it is experienced in, and how this shapes their experience.

Bibliography

Allen, Colin. 'Animal Pain'. *Nous* 38, no. 4 (December 2004): 617–43. https://doi.org/10.1111/j.0029-4624.2004.00486.x.

———. 'The Discovery of Animal Consciousness: An Optimistic Assessment'. *Journal of Agricultural and Environmental Ethics* 10, no. 3 (1997): 217–25. https://doi.org/10.1023/A:1007716908222.

Allen, Colin, and Marc Bekoff. 'Animal Minds, Cognitive Ethology, and Ethics'. *The Journal of Ethics* 11, no. 3 (September 2007): 299–317. https://doi.org/10.1007/s10892-007-9016-5.

———. Species of Mind: The Philosophy and Biology of Cognitive Ethology. Cambridge, MA; London: MIT, 1999.

Allen, Colin, and Mark Bekoff. 'Animal Consciousness'. In *The Blackwell Companion to Consciousness*, edited by Max Velmans and Susan Schneider, 58–71. Malden, MA, USA: Blackwell Publishing, 2007. https://doi.org/10.1002/9780470751466.ch5.

Allen, Colin, Perry N Fuchs, Adam Shriver, and Hilary D Wilson. '20 Deciphering Animal Pain'. In *Pain: New Essays on Its Nature and the Methodology of Its Study*, edited by Murat Aydede, 351–66. Cambridge, MA: MIT Press, 2005.

Baptista-de-Souza, Daniela, Ana C. Nunciato, Barbara C. Pereira, Gabriel Fachinni, Caroline R. Zaniboni, and Azair Canto-de-Souza. 'Mice Undergoing Neuropathic Pain Induce Anxiogenic-like Effects and Hypernociception in Cagemates'. *Behavioural Pharmacology* 26, no. 7 (October 2015): 664–72. https://doi.org/10.1097/FBP.0000000000000170.

Barrot, Michel, Eric Salvat, and Ipek Yalcin. 'Tests and Models to Study Pain in Animal-Based Translational Research'. In *Meanings of Pain*, edited by Simon van Rysewyk, 375–88. Cham: Springer International Publishing, 2016. https://doi.org/10.1007/978-3-319-49022-9_22.

Bateson, Patrick. 'Assessment of Pain in Animals'. *Animal Behaviour* 42, no. 5 (November 1991): 827–39. https://doi.org/10.1016/S0003-3472(05)80127-7.

Bekoff, Marc. 'The Question of Animal Emotions: An Ethological Perspective'. In *Mental Health and Well-Being in Animals*, edited by Franklin D. McMillan, 15–27. Oxford, UK: Blackwell Publishing Ltd, 2008. https://doi.org/10.1002/9780470384947.ch2.

Birch, Jonathan. 'Animal Sentience and the Precautionary Principle'. *Animal Sentience* 2, no. 16 (1 January 2017). https://doi.org/10.51291/2377-7478.1200.

Breed, Michael D. *Conceptual Breakthroughs in Ethology and Animal Behavior*. London, United Kingdom: San Diego, CA: Academic Press, an imprint of Elsevier, 2017.

Burghardt, Gordon M. 'Amending Tinbergen: A Fifth Aim for Ethology'. In *Anthropomorphism, Anecdotes, and Animals*, edited by Robert W. Mitchell, Nicholas S. Thompson, and H. Lyn Miles, 254–76. Albany, New York: State University of New York Press, 1997.

———. 'Animal Awareness: Current Perceptions and Historical Perspective'. *American Psychologist* 40, no. 8 (1985): 905–19. https://doi.org/10.1037/0003-066X.40.8.905.

Burkett, J. P., E. Andari, Z. V. Johnson, D. C. Curry, F. B. M. de Waal, and L. J. Young. 'Oxytocin-Dependent Consolation Behavior in Rodents'. *Science*

351, no. 6271 (22 January 2016): 375–78. https://doi.org/10.1126/science.aac4785.

Church, Russell M. 'Emotional Reactions of Rats to the Pain of Others.' *Journal of Comparative and Physiological Psychology* 52, no. 2 (1959): 132–34. https://doi.org/10.1037/h0043531.

Crook, R. J., and E. T. Walters. 'Nociceptive Behavior and Physiology of Molluscs: Animal Welfare Implications'. *ILAR Journal* 52, no. 2 (1 January 2011): 185–95. https://doi.org/10.1093/ilar.52.2.185.

Damasio, Antonio, and Hanna Damasio. 'Pain and Other Feelings in Humans and Animals'. *Animal Sentience* 1, no. 3 (1 January 2016). https://doi.org/10.51291/2377-7478.1064.

Dawkins, Marian Stamp. 'From an Animal's Point of View: Motivation, Fitness, and Animal Welfare'. *Behavioral and Brain Sciences* 13, no. 1 (March 1990): 1–9. https://doi.org/10.1017/S0140525X00077104.

Désiré, Lara, Alain Boissy, and Isabelle Veissier. 'Emotions in Farm Animals': *Behavioural Processes* 60, no. 2 (November 2002): 165–80. https://doi.org/10.1016/S0376-6357(02)00081-5.

Devor, Marshall. 'Where Is Pain in the Brain?' *Animal Sentience* 1, no. 3 (1 January 2016). https://doi.org/10.51291/2377-7478.1071.

Dolensek, Nate, Daniel A. Gehrlach, Alexandra S. Klein, and Nadine Gogolla. 'Facial Expressions of Emotion States and Their Neuronal Correlates in Mice'. *Science* 368, no. 6486 (3 April 2020): 89–94. https://doi.org/10.1126/science.aaz9468.

Eisenberger, Naomi I. 'The Pain of Social Disconnection: Examining the Shared Neural Underpinnings of Physical and Social Pain'. *Nature Reviews Neuroscience* 13, no. 6 (June 2012): 421–34. https://doi.org/10.1038/nrn3231.

Elwood, Robert W. 'Assessing Negative and Positive Evidence for Animal Pain'. *Animal Sentience* 2, no. 16 (1 January 2017). https://doi.org/10.51291/2377-7478.1283.

———. 'Discrimination between Nociceptive Reflexes and More Complex Responses Consistent with Pain in Crustaceans'. *Philosophical Transactions of the Royal Society B: Biological Sciences* 374, no. 1785 (11 November 2019): 20190368. https://doi.org/10.1098/rstb.2019.0368.

Fanselow, Michael S. 'Odors Released by Stressed Rats Produce Opioid Analgesia in Unstressed Rats.' *Behavioral Neuroscience* 99, no. 3 (1985): 589–600. https://doi.org/10.1037/0735-7044.99.3.589.

Gibbons, Matilda, Elisabetta Versace, Andrew Crump, Bartosz Baran, and Lars Chittka. 'Motivational Trade-Offs and Modulation of Nociception in Bumblebees'. Proceedings of the National Academy of Sciences 119, no. 31 (2 August 2022): e2205821119. https://doi.org/10.1073/pnas.2205821119.

Gregory, Neville G. *Physiology and Behaviour of Animal Suffering*. Oxford, UK: Blackwell Science, 2004.

Griffin, Donald R. 'Prospects for a Cognitive Ethology'. *Behavioral and Brain Sciences* 1, no. 4 (December 1978): 527–38. https://doi.org/10.1017/S0140525X00076524.

Heidegger, Martin. *On the Essence of Language: The Metaphysics of Language and the Essencing of the Word: Concerning Herder's Treatise on the Origin of Language*. Translated by Wanda Torres Gregory and Yvonne Unna. Albany: State University of New York Press, 2004.

———. 'The Question Concerning Technology'. In *Basic Writings*, edited by David Farrell Krell. New York: Harper & Row, 1977.

Husserl, Edmund. *Logical Investigations*. Edited by Dermot Moran. Translated by John N. Findlay. International Library of Philosophy. London; New York: Routledge, 2001.

Institute for Laboratory Animal Research (U.S.), ed. *Recognition and Alleviation of Pain in Laboratory Animals*. Washington, DC: National Academies Press, 2009.

Key, Brian. 'Why Fish Do Not Feel Pain'. *Animal Sentience* 1, no. 3 (1 January 2016). https://doi.org/10.51291/2377-7478.1011.

Kim, Eun Joo, Earnest S. Kim, Ellen Covey, and Jeansok J. Kim. 'Social Transmission of Fear in Rats: The Role of 22-KHz Ultrasonic Distress Vocalization'. Edited by Georges Chapouthier. *PLoS ONE* 5, no. 12 (1 December 2010): e15077. https://doi.org/10.1371/journal.pone.0015077.

Klune, Cassandra B., Amy E. Larkin, Vivian S.Y. Leung, and Daniel Pang. 'Comparing the Rat Grimace Scale and a Composite Behaviour Score in Rats'. Edited by Francesco Staffieri. *PLOS ONE* 14, no. 5 (31 May 2019): e0209467. https://doi.org/10.1371/journal.pone.0209467.

Kurejova, Martina, Ulrike Nattenmüller, Ullrich Hildebrandt, Deepitha Selvaraj, Sebastian Stösser, and Rohini Kuner. 'An Improved Behavioural Assay Demonstrates That Ultrasound Vocalizations Constitute a Reliable Indicator of Chronic Cancer Pain and Neuropathic Pain'. *Molecular Pain* 6 (1 January 2010): 1744-8069-6–18. https://doi.org/10.1186/1744-8069-6-18.

Langford, Dale J, Andrea L Bailey, Mona Lisa Chanda, Sarah E Clarke, Tanya E Drummond, Stephanie Echols, Sarah Glick, et al. 'Coding of Facial Expressions of Pain in the Laboratory Mouse'. *Nature Methods* 7, no. 6 (June 2010): 447–49. https://doi.org/10.1038/nmeth.1455.

Langford, Dale J., Sara E. Crager, Zarrar Shehzad, Shad B. Smith, Susana G. Sotocinal, Jeremy S. Levenstadt, Mona Lisa Chanda, Daniel J. Levitin, and Jeffrey S. Mogil. 'Social Modulation of Pain as Evidence for Empathy in Mice'. *Science* 312, no. 5782 (30 June 2006): 1967–70. https://doi.org/10.1126/science.1128322.

Li, Zhen, Yun-Fei Lu, Chun-Li Li, Yan Wang, Wei Sun, Ting He, Xue-Feng Chen, Xiao-Liang Wang, and Jun Chen. 'Social Interaction with a Cagemate in Pain Facilitates Subsequent Spinal Nociception via Activation of the Medial

Prefrontal Cortex in Rats'. *Pain* 155, no. 7 (July 2014): 1253–61. https://doi.org/10.1016/j.pain.2014.03.019.

Lockwood, Randall. 'Anthropomorphism Is Not a Four-Letter Word'. In *Advances in Animal Welfare Science 1985*, edited by Michael W. Fox and Linda D. Mickley, 185–99. Advances in Animal Welfare Science. Dordrecht: Springer Netherlands, 1986. https://doi.org/10.1007/978-94-009-4247-9_10.

Lu, Yun-Fei, Bo Ren, Bin-Fang Ling, Jing Zhang, Chen Xu, and Zhen Li. 'Social Interaction with a Cagemate in Pain Increases Allogrooming and Induces Pain Hypersensitivity in the Observer Rats'. *Neuroscience Letters* 662 (January 2018): 385–88. https://doi.org/10.1016/j.neulet.2017.10.063.

McMillan, Franklin D., ed. *Mental Health and Well-Being in Animals*. 2nd edition. Wallingford, Oxfordshire; Boston, MA: CABI, 2020.

———. 'The Psychobiology of Social Pain: Evidence for a Neurocognitive Overlap with Physical Pain and Welfare Implications for Social Animals with Special Attention to the Domestic Dog (*Canis familiaris*)'. *Physiology & Behavior* 167 (December 2016): 154–71. https://doi.org/10.1016/j.physbeh.2016.09.013.

Mitchell, Les. *Reading the Animal Text in the Landscape of the Damned*. First edition. Grahamstown, South Africa: NISC, 2019.

Mogil, Jeffrey S. 'The Measurement of Pain in the Laboratory Rodent'. In *The Oxford Handbook of the Neurobiology of Pain*, edited by John N. Wood, 1st ed., 28–60. Oxford University Press, 2019. https://doi.org/10.1093/oxfordhb/9780190860509.013.21.

Mogil, Jeffrey S., Daniel S.J. Pang, Gabrielle Guanaes Silva Dutra, and Christine T. Chambers. 'The Development and Use of Facial Grimace Scales for Pain Measurement in Animals'. *Neuroscience & Biobehavioral Reviews* 116 (September 2020): 480–93. https://doi.org/10.1016/j.neubiorev.2020.07.013.

Moscoso, Javier. *Pain*. London: Palgrave Macmillan UK, 2012. https://doi.org/10.1057/9781137284235.

Mota-Rojas, Daniel, Míriam Marcet-Rius, Asahi Ogi, Ismael Hernández-Ávalos, Chiara Mariti, Julio Martínez-Burnes, Patricia Mora-Medina, et al. 'Current Advances in Assessment of Dog's Emotions, Facial Expressions, and Their Use for Clinical Recognition of Pain'. *Animals* 11, no. 11 (22 November 2021): 3334. https://doi.org/10.3390/ani11113334.

Nakashima, Satoshi F., Masatoshi Ukezono, Hiroshi Nishida, Ryunosuke Sudo, and Yuji Takano. 'Receiving of Emotional Signal of Pain from Conspecifics in Laboratory Rats'. *Royal Society Open Science* 2, no. 4 (April 2015): 140381. https://doi.org/10.1098/rsos.140381.

Olivier, Abraham. 'On the Nature of Language—Heidegger and African Philosophy'. *South African Journal of Philosophy* 27, no. 4 (January 2008): 310–24. https://doi.org/10.4314/sajpem.v27i4.31520.

———. 'The Social Dimension of Pain'. *Phenomenology and the Cognitive Sciences*, 23 December 2022. https://doi.org/10.1007/s11097-022-09879-7.

Panksepp, Jaak. 'Cross-Species Affective Neuroscience Decoding of the Primal Affective Experiences of Humans and Related Animals'. Edited by Angela Sirigu. *PLoS ONE* 6, no. 9 (7 September 2011a): e21236. https://doi.org/10.1371/journal.pone.0021236.

———. 'Toward a Cross-Species Neuroscientific Understanding of the Affective Mind: Do Animals Have Emotional Feelings?' *American Journal of Primatology* 73, no. 6 (June 2011b): 545–61. https://doi.org/10.1002/ajp.20929.

Panksepp, Jaak, and Jeffrey Burgdorf. '50-KHz Chirping (Laughter?) In Response to Conditioned and Unconditioned Tickle-Induced Reward in Rats: Effects of Social Housing and Genetic Variables'. *Behavioural Brain Research* 115, no. 1 (October 2000): 25–38. https://doi.org/10.1016/S0166-4328(00)00238-2.

Park, Thomas J., Ying Lu, René Jüttner, Ewan St. J. Smith, Jing Hu, Antje Brand, Christiane Wetzel, et al. 'Selective Inflammatory Pain Insensitivity in the African Naked Mole-Rat (*Heterocephalus glaber*)'. Edited by Allan I Basbaum. *PLoS Biology* 6, no. 1 (29 January 2008): e13. https://doi.org/10.1371/journal.pbio.0060013.

Raja, Srinivasa N., Daniel B. Carr, Milton Cohen, Nanna B. Finnerup, Herta Flor, Stephen Gibson, Francis J. Keefe, et al. 'The Revised International Association for the Study of Pain Definition of Pain: Concepts, Challenges, and Compromises'. *Pain* 161, no. 9 (September 2020): 1976–82. https://doi.org/10.1097/j.pain.0000000000001939.

Rollin, Bernard E. 'Animal Pain: What It Is and Why It Matters'. *The Journal of Ethics* 15, no. 4 (December 2011): 425–37. https://doi.org/10.1007/s10892-010-9090-y.

Rose, J D, R Arlinghaus, S J Cooke, B K Diggles, W Sawynok, E D Stevens, and C D L Wynne. 'Can Fish Really Feel Pain?' *Fish and Fisheries* 15, no. 1 (March 2014): 97–133. https://doi.org/10.1111/faf.12010.

Smith, Monique L., Naoyuki Asada, and Robert C. Malenka. 'Anterior Cingulate Inputs to Nucleus Accumbens Control the Social Transfer of Pain and Analgesia'. *Science* 371, no. 6525 (8 January 2021): 153–59. https://doi.org/10.1126/science.abe3040.

Smith, Monique L, Caroline M Hostetler, Mary M Heinricher, and Andrey E Ryabinin. 'Social Transfer of Pain in Mice'. *SCIENCE ADVANCES*, 2016.

Sneddon, Lynne U. 'Comparative Physiology of Nociception and Pain'. *Physiology* 33, no. 1 (1 January 2018): 63–73. https://doi.org/10.1152/physiol.00022.2017.

———. 'Evolution of Nociception and Pain: Evidence from Fish Models'. *Philosophical Transactions of the Royal Society B: Biological Sciences* 374, no. 1785 (11 November 2019): 20190290. https://doi.org/10.1098/rstb.2019.0290.

Sneddon, Lynne U., Robert W. Elwood, Shelley A. Adamo, and Matthew C. Leach. 'Defining and Assessing Animal Pain'. *Animal Behaviour* 97 (November 2014): 201–12. https://doi.org/10.1016/j.anbehav.2014.09.007.

Suzuki, Toshitaka N. 'Animal Linguistics: Exploring Referentiality and Compositionality in Bird Calls'. *Ecological Research* 36, no. 2 (March 2021): 221–31. https://doi.org/10.1111/1440-1703.12200.

Waal, Frans B. M. de, and Stephanie D. Preston. 'Mammalian Empathy: Behavioural Manifestations and Neural Basis'. *Nature Reviews Neuroscience* 18, no. 8 (August 2017): 498–509. https://doi.org/10.1038/nrn.2017.72.

de Waal, Frans B.M. 'What Is an Animal Emotion?' *Annals of the New York Academy of Sciences* 1224, no. 1 (April 2011): 191–206. https://doi.org/10.1111/j.1749-6632.2010.05912.x.

Walters, Edgar T. 'Defining Pain and Painful Sentience in Animals'. *Animal Sentience* 3, no. 21 (1 January 2018a). https://doi.org/10.51291/2377-7478.1360.

———. 'Nociceptive Biology of Molluscs and Arthropods: Evolutionary Clues About Functions and Mechanisms Potentially Related to Pain'. *Frontiers in Physiology* 9 (3 August 2018b): 1049. https://doi.org/10.3389/fphys.2018.01049.

Walters, Edgar T., and Amanda C. de C. Williams. 'Evolution of Mechanisms and Behaviour Important for Pain'. *Philosophical Transactions of the Royal Society B: Biological Sciences* 374, no. 1785 (11 November 2019): 20190275. https://doi.org/10.1098/rstb.2019.0275.

Watanabe, Shigeru, and Kunihiko Ono. 'An Experimental Analysis of "Empathic" Response: Effects of Pain Reactions of Pigeon upon Other Pigeon's Operant Behavior.' *Behavioural Processes* 13, no. 3 (September 1986): 269–77. https://doi.org/10.1016/0376-6357(86)90089-6.

Watanabe, Shigeru, Kazutaka Shinozuka, and Takefumi Kikusui. 'Preference for and Discrimination of Videos of Conspecific Social Behavior in Mice'. *Animal Cognition* 19, no. 3 (May 2016): 523–31. https://doi.org/10.1007/s10071-016-0953-x.

Zimmermann, M. 'Physiologische Mechanismen Des Schmerzes Und Seiner Behandlung'. In *Schmerztherapie*, edited by Hans Bergmann, 32:1–19. Klinische Anästhesiologie Und Intensivtherapie. Berlin, Heidelberg: Springer Berlin Heidelberg, 1986. https://doi.org/10.1007/978-3-642-71333-0_1.

CHAPTER 4

Critical Animal Historiography, Experiential Subjectivity and Animal Standpoint Theory

Kai Horsthemke

Stories of the hunt will be stories of glory until the day when animals have their own historians.
—Zimbabwean proverb

Introduction: Mainstream and Critical Animal Studies

'Animal studies' (sometimes also referred to as 'human-animal studies') is a fairly young field within academic education and research that developed in part out of the animal liberation movement, following the publication

This essay contains certain ideas and arguments originally articulated in Kai Horsthemke, *Animal Rights Education*, Cham: Palgrave Macmillan, 2018: 197–216.

K. Horsthemke (✉)
Wits School of Education, University of the Witwatersrand, Johannesburg, South Africa

M. J. Glover, L. Mitchell (eds.), *Animals as Experiencing Entities*, The Palgrave Macmillan Animal Ethics Series,
https://doi.org/10.1007/978-3-031-46456-0_4

of seminal books like Peter Singer's *Animal Liberation*[1] and Tom Regan's *The Case for Animal Rights*.[2] Within this paradigm other-than-human animals are not viewed as mere historical referents, abstract cultural objects, symbols or patterns, but as "beings who live and suffer now",[3] as individual agents and bearers of inherent value.[4] Initially concerned largely with ethical and epistemological questions, animal studies signalled a growing movement that combined literary, philosophical, psychological, sociological and historical research on animals and human-animal relationships. It gave birth to a plethora of journals, books, academic seminars, courses, colloquia and conferences dedicated to discussing the 'animal turn' in literature, the visual arts, cultural studies, geography, history and anthropology, and the like.[5] The various contributors to animal studies are chiefly concerned with exploring the interaction and interplay between human and other animals. As Reingard Spannring points out,[6] academics have also begun to consider other animals as "living beings with their own experiences, perceptions and interests" in environmental education research, a field that has traditionally been concerned more with larger communities, conservation and ecological sustainability.

'Critical animal studies', as championed by Richard Kahn, Anthony Nocella II, Steven Best and others, emerged in higher education as a result of mainstream animal studies being perceived as having sold out, as having been domesticated "within the sterile, normalising, hierarchical, and repressive environment of academia",[7] "muzzled and neutralised by a corporate-bureaucratic machine and its own codes and logics",[8] colonised by organisations and "academics whose commitment to animals is strictly abstract, nothing more than an interesting topic of research and form of academic capital", individuals without any real or practical commitment to animal rights and liberation, to a vegan and generally anti-speciesist

[1] Peter Singer, *Animal Liberation* (New York: Random House, 1975).

[2] Tom Regan, *The Case for Animal Rights* (London: Routledge & Kegan Paul, 1983).

[3] Steven Best, 'The Rise of Critical Animal Studies: Putting Theory into Action and Animal Liberation into Higher Education', *Journal for Critical Animal Studies* 7, no. 1 (2009): 22.

[4] Reingard Spannring, 'Animals in Environmental Education Research', *Environmental Education Research* 23, no. 1 (2016).

[5] Kenneth Shapiro and Margo DeMello, 'The State of Human-Animal Studies', *Society and Animals* 18 (2010); Spannring (2016), op. cit.

[6] Spannring (2016), 63.

[7] Best (2009), 10.

[8] Ibid.

lifestyle, etc. The net effect has been that all kinds of scholars, even those more or less tacitly committed to (and benefiting from) so-called sustainable use of non-humans, a carnivorist lifestyle and also vivisection and cosmetics testing on animals, contribute to and make a career out of animal studies. Critics complain that work in animal studies has largely become a way of securing funding and academic recognition, getting published and engaging in theory-for-theory's sake across a wide range of disciplines within the natural and social sciences, without ultimately contributing to the amelioration of the plight of animals, let alone the cessation of abusive and oppressive practices, in any meaningful or profound way. An even more serious concern is that animal studies actually serves to legitimate the ongoing exploitation and subjugation of other animals, that the interests of those teaching and researching in animal studies more often than not run counter to the interests of those who are being so studied, and that it is precisely the continuing utilisation of animals that makes animal studies an academically viable and feasible domain. In other words, animals remain either worthwhile topics of purely theoretical concern or continue to be instrumentalised for largely human purposes, on the basis of empirical investigations. There are several examples of applied research pertaining to human-animal relationships that nonetheless retain a strongly anthropocentric focus on, and attitude towards, animals:

- animal-assisted therapy;
- animal welfare science "as a source of empirical findings on the needs of animals in a variety of settings", where it remains unclear whether "the impact of this literature will be limited to the promotion of enhanced well-being of animals and improving the conditions of their use, or will provide empirical grounds for the abolition of those uses";[9] and
- trans-species psychology, which "applies findings from human psychology to the understanding of other animals, reversing the ethically and scientifically contentious construction of animal models to understand human conditions". Whether "extrapolation from human to animal", which "implies similarity across species", will "reinforce the posthumanist idea of blurred categories, or will …

[9] Shapiro and DeMello (2010), 312.

buttress animal-model-based laboratory research, which perennially uses that argument to justify its enterprise", too, remains to be seen.[10]

A particularly crass example of how instrumentalising approaches have infiltrated the initially non-anthropocentric field of animal studies is provided by Helena Pedersen's[11] description of hunter education electives in social science-oriented schools. There are further examples of how everyday school activities reflect the commodification of animals. Mueller, Tippins and Stewart,[12] introducing their anthology on the role of animals in science education, speak of "the ways in which [animals] provide countless contributions to the health, beauty, and lifestyles comprising the status quo [,]...to the health and lifestyles we all enjoy". In the same volume, Alexandra West Jefferies[13] discusses solutions for "managing fish waste". After referring to the "importance of fish in human diet, coastal communities' economies, sport and leisure, recreation and tourism",[14] she states, "Not only can fish waste create water quality issues, smell and attract *pests* [such as gulls or bears!], but it can also create an aesthetic concern for visitors".[15] No concern at all is exhibited here for the fish themselves, only broadly "environmental concerns".[16] In Jimmy Karlan's[17] examination of the roles of rats and spiders in science education, "moral objections" by students are "welcomed", but the practice of animal use in science classrooms remains essentially unchallenged by the author. Similarly, although Eduardo Dopico and Eva Garcia-Vazquez[18] list a number of animal-free

[10] Ibid.

[11] Helena Pedersen, *Animals in Schools: Processes and Strategies in Human-Animal Education* (West Lafayette/IN: Purdue University Press, 2010): 93ff.

[12] Michael P. Mueller, Deborah J. Tippins and Arthur J. Stewart, 'Wild Awakedness and Animalistic Inquiry: Introducing a Book on the Role of Animals with/in Science Education', in *Animals and Science Education: Ethics, Curriculum and Pedagogy*, edited by Michael P. Mueller, Deborah J. Tippins and Anthony J. Stewart (Cham: Springer, 2017): x, xiii.

[13] Alexandra West Jefferies, 'Engineering a Solution for Managing Fish Waste', in *Animals and Science Education: Ethics, Curriculum and Pedagogy*, edited by Michael P. Mueller, Deborah J. Tippins and Anthony J. Stewart (Cham: Springer, 2017).

[14] Ibid.: 29.

[15] Ibid.: 30, 38; emphasis added.

[16] Ibid.: 37.

[17] Jimmy Karlan, 'Spiders, Rats, and Education', in *Animals and Science Education: Ethics, Curriculum and Pedagogy*, edited by Michael P. Mueller, Deborah J. Tippins and Anthony J. Stewart (Cham: Springer, 2017): 125–130.

[18] Eduardo Dopico and Eva Garcia-Vazquez, 'How Technology Can Replace Animals in Lab Practices', in *Animals and Science Education: Ethics, Curriculum and Pedagogy*, edited by Michael P. Mueller, Deborah J. Tippins and Anthony J. Stewart (Cham: Springer, 2017): 142.

teaching alternatives and "cruelty-free methodologies", they accept that some testing remains necessary. Finally, in his presentation of "a framework within which to determine how we should use animals in science education", Michael Reiss[19] suggests that animal use in science education is fairly negligible, that "animals are not included much in curricula". This is surely incorrect. If anything, it is the *acknowledgement* within the science education community that animal use is important in science education (let alone raises ethical questions) that is negligible. In Michael Matthews's 3-volume, 2532-page *International Handbook of Research in History, Philosophy and Science Teaching*,[20] merely 5 pages contain reference to animals. In Reiss's[21] discussion of dissection in schools, there is no principled objection on the author's part, nor to keeping animals in schools or bringing animals to schools for educational purposes:[22] "students themselves should consider the use of animals by humans", because such decision-making would benefit students in terms of heightening their "ethical sensitivity", increasing their "ethical knowledge", improving their "ethical judgement" and making them "better people in the sense of making them more virtuous or otherwise more likely to implement normatively right choices".[23] Apart from the real possibility that a "normatively right choice" may indeed be the essential instrumentalisation of animals, there is something quite repugnant at work here: consider mandating able or white or male students to make decisions about the school admitting, or including, disabled or black or female students, respectively.

Kenneth Shapiro and Margo DeMello[24] list several strategies for increasing the institutional base and influence of human-animal studies. These include, amongst other things, taking advantage of the bias toward anthropocentric interests and selecting for study "areas of investigation that connect directly with existing human values", such as the "study of the

[19] Michael J. Reiss, 'A Framework within which to Determine How We Should Use Animals in Science Education', in *Animals and Science Education: Ethics, Curriculum and Pedagogy*, edited by Michael P. Mueller, Deborah J. Tippins and Anthony J. Stewart (Cham: Springer, 2017): 243.

[20] Michael R. Matthews, ed. *International Handbook of Research in History, Philosophy and Science Teaching* (Dordrecht & Heidelberg: Springer, 2014).

[21] Reiss (2017), 255–256.

[22] Ibid.: 256.

[23] Ibid.: 257, 258.

[24] Shapiro and DeMello (2010), op. cit.

relationship between human violence and animal abuse [… as] a bridge to concern about domestic and street violence".[25] Other suggestions, too, have a distinct historical-educational value: selection for study "areas of investigation that are topical … (animals in natural disasters, or the appropriateness of using dolphins for 'therapy', or the plight of the polar bear)"; development of "working relationships with professionals working in shelters, zoos, sanctuaries, veterinary, and animal-assisted practices", "with the animal protection movement", and "with scholars whose study is related to other social justice movements", in particular those who work in women's studies and environmental studies.[26]

According to Kahn,[27] Best,[28] and Nocella, Sorenson et al.,[29] critical animal studies positions itself against the welfarism and utilitarianism pervading these suggestions. It repudiates alliances and working relationships with individuals or groups who remain, implicitly or explicitly, committed to the ontology and ethics of anthropocentric, species-based domination. It is 'critical' also in a second sense, in its commitment to overthrowing all systems of hierarchical subjugation and exploitation and in its rejection of global capitalism and neo-liberalism. It is not anti-theory but rather "uses theory as a means to the end of illuminating and eliminating domination".[30]

Critical Animal Historiography

'Critical animal historiography' arguably emanates from critical animal studies. It regards other-than-human animals as subjects of and as both agents and recipients in history, as individuals who impact and who are

[25] Ibid.: 315.

[26] Ibid.

[27] Richard Kahn, 'Afterword', in *The Educational Significance of Human and Non-human Animal Interactions: Blurring the Species Line*, edited by Suzanne Rice and A.G. Rud (Basingstoke & New York: Palgrave Macmillan, 2016): 218 and 222n.4; see also Richard Kahn, 'Towards an Animal Standpoint: Vegan Education and the Epistemology of Ignorance', in *Epistemologies of Ignorance in Education*, edited by Eric Malewski and Nathalia Jaramillo (Charlotte: Information Age Publishing, 2011).

[28] Best (2009), op. cit.

[29] Anthony J. Nocella II, John Sorenson, Kim Socha and Atsuko Matsuoka, eds. *Defining Critical Animal Studies: An Intersectional Social Justice Approach for Liberation* (New York: Peter Lang, 2014).

[30] Steven Best, 'Introduction', *Animal Liberation Philosophy and Policy Journal* 5, no. 1 (2007): 2.

impacted by historical processes.[31] In recent years, in other words, research guided both by cultural studies and by interdisciplinary approaches that focus on the practices, materiality and spatiality of human-animal connections has repeatedly emphasised the historicity of human-animal relationships. In Western philosophical and educational traditions, the historicity of non-human animals has all but been ignored. Animals, unlike humans, are seldom granted historical subject status, let alone agent status. Historical progress and change are seen in the canon as essentially determined by what human agents are and have negotiated and not in terms of the contributions (willing or unwilling) made by other animals. (Koko, Washoe and Lucy seem to have embraced being taught sign language, and apes are known not only to pass this knowledge on to other apes but also to work creatively within the framework of sign language, inventing new signs that either reflect or closely resemble established signs. But would Hannibal's elephants have crossed the Alps on their own accord? Laika is one of the few non-human beings whose tragic biography is captured at least in part on Wikipedia.) Historiography, the writing of history, has characteristically posited the ahistoricity of animals. In contrast with humans, animals continue to be perceived as the embodiment of the ahistorical 'other'. For example, 'cattle' have been referred to, albeit mistakenly,[32] as the oldest and most important domestic animals. Limited to their role as 'livestock' and defined exclusively in terms of their instrumental value within human society,[33] they have no history—and no identity. Without a notion of animals' historicity and different forms of

[31] See Steven Best, 'The Revolutionary Implications of Animal Standpoint Theory' (2012): 9.https://drstevebest.wordpress.com/2012/05/14/the-revolutionary-implications-of-animal-standpoint-theory-2/.

[32] Evidence suggests that the first companion animal was the dog, whose free-roaming ancestor is the wolf. Domestication of the dog took place in the Upper Palaeolithic Period (see Valeska Becker, 'Auf den Schultern von Riesen. Zum Umgang mit Tieren aus Sicht der Ur- und Frühgeschichte', in *Interspezies Lernen: Grundlinien interdisziplinärer Tierschutz- und Tierrechtsbildung*, edited by Simone Horstmann, 2021: 35–37).

[33] August Renner, Gustav Feddeler, J.F. Hüttmann, Heinrich Jastram and Adolf Marten, eds. *Weltkunde: Leitfaden der Geographie, Geschichte, Naturgeschichte, Physik und Chemie für Mitteschulen und mehrklassige Volksschulen* (Hannover: Helwing, 1896); see also Andreas Hübner, '"Mißhandle und mißbrauche nie ein Tier!" Tierschutz- und Tierrechtsbildung in geschichtsdidaktischer Perspektive', in *Interspezies Lernen: Grundlinien interdisziplinärer Tierschutz- und Tierrechtsbildung*, edited by Simone Horstmann (Bielefeld: Transcript Verlag, 2021): 91–92.

agency,[34] animal welfare education and animal rights education will be seriously deficient. The assumption that animals, too, have a history needs to be foregrounded in didactic and pedagogical considerations, just as biological reductionism and the cultural uniqueness of humans needs to be questioned. Human and non-human 'co-history' involves recognition and analysis of 'co-constitutive relationships'[35] and conceptualises animals as historical and experiential subjects capable of being protected via an appeal to rights.

Best, Kahn and Nocella collaborated on establishing the principles of critical animal studies as a critical and radical platform for transformation, of which the following are of particular educational and pedagogical relevance:

- [Critical animal studies rejects] pseudo-objective academic analysis by explicitly clarifying its normative values and political commitments, such that there are no positivist illusions whatsoever that theory is disinterested or writing and research is non-political.[36]

[34] Given the heterogeneity of nonhuman animals, Andreas Hübner ('"Mißhandle und mißbrauche nie ein Tier!". Tierschutz- und Tierrechtsbildung in geschichtsdidaktischer Perspektive', in *Interspezies Lernen: Grundlinien interdisziplinärer Tierschutz- und Tierrechtsbildung*, edited by Simone Horstmann, 2021: 92) distinguishes between three different conceptions of agency: (1) *relational agency*, which concerns interactions and their impact on human-animal relationships; (2) *entangled agency*, the interweaving of actants manifested in networks, and (3) *embodied agency*, which refers to the corporeality of human-animal relationships and, as a praxeological approach, seems particularly promising for animal welfare and animal rights education.

[35] Donna Haraway, *The Companion Species Manifesto: Dogs, People and Significant Otherness* (Chicago: Prickly Paradigm Press, 2003).

[36] See also Cori Jakubiak ('Tracing the Anthrozoological Landscape of Central Iowa: Place and Pedagogical Possibilities', in *Animals and Science Education: Ethics, Curriculum and Pedagogy*, edited by Michael P. Mueller, Deborah J. Tippins and Anthony J. Stewart, 2017: 53): "Politically engaged social science—and by extension politically engaged classroom teaching—must pay heed to human-animal relations", which "provides insight on how power works in society" and "also draws our attention to potential vectors of change". Teresa Lloro-Bidart and Constance Russell ('Learning Science in Aquariums and on Whalewatching Boats: The Hidden Curriculum on the Deployment of Other Animals', in *Animals and Science Education: Ethics, Curriculum and Pedagogy*, edited by Michael P. Mueller, Deborah J. Tippins and Anthony J. Stewart, 2017: 47), too, make a normative rather than an empirical point about teaching. They consider "pedagogies that rely on transmission of depoliticised facts" to be problematic.

- [Critical animal studies eschews] narrow academic viewpoints and the debilitating theory-for-theory's sake position in order to link theory to practice, analysis to politics, and the academy to the community.
- Through—and only through—new paradigms of ecopedagogy, bridge-building with other social movements, and a solidarity-based alliance politics, is it possible to build the new forms of consciousness, knowledge, social institutions that are necessary to dissolve the hierarchical society that has enslaved this planet for the last ten thousand years.[37]

This new conception of animal studies thus critiques the apolitical and supposedly objective, value-free orientation and the compartmentalisation of academic perspectives and analysis.

Best's analysis of education seems to reveal certain contradictions. On the one hand, he argues[38] that education, like other Western academic disciplines, has been an anthropocentric and speciesist discipline since its beginnings. Education began by defining the human against the animal—rather than as a differently endowed kind of animal embedded in relations with other animals—and developed along the dualist path of divorcing culture from nature. Humanism and its academic representatives were progressive in educating the human mind, but they never questioned the anthropocentric and speciesist logic contained in educational theory and pedagogical practice, and the erroneous understanding of humans, animals and the natural world. This implies that education is inherently problematic. On the other hand, there is the view that education is under some kind of external threat. In Best's[39] quotation from the *Uncut Conscience* website, "Education is under assault because it provides access to the historical truths, critical thinking, and alternative perspectives that lay the groundwork for structural change". An intermediate position, one that

[37] Best (2009), 24, 26; Steven Best, Anthony J. Nocella II, Richard Kahn, Carol Gigliotti and Lisa Kemmerer, 'Introducing Critical Animal Studies', *Animal Liberation Philosophy and Policy Journal* 5, no.1 (2007): 4, 5.

[38] Steven Best, 'Education in the Time of Eco-crisis', presentation in the Wits School of Education, University of the Witwatersrand (2 October, 2008).

[39] Steven Best, 'The Rise (and Fall) of Critical Animal Studies', *Liberazioni* (n.d.): 5. http://www.liberazioni.org/articoli/BestS-TheRise%28and%20Fall%29ofCriticalAnimalStudies.pdf.

articulates both the shortcomings and the promise of education, is arguably held by historian and political scientist Howard Zinn:[40]

> Unfortunately, our education system is geared to prepare young people to become successful within the confines of the present society. It doesn't prepare them to question this present society, to ask if fundamental change is needed. And so I believe the most important thing education can do is to take the students out of this narrow concern with learning what they need to be successful in their profession and make them aware that the most important thing they can do in their lives is to play a role in creating a better society, whether it's stopping war, or ending racial inequality, or ending economic inequality. This is the most important thing that education can do. And I think our most wise of educators—our philosophers of education, like John Dewey—have recognised this as the critical problem of education.

Kahn,[41] too, is aware of the potential of both education and miseducation:

> To take up the militant defence of animals in education is to actualise intensely a key political contradiction and to challenge the institutionalisation of speciesism quite openly. But since we live in a speciesist society, and schools remain institutions that legitimate speciesism,[42] to try to advocate for animals through one's research therein immediately subjects one to all manner of microaggressions from one's colleagues and the everyday functioning of the organisational structure of the school proper.

In Best's[43] analysis of higher education in particular,

> With the potential for enlightenment and edification, books, research, writers, and professors instead perpetuate ignorance, egoism, and apathy. Erudite professors train their students in their methods of abstraction,

[40] Howard Zinn, 'Science, the Public, and Social Change', interviewed by G.M. Schivone (2008): 11. https://www.howardzinn.org/the-citizens-among-us/.

[41] Kahn (2016), 217–218.

[42] Richard Kahn ('Afterword', in *The Educational Significance of Human and Non-human Animal Interactions: Blurring the Species Line*, edited by Suzanne Rice and A.G. Rud, 2016: 222n.2) quotes from Veganism.com, 2011: "From our education and social experiences we learn to see human characteristics and abilities as the ideal standard against which all others are measured".

[43] Best (2009), 33.

> obfuscation, and alienated detachment, as the disease of intellectualism spreads from generation to generation.

In other words, "Still brimming with potential for knowledge, enlightenment, autonomy, and provoking radical change, higher education more often promotes ignorance, conformism, egoism, and apathy".[44]

Critical animal studies, then, aims to revitalise education as a personally and politically empowering phenomenon that aims at both critical conscientisation and the development of a sense of practical purpose, as well as structural change through personal agency. Lauren Corman and Tereza Vandrovcová,[45] in their discussion of critical animal studies pedagogy, argue that animals, like humans, ought to be recognised as potential teachers. The authors also seek to counter what they deem to be an over-representation of animals as victims, by emphasising the need to represent animal subjectivities in theory, teaching and activism. They point out that critical animal studies pedagogy can help us appreciate the subjectivities not only of animals in our immediate (domestic) contexts but also of those in factory farms and vivisection laboratories. Among the wide array of practical concerns they address is also the issue of whether or not school teaching should include graphic imagery of exploitation and violence perpetrated against animals.

However, the novel domain of critical animal studies is not without problems. As Best himself acknowledges,[46] critical animal studies is—like mainstream animal studies—under threat of co-option, appropriation and domestication by academics and "university bureaucrats", "opportunists eager to exploit novel discourses to survive in a cutthroat environment, and by a moribund publishing industry dependent on an endless succession of transitory celebrities and [ephemeral] fads", and absorption "into personal agendas and conservative paradigms and programs". But even without considering the bandwagon syndrome and the willingness of many contributors "to bear the heavy costs of betrayal, selling-out, betraying animals and militant activists, and irreparable loss of integrity and credibility", as diagnosed by Best, it is unclear to what extent critical

[44] Best (n.d.), 19.

[45] Lauren Corman and Tereza Vandrovcová, 'Radical Humility: Toward a More Holistic Critical Animal Studies Pedagogy', in *Defining Critical Animal Studies: An Intersectional Social Justice Approach for Liberation*, edited by Anthony J. Nocella II, John Sorenson, Kim Socha and Atsuko Matsuoka (New York: Peter Lang, 2014).

[46] Best (n.d.), 29.

animal studies manages to avoid mainstream animal studies' frequent "elitist and inscrutable discourse".[47] Many of the sources it draws on, the Frankfurt School, latter-day Marxist writings and post-modernist affiliates of critical pedagogy, could be accused of peddling such "elitist and inscrutable discourse" and as constituting mere "theories divorced from practice, political struggle, and social transformation".[48] They are not exactly known for their accessible language or endorsement of "militant direct action", such as proposed by Best.[49] Whether or not the resultant theory and research are intentionally obscure, many of their virtues have arguably "been lost ... in the elitist pomposity of cultural studies, postmodernism" and critical theory, in "trendy Continental theory-babble":[50] in other words, "the concrete realities of animal suffering, violence and exploitation, economic crisis and social power, and the rapidly worsening planetary ecological catastrophe are entirely muted and virtually barred from the hermetically-sealed chambers of theory-babble".[51]

Finally, Best's[52] own dismissal of individualism is both puzzling and unfortunate for someone who presumably does not care for animals as groups or species but as individuals.[53] Animals are "sentient beings who live and die in the most sadistic, barbaric, and wretched cages of techno-hell that humanity has been able to devise, the better to exploit them for all they are worth",[54] "beings who live and suffer now".[55] If "Speciesism strips animals of all intrinsic value to reduce them to instrumental value, to mere tools and objects whose cosmic purpose is to satisfy human purposes",[56] then why lambaste individualism—which acknowledges and prizes this intrinsic value? How would one make sense of Best's[57] challenge, "We must revolutionise both our psychologies (in post-

[47] Ibid.: 4.

[48] Ibid.

[49] Ibid.: 30; Best (2009), 48n.22.

[50] Best (n.d.), 25, 30.

[51] Best (2009), 32.

[52] Ibid.: 32–33n.5

[53] See also Steven Best, Peter McLaren and Anthony J. Nocella II, Anthony J., 'Revolutionary Peacemaking: Using a Critical Pedagogy Approach for Peacemaking with "Terrorists"', *Journal for Critical Education Policy Studies* 5, no. 2 (2007): 515; Best (n.d.), 7, 11, 27.

[54] Best, Nocella II et al. (2007), 4.

[55] Best (2009), 22.

[56] Best (n.d.), 7.

[57] Ibid.: 27.

anthropocentric, post-speciesist, and post-discriminatory form) and our institutions (in post-capitalist form that promotes autonomy, self-determination, decentralisation, and radical direct democracy)", while disavowing individualism? Best[58] later qualifies his dismissal, by targeting "individualist pursuits of 'the good life', as defined apart from what is good for animals and the land". In other words, it is only anthropocentric and speciesist individualism that ought to be rejected.

The fact that animals, too, are experiencing entities, conscious individuals, subjects of a life that matters *to them*, that can be better or worse *for them*, who can be benefited and harmed, individuals with needs, beliefs, desires, preferences and expectations, best accounts for the initial intuitions behind animal standpoint theory. There is something that it is like to be a particular animal, who experiences life and the world from a particular perspective, vantage- or standpoint.

Standpoint Theory

The classical model for standpoint theory is Marxist standpoint theory, which was inspired by Georg Wilhelm Friedrich Hegel's analysis of the master-slave relationship and developed by Karl Marx and Friedrich Engels to emphasise the proletarian standpoint.[59] During the last two decades of the previous century, it led to the formulation of feminist standpoint theory,[60] designed to draw attention to a more credible alternative to the patriarchal power systems that pervasively influenced society and history.

[58] Best (2012), 21.

[59] Karl Marx and Friedrich Engels, *The German Ideology*, edited by C.J. Arthur (New York: International Publishers, 1970); Karl Marx, *Economic and Philosophic Manuscripts of 1844*, edited by D. Struik (New York: International Publishers, 1984); Friedrich Engels, 'Socialism: Utopian and Scientific', in *The Marx and Engels Reader*, edited by Robert C. Tucker (New York: Norton, 1972); Georg Lukács, 'Reification and the Consciousness of the Proletariat', in *History and Class Consciousness*, translated by Rodney Livingstone (Cambridge: MIT Press, 1971).

[60] Nancy Hartsock, 'The Feminist Standpoint: Developing the Ground for a Specifically Feminist Historical Materialism', in *Discovering Reality: Feminist Perspectives on Epistemology, Metaphysics, Methodology and Philosophy of Science*, edited by Sandra Harding and Merrill B. Hintikka (Dordrecht, London & Boston: D. Reidel, 1983); Sandra G. Harding, *The Feminist Standpoint Theory Reader: Intellectual and Political Controversies* (New York: Routledge, 2004); Donna Haraway, 'Situated Knowledges', *Feminist Studies* 3 (1988).

Animal standpoint theory[61] is "an extension of feminist standpoint theory, which was developed to illuminate patriarchal domination and its debilitating impact on women and humanity as a whole".[62]

According to Kahn,[63] following both Paulo Freire and Sandra Harding, "emancipatory science (knowledge) is never neutral but instead always produced from and serves traditionally silenced movement standpoints". A standpoint is an individual's unique perspective from which the world is experienced and perceived. A standpoint theory is a representation of the world from a particular socially situated perspective that can lay claim to epistemic advantage or authority. Many claims to epistemic advantage on behalf of particular perspectives with regard to certain issues are commonplace and generally uncontroversial (e.g., the claims by specialist professionals, subject methodologists, even education policy analysts, etc.). According to Harding,[64] "human activity not only structures but also sets limits on understanding". If

> knowledge is supposed to be based on experience[, then] the reason the feminist claims can turn out to be scientifically preferable is that they originate in, and are tested against, a more complete and less distorting kind of social experience. The experiences arising from the activities assigned to women, seen through feminist theory, provide a grounding for potentially more complete and less distorted knowledge claims than do men's experiences.

Lorraine Code,[65] similarly, refers to a theory of knowledge that is informed by women's experiences and will focus on understanding rather than on justification, verification and control and will find value in first person experiential stories. Feminist standpoint epistemology claims, asserts Harding, that the point of view of women is a privileged one for knowledge.

[61] Haraway (2003), op. cit.; Josephine Donovan, 'Animal Rights and Feminist Theory', in *Ecofeminism: Women, Animals, Nature*, edited by Greta Gaard (Philadelphia: Temple University Press, 1993); Kahn (2011), op. cit.; Best (2009), op. cit. Best (2012), op. cit.; Steven Best, *The Politics of Total Liberation: Revolution for the 21st Century* (New York: Palgrave Macmillan, 2014).

[62] Best (2012), 3; Best (2014), 2–3.

[63] Kahn (2016), 216.

[64] Sandra G. Harding, 'Feminist Justificatory Strategies', in *Women, Knowledge and Reality: Explorations in Feminist Philosophy*, edited by Ann Garry and Marilyn Pearsall (New York & London: Routledge, 1992): 194.

[65] Lorraine Code, 'Feminist Epistemology', in *A Companion to Epistemology*, edited by Jonathan Dancy and Ernest Sosa (Oxford: Blackwell, 1992).

Standpoint theories become controversial when they lay claim to epistemic and cognitive advantage over socio-politically contentious topics and issues on behalf of the perspectives of systematically *dis*advantaged, historically *under*privileged social groups, relative to the perspectives of the dominant group/s. Harding is arguably also guilty of committing what might be called 'the fallacy of the collective singular' here (invoking a singular point of view when there is actually a multitude). Kahn[66] seems to be cognisant of this charge when he explains that "the animal standpoint actually seeks to understand the world from multiple evolving locations, and so there are at present a multitude of heterogeneous and contradictory animal standpoint situations, not a singular universal standpoint that can be utilised like a cryptographic key for a theory of everything". Nevertheless, he endorses Harding's[67] argument that although "this form of subjugated knowledge may be unable to escape being pluralist and partial in nature, it can thereby serve positively as a powerful resource to increase our objective understanding of society and provide for a more robustly democratic public sphere beyond majoritarian accounts".[68,69] While this argument is noteworthy for its mention of values

[66] Kahn (2011), 4.

[67] Harding (2004), 127–138.

[68] Kahn (2011), 4.

[69] Sandra Harding's (and Haraway's) ideas about "situated knowledges" and "partial perspectives" are also picked up by Brandy Humes ('Moving Toward a Liberatory Pedagogy for All Species: Mapping the Need for Dialogue Between Humane and Anti-oppressive Education', *Green Theory & Praxis: The Journal of Ecopedagogy* 4, no. 1, 2008: 77–78), in her attempt to merge humane and non-oppressive education into a "liberatory pedagogy":

> By disrupting grand or meta-narratives, what is considered knowledge, common sense, 'normal' and 'abnormal' are all brought into a contested terrain and examined, scrutinised and disrupted allowing for the 'unknowable' to surface. … Lessons about oppression include learning to resist the desire to know it, to reject searching for the 'Truth' as the end point of knowing. The goal is not a state of final or complete knowledge, a final answer, or the satisfaction that comes with obtaining those, but rather is partial (or 'situated') knowledge, the disruption of existing knowledge and the discomfort in not knowing, and desire for more change and for not closing off further learning opportunities….

But does "anti-oppression" not constitute just such a "grand or meta-narrative", and rightfully so? And is liberation from abuse and exploitation not the "final answer" that sincere defenders of all social causes have set as their goal? I find it puzzling that defenders of a strong, unequivocal ethical stance can be so taken by dodgy epistemology, and tacitly embrace relativism about both truth and knowledge.

like objective understanding and democratic engagement, it remains unclear what establishes something as knowledge, and what distinguishes knowledge from, say, prejudice or bias.

A problem with standpoint epistemology in particular is a fundamental confusion of issues in epistemology with issues in social justice. This is not just an unintentional category mistake, the pointing out of which would resolve the dispute. Kahn appears to be guilty of treating epistemological matters as necessarily inseparable from matters of social justice when he notes that "a primary concern of the animal standpoint is to provide counter-histories" to what might be called anthropocentric certainties and convictions, counter-histories that "can help to illuminate profound silences on the animal standpoint in the socio-historical record as being often non-accidental, and institutionally perpetrated and organised, in order to legitimate *hegemonic regimes of truth and ways of knowing* that are foundational to our [present certainties]".[70] What we are dealing with here is not truth or knowledge, but beliefs.[71]

The so-called hierarchy of knowledge[72] favours not androcentric, Eurocentric and anthropocentric ways of *knowing*, or *knowledge*, but rather androcentric, Eurocentric and anthropocentric knowledge *claims*, i.e. ways of *claiming to know*—which are, indeed, governed by hegemonic interests and power relations. Once such a distinction is made between knowledge and knowledge claims, the tension between objectivity and subjectivity disappears, and considerations of social justice and privilege can be foregrounded without invoking a problematic constructivist or relativist view of knowledge. It is knowledge claims, but not knowledge as such, that may result from social relations and that are often ideologically biased. Knowledge claims may be mistaken, involve false beliefs, or be only

[70] Kahn (2011), 3; emphasis added.

[71] See also Helena Pedersen (*Animals in Schools: Processes and Strategies in Human-Animal Education*, West Lafayette/IN: Purdue University Press, 2010: 23), who problematises the idea that "scientific" ways of thinking (for example, about animals) are necessarily considered to be "legitimate knowledge". This indicates both a tautology and a conceptual error. Knowledge comprises not only belief and truth but also justification and is, therefore, necessarily legitimate. There may be illegitimate and unethical ways of acquiring knowledge (animal experimentation, although still legal, is a prime example of the latter), but this is a matter of the ethics of practice and not of epistemology.

[72] See Spannring (2016), op. cit.

minimally justified. And while it may be plausible to consider some points of view and knowledge claims privileged (for example, those of specialists and authorities within a particular discipline), they are not therefore infallible. If they are genuine claims to knowledge, they must be checked against the facts or prove their mettle in the give-and-take of democratic deliberation and reasoned, scholarly argumentation. Some will survive such fact checking and disputation, and some will not. I will return to these and similar ideas in my discussion of Best's contribution to animal standpoint theory.

Animal Standpoint Theory

According to Best,[73]

> The complexity of historical and social dynamics can only be adequately grasped from a *plurality* of perspectives. These include critical approaches such as Marxism, feminism, and critical race/postcolonial studies, and environmental determinism. To determinants such as class, gender, race, and geography and climate we must also add the decisive influence of animals on human history—or, rather, human-animal relations and interactions—such as become clear through use of *the animal standpoint.*

What, then, is animal standpoint theory? If history is examined "from the animal standpoint, that is, from the crucial role that animals have played in human evolution and the consequences of human domination of non-human animals, we can glean new and invaluable insights into psychological, social, historical, and ecological phenomena, problems, and crises".[74] Animal standpoint theory, as Best characterises it, investigates "the fundamental role animals play in sustaining the natural world and shaping the human world in co-evolutionary relations". While animals have played a crucial positive role in shaping human existence, psychology and social life, "they have seldom been willing partners". The central thesis of animal standpoint theory is that "animals have been key determining forces of human psychology, social life, and history overall", and that in fundamental ways "the domination of humans over non-human animals underpins the domination of humans over one another and over the natural world".[75]

[73] Best (2012), 1.
[74] Best (2014), 1.
[75] Ibid.

The animal standpoint is vital to a "total liberation politics that promotes human, animal, and earth liberation as inseparably related struggles that need to unite against common enemies"[76] such as neo-liberalism and global capitalism: "It advocates an alliance politics in which different radical movements work together toward the positive goal of shifting the dominant paradigms from hierarchy to equality, from growth to sustainability, from alienation to harmony, and from violence to peace".[77]

The Cartesian or behaviourist sceptic's objection that animals do not have a standpoint is quickly dispensed with. As I have shown elsewhere,[78] there is an impressive, steadily expanding array of empirical evidence available for establishing that animals have rich conscious and conative lives, experiences and abilities that are often on a par with those of human beings—and sometimes indeed transcend these. The more serious question would be, however, what animals' perspectives actually are, what they consist in? How do we know what "the animal standpoint" actually comprises? How can we speak with any authority on the plurality of perspectives that make up the animal standpoint? The contention is that we cannot, as humans, presume to make claims or demands from a standpoint that is necessarily, and essentially, not our own. Before I respond to these queries I want to mention a related concern that critics have raised with regard to animal standpoint theory, one that pertains to all standpoint theories. If there is *one* such standpoint, there are *many*. Animals, like women, workers and indigenous people, are too diverse in their experiences to generate a single framework. In addition, the realms and fields of inquiry (like the social and natural sciences) are too diverse to give rise to great confidence that they might be equally transformed by such a framework. I do not think that the problems raised so far are insurmountable. Despite the enormous diversity of animals, in their needs, abilities and experiences, it is arguably easier to establish something like an 'animal standpoint' than it may be to determine 'the' standpoint of women, workers or indigenous people. We can ask the questions: What are the "interests" of animals "independent from those of exploiters or humans generally"? "What would they want, to live or to die?" To flourish, to "exist intact and in peace or to be violently assaulted, … and destroyed by greedy human industries and the increasing demands of a rapidly growing

[76] Best (2012), 2.
[77] Ibid.
[78] Kai Horsthemke, *Animal Rights Education* (Cham: Palgrave Macmillan, 2018), Part I.

human population?"[79] Perhaps a little more misleading are the questions, "What would oppressed and tortured animals want us to do? What courses of action would they approve and which would they condemn as inadequate and a betrayal?",[80] insofar as this type of inquiry presupposes animals' ability to differentiate between militant direct, liberatory action and strategies of education, patient lobbying, and the like. Yet, with a modicum of imagination and empathy we can arrive at conclusive answers, without needing to consult past histories and enter into discourses of reparation, restorative justice and affirmative action. The morally significant aspects of the animal standpoint are not inaccessible. They concern the need and ability to live in peace, without being subjected to physical and psychological discomfort, stress, distress and trauma, and without their lives being prematurely terminated. These needs and preferences are essential features of the animal standpoint that are easily determinable and that lend themselves to being taught to children and adult learners, to engage with, to understand and to value.

There are further concerns, however. The referential proximity of animal standpoint theory to sustainability discourse[81] is worrying. While arguably more legitimate than proximity to growth discourse, Best would arguably do well to address directly concerns about sustainability talk, especially the largely instrumental conception of animals its conservative variant involves. That he does not in any way subscribe to this conception is abundantly clear—which makes his passing, uncritical references to such discourse all the more puzzling. Even more problematic is Best's implicit endorsement of epistemological and moral relativism. Thus, he maintains[82] that animal standpoint theory is substantially influenced by Friedrich Nietzsche's "perspectivalism", the view that perception and cognition are always perspectival, that "there are no explanations, only interpretations, and science itself is interpretation". No one, and this includes scientists, "has privileged access to reality expressed in 'objective' knowledge and truths". "Individuals always come to any type of knowing or inquiry already burdened by a host of presuppositions, biases, and limitations".[83] Best explicitly rejects "absolutist truths, universal values, and reductionist

[79] Best (2012), 21; Best (2014), 16.
[80] Best (2014), 15.
[81] Best (2012), 2; Best (2014), 160.
[82] Best (2012), 3; Best (2014), 2.
[83] Best (2014), 2.

models".[84] This is problematic for several reasons. For one thing, the rejection of objective, absolutist knowledge and 'truths' is contradicted by Best's own knowledge and truth claims—for, if they are also perspectival, subjective, with no objective and general explanatory purchase, it would be difficult to understand why he bothers advancing them in the first place. Moreover, the rejection of universal values is glaringly at odds with Best's anti-speciesism, revolutionary environmentalism and "politics of total liberation"—all of which express values that are hardly anything short of universal. To put it another way, it is certainly remarkable for someone who rejects absolutes, notably moral absolutes, to evoke a "politics of *total* liberation". It may be correct that our inquiry and knowledge acquisition are necessarily limited by certain assumptions and proclivities, but this does not mean that they cannot also be based on good reasons and outflank, or be better justified than, competing arguments.

Consider what Best himself refers to as "an unpalatable truth", namely that despite "the efforts of social, environmental, and animal advocacy movements over the last two centuries, we are nevertheless losing ground in the battle for democracy, equality, autonomy, ecology, peace, and sustainability".[85] If this truth can lay no claim to objectivity and absolutism, then there is presumably little to worry about. Consider also the following telling passage:

> Articulating the connections between human, animal, and earth liberation movements will be challenging, to say the least. Given the advanced nature of the crisis, the feeble and fragmented resistance to total war, the tribalist tendency of humans to divide rather than unite, and the ferocity of repression one can expect to the degree this struggle could advance, we must acknowledge this truth.[86]

Again, what is the status of this truth? Moreover, given that we *must* acknowledge it, is this a relative, situated obligation? And does "the tribalist tendency of humans to divide rather than unite" not emanate from a logically problematic and practically incoherent adherence to perspectivalism and relativism? It should be clear that any such allegiance is inconsistent with Best's own intellectual and practical pursuits and commitments.

[84] Ibid.: 52.
[85] Ibid.: 159–160.
[86] Ibid.: 164.

An additional point of contention is the conceptual and hermeneutical overloading of the animal standpoint:

> The animal standpoint seeks generally to illuminate human biological and social evolution in important new ways, such as revealing the origins, dynamics, and development of dominator cultures, social hierarchies, economic and political inequalities, and asymmetrical systems of power. Through the animal standpoint, we can glean important lessons regarding the origins of hatred, hierarchy, violence, war, genocide, slavery, colonialism, racism, and hierarchy. Providing perspectives and insights unattainable through other historical approaches, the animal standpoint analyses how the domination of humans over animals is intimately linked to the domination of humans over one another, as it also brings to light the environmental impact of large-scale animal slaughter and exploitation.
>
> According to the animal standpoint, speciesism was the first form of hierarchy and domination, and laid the groundwork for other forms of oppression, power, and power systems, one is tempted to say, paraphrasing Marx, that animal standpoint theory, not communism, is the solution to the 'riddle of history'. The animal standpoint casts a brilliant light on problems that one cannot even see or identify through the opaque lens of humanism or its theoretical offshoots.[87]

In Best's[88] characterisation, it is difficult (if not impossible) to establish whether the insights and imperatives emerge from a genuinely 'animal' standpoint, rather than from concerned humans' standpoint and interpretation:

> As feminist and critical race standpoints yield crucial insights into the power pathologies of modernity and more, so animal standpoint theory, interpreting history from the perspective of human-non-human interactions, shows how speciesism and the exploitation of animals has had, is having, and increasingly will have momentous and disastrous consequences on all levels. The animal standpoint provides crucial insight for understanding the evolution and dynamics of violence, power, hierarchical domination, and dysfunctional and unsustainable societies.

The notable difference is, of course, that women and members of different race groups articulate their insights, anger, and condemnation. The animal

[87] Ibid.: 17–18.
[88] Best (2012), 5.

standpoint requires interpretation on our part. It is not the animal standpoint that provides "crucial insight" but rather the anti-speciesist, "total liberation standpoint" of Best and other like-minded (human) theorists. It is not the animal standpoint that "examines the origins and development of societies through the dynamic, symbiotic interrelationship between human and non-human animals" but rather the enlightened human perspective that "interprets history not from an evolutionary position that reifies human agency as the *sui generis* of all things, nor as the autonomous actions of a Promethean species, but rather from a *co-evolutionary* optic that sees animals as an inseparable part of human history and as autonomous subjects and moral agents in their own right".[89] Best acknowledges that "animals cannot tell us in human language what they really think about us", but how can he be "sure we would wither from their righteous anger, and profane diatribes"[90]? "Righteous anger" and "profane diatribes" inform Best's anti-speciesist standpoint, quite understandably. When he includes a photograph of a polar bear clinging to a tiny, ever-diminishing block of ice, a photo that contains the caption "Fuck you humans ... Fuck you.",[91] this is not the polar bear's profane fulmination or expression of anger. If anything, the polar bear's standpoint is one of bewilderment, of helplessness and perplexity in the face of nearly insurmountable odds. It is easy, and natural, to feel both a profound sadness and real anger when one witnesses a dog short-chained to a post without being able to seek shelter from the sun or rain, or a horse being beaten, or an emaciated donkey pulling an overloaded cart, when one sees footage of seals being clubbed to death, dolphins and whales being massacred, pigs being assaulted with blowtorches or having their noses smashed with bricks in abattoirs, a drugged lioness being shot by a foreign hunter who has paid a considerable sum for his 'trophy', 'fur' animals being skinned alive—with as little damage as possible to their coats, but their bodies a bleeding, writhing mess, their eyes betraying both the excruciating pain they must be feeling and the utter failure to comprehend what has just happened to them. *This*, the pain and the failure to make sense of a situation that is—more often than not—inescapable, desperate, is part of what constitutes the animal standpoint, and this is something we can understand, learn and teach about, on the basis of our imagination and empathy.

[89] Ibid.: 9.
[90] Best (2014), 11.
[91] Best (2012), 20.

There is also a potential weakening of the relevant concepts here, as when Best speaks of "the standpoint of the earth".[92] Does (or could) the earth have a standpoint? It is difficult, if not impossible, to make sense of these kinds of claims in the absence of, or without reference to, conscious individuals who are the subjects of a life that is experienced by them. While trees and plants in general have a physical standpoint, insofar as they are physically located and locatable organisms, they lack the central nervous system and cognitive processing mechanisms necessary for viewing and perceiving the world. To speak of a standpoint in the relevant sense here would involve unnecessary dilution of the concept—unnecessary, because there are more appropriate ways of caring and acting for these organisms and, indeed, for the earth. It would also render opaque what is special and morally pertinent about individual, subjective experience.

How, then, can we make sense of animal standpoint theory? To what extent can and does representation of the world from animals' particular, socially (or other) situated perspectives lay claim to epistemic advantage or authority? The experiences of other-than-human animals, understood on the basis of evolutionary theory, ethology and the like, provide a basis for potentially more complete and less distorted knowledge claims than do anthropocentric orientations. Animals manifest complex forms of communication, intelligence, emotions, thought and purposive behaviour. They have obvious needs and preferences, like a desire for freedom. They will attempt to escape, avoid noxious stimuli, and have been known to attack their tormentors. This also takes us back to Best's statement[93] about animals being seen as "an inseparable part of human history and as autonomous subjects and moral agents in their own right". There seems to be nothing implausible or incoherent about the idea that moral agency, self-determination, and autonomy are matters of degree. Animals may not attain more sophisticated levels of agency. Like virtuous humans, however, they are motivated by tenderness, affection, prudence, deference, altruism, aggression and dominance. They certainly possess the rudiments of agency, in varying degrees. All this is pivotal in animal standpoint theory. That there is something that it is like to be a particular animal, that there is a particular vantage—or standpoint from which life and the world are perceived and experienced, a perspective that is communicated in a variety of ways and that is informed by more or less complex needs, beliefs, preferences, desires and emotions, also constitutes a suitable range of topics for education.

[92] Ibid.: 21; Best (2014), 16, 104, 165.
[93] Best (2012), 9.

Bibliography

Becker, Valeska. 'Auf den Schultern von Riesen. Zum Umgang mit Tieren aus Sicht der Ur- und Frühgeschichte'. In *Interspezies Lernen: Grundlinien interdisziplinärer Tierschutz- und Tierrechtsbildung*, edited by Simone Horstmann, 29–60. Bielefeld: Transcript Verlag, 2021.

Best, Steven. 'Introduction', *Animal Liberation Philosophy and Policy Journal* 5, no. 1 (2007): 2–3.

Best, Steven. 'Education in the Time of Eco-crisis'. Presentation in the Wits School of Education, University of the Witwatersrand, 2 October (2008).

Best, Steven. 'The Rise of Critical Animal Studies: Putting Theory into Action and Animal Liberation into Higher Education'. *Journal for Critical Animal Studies* 7, no. 1 (2009): 9–52.

Best, Steven. 'The Revolutionary Implications of Animal Standpoint Theory' (2012). https://drstevebest.wordpress.com/2012/05/14/the-revolutionary-implications-of-animal-standpoint-theory-2/. Accessed 23 March 2018.

Best, Steven. *The Politics of Total Liberation: Revolution for the 21st Century.* New York: Palgrave Macmillan, 2014.

Best, Steven. 'The Rise (and Fall) of Critical Animal Studies'. *Liberazioni* (n.d.): 1–35 http://www.liberazioni.org/articoli/BestS-TheRise%28and%20Fall%29ofCriticalAnimalStudies.pdf. Accessed 22 March 2018.

Best, Steven, Peter McLaren and Anthony J. Nocella II, 'Revolutionary Peacemaking: Using a Critical Pedagogy Approach for Peacemaking with "Terrorists"'. *Journal for Critical Education Policy Studies* 5, no. 2 (2007a): 494–528.

Best, Steven, Anthony J. Nocella II, Richard Kahn, Carol Gigliotti and Lisa Kemmerer. 'Introducing Critical Animal Studies'. *Animal Liberation Philosophy and Policy Journal* 5, no. 1 (2007b): 4–5.

Code, Lorraine. 'Feminist Epistemology'. In *A Companion to Epistemology*, edited by Jonathan Dancy and Ernest Sosa, 138–142. Oxford: Blackwell, 1992.

Corman, Lauren and Tereza Vandrovcová. 'Radical Humility: Toward a More Holistic Critical Animal Studies Pedagogy'. In *Defining Critical Animal Studies: An Intersectional Social Justice Approach for Liberation*, edited by Anthony J. Nocella II, John Sorenson, Kim Socha and Atsuko Matsuoka, 135–157. New York: Peter Lang, 2014.

Donovan, Josephine. 'Animal Rights and Feminist Theory'. In *Ecofeminism: Women, Animals, Nature*, edited by Greta Gaard, 167–194. Philadelphia: Temple University Press, 1993.

Dopico, Eduardo and Garcia-Vazquez, Eva. 'How Technology Can Replace Animals in Lab Practices'. In *Animals and Science Education: Ethics, Curriculum and Pedagogy*, edited by Michael P. Mueller, Deborah J. Tippins and Anthony J. Stewart, 135–144. Cham: Springer, 2017.

Engels, Friedrich. 'Socialism: Utopian and Scientific'. In *The Marx and Engels Reader*, edited by Robert C. Tucker: 683–717. New York: Norton, 1972.

Haraway, Donna. 'Situated Knowledges'. *Feminist Studies* 3 (1988): 575–599.

Haraway, Donna. *The Companion Species Manifesto: Dogs, People and Significant Otherness.* Chicago: Prickly Paradigm Press, 2003.

Harding, Sandra G. 'Feminist Justificatory Strategies'. In *Women, Knowledge and Reality: Explorations in Feminist Philosophy*, edited by Ann Garry and Marilyn Pearsall, 189–201. New York & London: Routledge, 1992.

Harding, Sandra G. *The Feminist Standpoint Theory Reader: Intellectual and Political Controversies.* New York: Routledge, 2004.

Hartsock, Nancy. 'The Feminist Standpoint: Developing the Ground for a Specifically Feminist Historical Materialism'. In *Discovering Reality: Feminist Perspectives on Epistemology, Metaphysics, Methodology and Philosophy of Science*, edited by Sandra Harding and Merrill B. Hintikka, 283–310. Dordrecht, London & Boston: D. Reidel, 1983.

Horsthemke, Kai. *Animal Rights Education.* Cham: Palgrave Macmillan, 2018.

Hübner, Andreas. '"Mißhandle und mißbrauche nie ein Tier!" Tierschutz- und Tierrechtsbildung in geschichtsdidaktischer Perspektive'. In *Interspezies Lernen: Grundlinien interdisziplinärer Tierschutz- und Tierrechtsbildung*, edited by Simone Horstmann, 89–106. Bielefeld: Transcript Verlag, 2021.

Humes, Brandy. 'Moving Toward a Liberatory Pedagogy for All Species: Mapping the Need for Dialogue between Humane and Anti-oppressive Education'. *Green Theory & Praxis: The Journal of Ecopedagogy* 4, no. 1 (2008): 65–85.

Jakubiak, Cori. 'Tracing the Anthrozoological Landscape of Central Iowa: Place and Pedagogical Possibilities'. In *Animals and Science Education: Ethics, Curriculum and Pedagogy*, edited by Michael P. Mueller, Deborah J. Tippins and Anthony J. Stewart, 51–68. Cham: Springer, 2017.

Kahn, Richard. 'Towards an Animal Standpoint: Vegan Education and the Epistemology of Ignorance'. In *Epistemologies of Ignorance in Education*, edited by Eric Malewski and Nathalia Jaramillo, 53–70. Charlotte: Information Age Publishing, 2011.

Kahn, Richard. 'Afterword'. In *The Educational Significance of Human and Non-human Animal Interactions: Blurring the Species Line*, edited by Suzanne Rice and A.G. Rud, 215–226. Basingstoke & New York: Palgrave Macmillan, 2016.

Karlan, Jimmy. 'Spiders, Rats, and Education'. In *Animals and Science Education: Ethics, Curriculum and Pedagogy*, edited by Michael P. Mueller, Deborah J. Tippins and Anthony J. Stewart, 123–134. Cham: Springer, 2017.

Lloro-Bidart, Teresa and Russell, Constance. 'Learning Science in Aquariums and on Whalewatching Boats: The Hidden Curriculum on the Deployment of Other Animals'. In *Animals and Science Education: Ethics, Curriculum and Pedagogy*, edited by Michael P. Mueller, Deborah J. Tippins and Anthony J. Stewart, 41–50. Cham: Springer, 2017.

Lukács, Georg. 'Reification and the Consciousness of the Proletariat'. In *History and Class Consciousness.* Translated by Rodney Livingstone. Cambridge: MIT Press, 1971.

Marx, Karl and Friedrich Engels. *The German Ideology*. Edited by C.J. Arthur. New York: International Publishers, 1970.

Marx, Karl. *Economic and Philosophic Manuscripts of 1844*. Edited by D. Struik. New York: International Publishers, 1984.

Matthews, Michael R., ed. *International Handbook of Research in History, Philosophy and Science Teaching*. Dordrecht & Heidelberg: Springer, 2014.

Mueller, Michael P., Tippins, Deborah J. and Stewart, Arthur J. 'Wild Awakedness and Animalistic Inquiry: Introducing a Book on the Role of Animals with/in Science Education'. In *Animals and Science Education: Ethics, Curriculum and Pedagogy*, edited by Michael P. Mueller, Deborah J. Tippins and Anthony J. Stewart, v–xii. Cham: Springer, 2017.

Nocella II, Anthony J., John Sorenson, Kim Socha and Atsuko Matsuoka, eds. *Defining Critical Animal Studies: An Intersectional Social Justice Approach for Liberation*. New York: Peter Lang, 2014.

Pedersen, Helena. *Animals in Schools: Processes and Strategies in Human-Animal Education*. West Lafayette, IN: Purdue University Press, 2010.

Regan, Tom. *The Case for Animal Rights*. London: Routledge & Kegan Paul, 1983.

Reiss, Michael J. 'A Framework within which to Determine how We Should Use Animals in Science Education'. In *Animals and Science Education: Ethics, Curriculum and Pedagogy*, edited by Michael P. Mueller, Deborah J. Tippins and Anthony J. Stewart, 243–259. Cham: Springer, 2017.

Renner, August, Gustav Feddeler, J.F. Hüttmann, Heinrich Jastram and Adolf Marten, eds. *Weltkunde: Leitfaden der Geographie, Geschichte, Naturgeschichte, Physik und Chemie für Mitteschulen und mehrklassige Volksschulen*. Hannover: Helwing, 1896.

Shapiro, Kenneth and DeMello, Margo. 'The State of Human-Animal Studies'. *Society and Animals* 18 (2010): 307–318.

Singer, Peter. *Animal Liberation*. New York: Random House, 1975.

Spannring, Reingard. 'Animals in Environmental Education Research'. *Environmental Education Research* 23, no. 1 (2016): 63–74.

West Jefferies, Alexandra. 'Engineering a Solution for Managing Fish Waste'. In *Animals and Science Education: Ethics, Curriculum and Pedagogy*, edited by Michael P. Mueller, Deborah J. Tippins and Anthony J. Stewart, 29–39. Cham: Springer, 2017.

Zinn, Howard. 'Science, the Public, and Social Change', interviewed by G.M. Schivone (2008). https://www.howardzinn.org/the-citizens-among-us/. Accessed 1 April 2018.

CHAPTER 5

Sensing Life: Intersections of Animal and Sensory Histories

Andrew Flack and Sandra Swart

Introduction

'Life', in its astonishing diversity, escapes a decisive definition. From philosophers to biologists, we struggle to delimit it. It can be defined in many ways; Aristotle (384–322 BC) had it as 'nothing is nourished which does not have a share in life'.[1] Frequently, one is offered a roll call of what life *does* rather than *is*—it can change, reproduce, adapt, and die. Very significantly, a living being is usually defined as something with the capacity to 'respond' to the fluctuating conditions of its wider environment and, by extension, the other living beings that inhabit those natural worlds. In

[1] Aristotle, *De Anima* ii 4, 415b27–415b28.

A. Flack (✉)
School of Humanities, University of Bristol, Bristol, UK
e-mail: Andrew.flack@bristol.ac.uk

S. Swart
Department of History, Stellenbosch University, Stellenbosch, South Africa
e-mail: sss@sun.ac.za

M. J. Glover, L. Mitchell (eds.), *Animals as Experiencing Entities*, The Palgrave Macmillan Animal Ethics Series,
https://doi.org/10.1007/978-3-031-46456-0_5

fact, it is this 'response' that helps us get closer to a definition: Adaptability is a function—an *ability*—shared by all life on Earth. Indeed, NASA used it, a generation ago, as the very basis for their endeavours to identify life on other planets: *Life is a self-sustaining chemical system capable of Darwinian evolution*; of adapting bodies and behaviours in response to the wider environment.[2]

The nature of these physical responses is critical to the building of relationships both within and between species, and between creatures and their wider worlds. The character of such responses reveals much about what the world *means* in a material sense to those who dwell in it. Of course, the world is always co-authored; the sensory systems of humans and other animals overlie and intersect in critical ways.[3] And yet, even when thinking about the 'sensory ecology' of other animals (their sensory systems and how these systems affect their perception of and responses to the world), humans often concentrate on the classical senses that culturally and historically make 'sense' to us: vision, smell, audition, touch, and taste. Significantly, however, new research, in the field of sensory ecology, reveals an extraordinary range of sensory abilities relating to bodily movement, vibration, chemical responses, electric and magnetic field detection. These more-than-human sensory capacities allow animals to function effectively in the world: to compete for resources[4] or mate,[5] find food,[6] or escape becoming food for someone else. Let us provide a couple of illustrative examples. For instance, desert ants in the hostile saltpans of Tunisia build cognitive maps based on smell, deploying their antennae 'in stereo' to process sensory data to achieve this olfactory cartography.[7] Elephants,

[2] Steven Benner, "Defining life", *Astrobiology,* no. 10 (2010): 1021–1030; Edward N. Trifonov, "Vocabulary of Definitions of Life Suggests a Definition", *Journal of Biomolecular Structure and Dynamics*, no. 29 (2011): 259–266.

[3] Stephen Burnett, "Perceptual Worlds and Sensory Ecology", *Nature Education Knowledge*, no. 3 (2011): 75.

[4] R. T. Mason and R.M. Parker, "Social behaviour and pheromonal communication in reptiles", *Journal of Comparative Physiology A*, no. 196 (2010): 729–749.

[5] A.S. Feng et al. "Ultrasonic communication in frogs", *Nature,* no. 440 (2006): 333–336.

[6] G. P. Bell, "The sensory basis of prey location by the California leaf-nosed bat *Macrotus californicus* (Chiroptera: Phyllostomidae)", *Behavioral Ecology and Sociobiology,* no. 16 (1985): 343–347; J.U. van Dyke and M.S. Grace, "The role of thermal contrast in infrared-based defensive targeting by the copperhead, *Agkistrodon contortrix*", *Animal Behaviour,* no. 79 (2010): 993–999.

[7] Kathrin Steck, Markus Knaden and Bill Hansson, "Do desert ants smell the scenery in stereo?" *Animal Behaviour*, no. 79 (2010): 939–945.

meanwhile, use low-frequency sounds or 'infrasounds' (below 20 Hz) as a form of long-distance—yet private—communication, allowing herds to maintain contact and convey important information about their world over tens of kilometres.[8] Thus, it is not just a matter of some creatures possessing 'better senses' than others, but about the existence of inherently diverse ways of experiencing the world. These are ways of sensing—ways of *perceiving and being*—far beyond our own.

That this is now so clearly the case makes it rather surprising that animal historians have not yet done more to think about multi-species pasts in the context of sensory entanglements. Certainly, environmental historians have explored a partial sensory turn in recent decades; scholars such as Alice Would, Peter Coates and Melanie A. Kiechle are notable for their attention to historical touchscapes, soundscapes, and smellscapes, respectively.[9] Some historians, including Erica Fudge, Christopher Plumb, Jonathan Saha, and Brett Walker, have deployed an engagement with the senses as a means of thinking through that doggedly addictive issue in animal and environmental history: agency, or the capacities of more-than-human beings to influence the material world with (or without) some degree of intentionality.[10] Meanwhile, scholars of human-animal relations beyond the historical have begun to think with the senses. The British Animal Studies Network, for instance, has gathered scholars together who work on particular senses, between 2013 and 2017.[11]

[8] J.H. Poole, et al. "The social contexts of some very low frequency calls of African elephants", *Behavioural Ecology and Sociobiology*, no. 22 (1998): 385–392; K. McComb et al. "Long-distance communication of acoustic cues to social identity in African elephants", *Animal Behaviour*, no. 65 (2003): 317–329.

[9] Peter A. Coates, "The Strange Stillness of the Past: Towards an Environmental History of Sound and Noise", *Environmental History*, no. 10 (2005): 636–665; Melanie A. Kiechle, *Smell Detectives: An Olfactory History of Urban America* (Seattle: University of Washington Press, 2019); Alice Would, "Fleshing Out the Animals of British Taxidermy in the long Nineteenth Century (1820–1914)", Doctoral dissertation, University of Bristol, 2021.

[10] Erica Fudge, "Milking other men's beasts", *History and Theory*, no. 52 (2013): 13–28; Christopher Plumb, "Reading menageries: using eighteenth-century print sources to historicise the sensorium of menagerie spectators and their encounters with exotic animals", *European Review of History: Revue européenne d'histoire*, no. 17 (2010): 265–286; Jonathan Saha, "Colonizing elephants: animal agency, undead capital and Imperial science in British Burma", *BJHS*, Themes 2 (2017): 2–17 and Brett L. Walker, "Animals and the Intimacy of History", *History and Theory*, no. 52 (2013): 45–67; 129–189.

[11] British Animal Studies Network, https://www.britishanimalstudiesnetwork.org.uk. They provide audio recordings of many of the papers.

Just over a decade ago, Sandra Swart followed Lucien Febvre's call for a sensory history by essaying a cursory sensory history of horses, proposing that they changed human history not only on the macro-level, but significantly in the small, intimate arena of the body.[12] Swart drew attention to a hippocentric sensory history, suffused in the horses' physical memory, intense fear and cyclical sexuality and fecundity, and their strongest traits (as grass-eating prey herbivores with a fatal tendency towards over-eating and over-heating). Such a history, she argued, might be a story of running, smelling for predators, smelling the chemical messages left in dung by the herd, listening, seeing, tasting pheromones, foals, blood, sex, pain, fear—perhaps mainly the heady pleasures of munching grass.[13] But critically, she suggested that it remained a Rorschach test, revealing more about the historian (and her own epoch or sensory imagination) than about horses, just as Clifford Geertz was sceptical of empathy in understanding other cultures lest the researcher project their own cultural norms under cover of purported empathy.[14] This challenge is right at the core of doing sensory history at the human-animal intersection.

The combination of expansive opportunity and irresistible methodological challenge motivates us to think that there is so much more to be investigated at the nexus of animal and sensory history. The theories, concepts, and methods developed by pioneers in the history of the senses offer exciting ways forward for scholars thinking about past engagements between animals and people across times and places. The field of sensory history has been shaped by scholars who wanted to trace both the physical operation and socio-cultural understanding of the senses in different historical contexts.[15] Constance Classen's, Mark Smith's, and Alain Corbin's contributions to the cultural deconstruction of the senses illustrate the ways in which senses and sensations are not only biochemical phenomena,

[12] L. Febvre, "Smells, Tastes, and Sounds", in *idem*, *The Problem of Unbelief in the Sixteenth Century* (Cambridge: CUP, 1982); Sandra Swart, "The World the Horses Made – A South African Case Study of Writing Animals into Social History", *International Review of Social History*, no. 55 (2010); Sandra Swart, *Riding High: Horses, Humans and History in South Africa* (Johannesburg: Witwatersrand University Press, 2010).

[13] This section draws on Swart, *Riding High*, 217.

[14] Clifford Geertz, "'From the Native's Point of View': On the Nature of Anthropological Understanding in Culture Theory", in Richard Shweder and Robert LeVine (eds), *Culture. Theory: Essays on Mind, Self, and Emotion* (Cambridge: CUP, 1984); Douglas Hollan and C. Jason Throop, "Whatever Happened to Empathy?: Introduction", *Ethos*, no. 36 (2008): 385–401, 385.

[15] Sensory history is sometimes also labelled the 'history of the senses', or sensate history.

but also significantly contoured by diverse and historically contingent cultural conditions. The theoretical, conceptual, and methodological richness of this work strongly suggests that animal historians have much work to do in understanding the sensory regimes inherent in historical phenomena such as pet-keeping, animal feeding, meat-eating, and the study of natural history (to name just a few potential topics).[16] Embracing a radical 'sensory turn' may provide fruitful avenues for deepening appreciation of the historical dimensions of the ways in which human-animal relations emerge, develop, and, indeed, come to an end.

In this chapter, we want to illustrate the significant potential of an animal-orientated sensory turn. We want to make the case that it is substantially through the senses—by thinking about what sensory experiences were and what they meant in the context of human-animal relations—that we can draw closer to the lived realities of past animals—and their people. This chapter has four parts; in the introduction, we have so far briefly traced the dimensions of existing sensory scholarship. Next, we will connect it to an ethical impulse and imperative surrounding the diverse lives of animals in historical and contemporary entanglement with people. It then illustrates these themes with reference to two distinct but, in the end, remarkably connected case studies. The first—'Anthropocene Nights'—relates to the history of darkness and the emergence of human-animal relations in nocturnal contexts where human infrastructures have impacted on sensory realities for nocturnal creatures. Here we highlight the sensory chasms separating diurnal from nocturnal species and argue that attention to the idea of 'disablement' in sensory contexts may help us to better understand and empathize with the real consequences of our increasing illumination of the night-time. The second—'Sensory Horstory'—emphasizes the intense intimacy of the relationships between people and horses, considering the rich potential of delving deeply into the sensory dimensions of these embodied and often deeply empathetic relationships at the micro-level. Over the course of the two case studies, we bring diverse approaches—not least from animal history and the history of the senses (very broadly conceived)—in what we hope might be generative new

[16] Constance Classen, *The Color of Angels: Cosmology, Gender and the Aesthetic Imagination* (London: Routledge, 1998); Alain Corbin, *The Foul and the Fragrant: Odor and the French Social Imagination* (Leamington Spa: Berg, 1986); Elizabeth D. Harvey, "The Portal of Touch", *American Historical Review*, no. 116 (2011): 385–400; and Mark M. Smith, *Sensing the Past: Seeing, Hearing, Smelling, Tasting and Touching in History* (Berkeley: University of California Press, 2007).

ways. We close by reflecting on the mouth-watering theoretical and methodological challenges inherent in engagement with more-than-human sensoria, while motivating for an expansive and intellectually ambitious 'sensory return' for historians interested in the experiential dimensions of human-animal relations.

Indeed, there are ethical imperatives underpinning the kind of sensory thinking that we want to see among animal historians and animal studies scholars more broadly. There is a deep connection between sensory histories and attempts to generate 'affective empathy'—'the ability to resonate with the emotions of others while retaining a clear self-other distinction' with the beings who inhabited—indeed *made*—the world. But these efforts are usually anthropocentric.[17] Here—and especially in the context of the human-horse relationship—we build on Despret's "interspecies empathy", which is not one that operates through analogy or identification, but an "embodied empathy"—the direct perception of others' subjectivity in the world—that operates as a mode of thinking *together with* rather than merely "about" the animal.[18] Our approach draws on Lori Gruen's idea of "entangled empathy"—the focussed sensitivity to the experiences of another, which therefore involves a relationship and responsibility to them.[19] In the process, a focus on the sensory troubles the

[17] For example, 'Historical empathy is the process of cognitive and affective engagement with historical figures to better understand how *people* from the past thought, felt, made decisions, acted, and faced consequences within a specific historical and social context' [emphasis added]: Jason L. Endacott, "Negotiating the Process of Historical Empathy", *Theory & Research in Social Education*, no. 42 (2014): 4–34. On 'affective empathy', see Elisa Aaltola, *Varieties of Empathy: Moral Psychology and Animal Ethics* (Lanham, MD: Rowman & Littlefield International, 2018).

[18] Vinciane Despret, "Responding Bodies and Partial Affinities in Human–Animal Worlds", *Theory, Culture & Society*, no. 30 (2013): 51–76. See also Aaltola, *Varieties of Empathy*.

[19] Gruen's entangled empathy involving 'reflectively imagin[ing] ourselves in the position of the other', and then to making 'judgment[s] about how the conditions that she finds herself in may contribute to her perceptions or state of mind and impact her interests'. Lori Gruen, "Navigating Difference (again): Animal Ethics and Entangled Empathy", *Strangers to Nature: Animal Lives and Human Ethics* (Lanham: Rowman and Littlefield, 2012: 213–233. Some go so far as to argue that empathy requires the sensory, especially 'touch'. Betty Stoneman, "The Dichotomy of Sight vs. Touch: The Harm of the Primacy of Sight for Empathy", *Sloth: A Journal of Emerging Voices in Human-Animal Studies*, no. 2 (2016). For an overview of the disputes over embodiment's links to empathy, see Nivedita Gangopadhyay, "Introduction: Embodiment and Empathy, Current Debates in Social Cognition", *Topoi*, no. 33 (2014): 117–127; Alvin Goldman and Frederique de Vignemont, "Is Social Cognition Embodied?", *Trends in Cognitive Sciences*, no. 13 (2009): 154–159.

overwhelming ascendancy of human rationality in how we historians approach our job, with an emphasis on the sensory so long misclassified as the province of the other animals.

We recognize, however, that such an approach to the sensory lives of historical others is fraught with problems that are themselves rooted in our *own* sensory lives. Humans are thoroughly limited by their sensory systems, which, in most circumstances, are adapted to diurnal or daylight living, where vision is the principal means of gathering knowledge of the environment.[20] This is not the case for all living things, however.[21] In fact, even within the context of vision, recent studies—coalescing in the emergent field of 'Vision Ecology'—reveal that 'seeing' means different things in different species contexts.[22] It is perhaps an astonishing and humbling realization that the human world is not the only world and that our world is thoroughly shaped by the constraints of our sensory apparatus. For other living things, 'reality' looks, feels, smells, sounds, and tastes quite differently. Every animal inhabits *our* world and *yet also their own world*—the totality of all the input sensed by their own nervous system.[23] An animal's sensory ecology needs to be grasped before their wider worldly experiences can be even approximated. It is this that makes it so hard to write animal history with the usual empathy required of an historian: it is impossible to 'see through the eyes of a horse', for example—our view has been staring down from the saddle or gazing up at the horse-human dyad from the ground. This challenge has been of enduring concern to both natural scientists like Donald R. Griffin who considered bats in 1958[24] and philosophers like Thomas Nagel who contemplated the same creatures some two decades later. Nagel argued that it would be impossible to know what it was like to be a bat; their sensory lives are, he thought, utterly alien

[20] For a history of vision in one specific context, see Martin Jay, *Downcast Eyes: The Denigration of Vision in Twentieth Century French Thought* (Berkeley: University of California Press, 1994).

[21] On more-than-human sensory adaptations, see Kenneth Katana, *Great Adaptations: Star-Nosed Moles, Electric Eels & Other Tales of Evolution's Mysteries Solved* (Princeton: Princeton University Press, 2020).

[22] Thomas W. Cronin, Sönke Johnson, N. Justin Marshall and Eric J. Warrant, *Visual Ecology* (Princeton; Princeton University Press, 2014); Michael F. Land and Dan-Eric Nilsson, *Animal Eyes* (Oxford: Oxford University Press, 2012).

[23] O. Dangles et al., "Variability in sensory ecology: Expanding the bridge between physiology and evolutionary biology", *Quarterly Review of Biology* no. 84 (2009): 51–74.

[24] Donald R. Griffin, *Listening in the Dark* (New Haven: Yale University Press, 1958).

to us, and this erects an impenetrable barrier to our ability to *be batty*.[25] Whether this sensory and experiential barrier—this chasm between species—is indeed impenetrable or not, the more significant question is whether it remains worth trying to get close to the realities of the lives of animal others. We think that it is.

Certainly, empathy, like the discipline of history itself, is modulated by context.[26] It permits individuals to share or at least comprehend the affective states of others and encourages prosocial behaviour. A kind of radically embodied empathy led to crossing the membrane between ours and other species, over time. Historians have shown how the recognition of the superiority of the other animals' senses led to past efforts to conscript animals to boost our human sensorium. Early treatises on natural history, for example, are replete with recipes for acquiring animal senses (*The Book of Secrets* accredited to Albertus Magnus suggested that if the heart of a weasel "be eaten yet quaking it meketh a man to know thinges to come", while the twelfth-century Abbess Hildegard contended that the same organ be "dried and placed with wax in the ear" to heal deafness). Much later, a self-consciously scientific sort of shapeshifting took place in the ethnological fieldwork of Nobel Prize winner Konrad Lorenz. From the 1920s, his imprinting and living with geese and jackdaws involved his own body in creating embodied relationships with birds.[27] One of his jackdaws tried to feed him with worms (it was sacrifice as Lorenz could not pretend to like the taste of finely minced worms, generously mixed with jackdaw saliva).[28] His body became a human-body-with-a-jackdaw, as Despret put it.[29]

Animals as animated sensory "prostheses" or "augmenters" entangled in affective sensory encounters continue to be vital in, for example, landmine snuffling giant rats, sniffer, "hearing" and "seeing-eye" dogs,[30] capuchin service monkeys for mobility impairments (or a mare—for an historian

[25] Thomas Nagel, "What is it like to be a bat?", *Philosophical Review*, no. 83 (1974): 435–450.

[26] Cristina Gonzalez-Liencres, Simone G. Shamay-Tsoory and Martin Brüne, "Towards a neuroscience of empathy: Ontogeny, phylogeny, brain mechanisms, context and psychopathology", *Neuroscience & Biobehavioral Reviews*, no. 37 (2013): 1537–1548.

[27] Konrad Lorenz, *King Solomon's Ring* (London: Routledge, 1952, 2002).

[28] Lorenz, *King Solomon's Ring*, 129.

[29] Despret, "Responding Bodies", 71.

[30] Monika Baár, "Prosthesis for the body and for the soul: the origins of guide dog provision for blind veterans in interwar Germany", *First World War Studies*, no. 6 (2015): 81–98.

on a night ride through a forest, as we shall see).[31] Animal others have thus, historically, provided a sensory reservoir, empowering humans to augment our own pool of sensory faculties through various forms of shape-shifting, directly using their bodies or by drawing on their example to invent or refine new kinds of acuity through technology (as in echolocation).

It is clear that sensory thinking presents opportunities for important scholarship at the intersection of animal history and histories of the senses that grapple hard with the tensions elucidated above. From the radical sensory chasms separating species and the fragile sensory configurations that perhaps 'disable' people and animals in fluxing nocturnal worlds to the embodied engagements between people and horses, we now proceed to argue that the intersection of animal history and sensory studies offers critical if not uncomplicated—but certainly as yet untapped—means of historicizing the lives of human and nonhuman animals, past and present.

Anthropocene Nights

The context of darkness is perhaps one of the most compelling illustrative examples of the ways in which radical divergences in sensoria across species could shape physical and imaginary interactions among living things. By thinking through and with the darkness, we might consider the ways in which bodies—with all their 'abilities' and concurrent 'limitations'—cohere (or not) with their environments. Indeed, *sensory negotiations* are central to the emergence and development of nocturnal human-animal relations. Not least, this is because people have historically been extremely maladapted to living in the darkness (or accessing it at all), whether that be during the fluxing darkness of the prosaic and rhythmic night or the deep and stable darkness of the subterranean. We might go as far as to say that, for most of human history, people have effectively been 'disabled' in the darkness, where their vision has rendered them vulnerable to attack or accident and has limited their capacity to function in the way that they do during the day. Other animals—many of which are primarily nocturnal in behaviour—were profoundly advantaged by their sensory apparatus. In the context of wider animal histories, where human power over the rest of the animal world has been emphasized, this apparent inversion seems to be

[31] Steven Connor, "The Menagerie of the Senses", *Senses and Society*, no. 1 (2006): 9–26, 15, 23.

important. Darkness—and the sensory reconfigurations that were necessary and/or inevitable in the shadows—offers a context for sensory thinking in which the traditional human-animal power dynamics were reconfigured. Like Val Plumwood in her visceral account of 'Being Prey', darkness—especially in pre-modern contexts, and those places where anthropogenic light is not pervasive—has operated across most of human history as a site of perceived vulnerability, fragility, where predators can become prey, or at least where people cannot always have things their own way.[32]

While principally emergent and considered in the context of capitalist systems of production, the notion of 'disability' might be useful here if we think about how sensory and behavioural ability and impairment operate beyond the human context. In particular, the 'social model' of disability emphasizes the emergence of disability when barriers erected by the normatively abled impact on the extent to which marginal bodies and the wider environment are able to cohere.[33] Disability, of course, emerged as a concept after the middle of the nineteenth century, when the notion of the 'normal' became more potent within Western societies increasingly driven by standardization, efficiency, and perpetual 'improvement'—of both bodies and social groups.[34] At its heart, though, disability is really about the extent to which beings can 'access' the world in ways that suit the bodies that they have. The context of nocturnal human-animal relations, with all their sensory tensions, really illuminates a history of adaptive bodies and reconfiguring sensoria.

In the first instance, nocturnal human-animal encounters in the West have increasingly been about the development of prostheses to enable visually limited people access to a shrouded world. In her 1985 'Cyborg Manifesto', philosopher Donna Haraway argues that during the twentieth century the boundary between living being and machine has become blurred, generating hybrids—cyborgs—in the process.[35] The sensory

[32] Val Plumwood, "Being Prey", *Kurungabaa,* no. 4 (18 January 2011), https://kurungabaa.wordpress.com/2011/01/18/being-prey-by-val-plumwood/.

[33] For a basic introduction to key concepts in Disability Studies, see Tom Shakespeare, *Disability: The Basics* (London: Routledge, 2018).

[34] Lennard Davis, *The End of Normal: Identity in a Biocultural Era* (Ann Arbor: University of Michigan Press, 2013).

[35] Donna Haraway, '"A Cyborg Manifesto: Science, Technology, and Socialist-Feminism in the Late Twentieth Century", in *Simians, Cyborgs and Women: The Reinvention of Nature* (New York; Routledge, 1991): 149–181.

prostheses developed by people in search of access to the darkness during the twentieth century illustrates several of the ways in which the limitations of vision could be addressed, and access granted to environments which had historically been 'exclusive'. Electric lighting technologies were developed and widely installed across the century, illuminating dark environments. Cities were increasingly bathed in light, rendering them ever more productive locales, while the lighting of automobile infrastructures established lines of lights crisscrossing previously dark domains. There is a growing body of literature that highlights the extent to which vast areas of the Global North—in particular—have been illuminated.[36] In the same vein, visual technologies such as infra-red thermal imaging devices—available after the Second World War—and the very recent innovation of moonlight cameras have increasingly operated as surveillance prostheses, permitting deep and intrusive vision for people, even in contexts—like places where wildlife and people share habitat—where travelling in the darkness might prove dangerous.[37]

This history of the adoption of prosthetics to allow human access to environments where the limitations of their bodies prevented access runs alongside a parallel process of more-than-human marginalization. For, in the transformation of night-time environments, and their modification for beings maladapted to dark-dwelling, they were also rendered increasingly inaccessible to beings whose senses—and wider ways of life—were adapted to the rhythms of light and dark provided by the interaction of celestial lights—the moon and stars—with atmospheric conditions like the movement of clouds across the night sky. As scholars of disability know, disablement emerges when systems are adapted to suit 'normative' bodies. As people increasingly used prosthetics to access dark worlds they considered

[36] Christopher Kyba, Sara Pritchard et al., "Night Matters—Why the Interdisciplinary Field of 'Night Studies' Is Needed", *Multidisciplinary Scientific Journal*, no. 3 (2020): 1–6; John Jackle, *City Lights: Illuminating the American Night* (Baltimore: Johns Hopkins University Press, 2001); Wolfgang Schivelbusch, *Disenchanted Night: The Industrialisation of Light in the Nineteenth Century* (Berkeley: University of California Press, 1995); Jeremy Zallen, *American Lucifers: The Dark History of Artificial Light, 1750–1865* (Chapel Hill: University of North Carolina Press, 2019). For environmental histories of darkness, light and night, see Andrew Flack, "Dark Trails: Animal histories beyond the light of day", *Environmental History*, no. 27 (2022): 215–241; Andrew Flack, "Dark Denegerations: Life, Light and Transformation Beneath the Earth, 1840–c. 1900", *ISIS*, no. 113 (2022): 331–351.

[37] See *Earth at Night* (Apple, 2020); *Night on Earth* (Netflix, 2020).

themselves (in a typical act of egotism) the mainstream occupants of those environments, marginalizing other beings in the process.

A growing scientific literature emphasises the distressing array of ways in which 'ecological light pollution' impacts on the bodies and behaviours of dark-adapted species.[38] The potential impacts of illumination on non-human animals have been becoming clearer for more than a century now. A 1900 *Country Life* article referred to eighteen years' worth of observations of high levels of mortality among nocturnal avian migrants, their lifeless bodies gathering around the bases of Irish lighthouses.[39] Later, in 1918, Squires and Hanson reported rumours of high levels of migrating bird mortality around lighthouses on the Californian pacific coast, implying that these had been evident for several decades, at the very least. Death serves as an extreme consequence of the illumination of—in this case—migratory highways that are in all probability at least many hundreds of thousands of years old. We know now that light pollution such as this profoundly impacts on avian sensory experience. A lighthouse as a single light in an otherwise gloomy environment can attract birds for whom the light forms a single point of interest in the night. This light, intense as it is, can also effectively blind and disorientate birds as they fly closer to it. At the same time, the light attracts insects, their sensoria stimulated too by the disturbance of the darkness. An abundance of insects presents an added stimuli for birds attracted in the first instance by the light source. As their attention is held by the blinding light, migratory birds quickly tire, rendering them vulnerable to accidents in the vicinity of the lighthouse.[40]

Fundamentally, the illumination from lighthouses illustrates the ways in which a human prosthetic device allowed access to the darkness for people while simultaneously disorienting avian senses associated with homing and migration.[41] In more recent decades, reports have increasingly noted the capacity of anthropogenic light to confuse animal senses. Polecats are found dead in huge numbers on the roadside leading North from Cape Town, South Africa, while disorientated birds have been observed

[38] See, for instance, Travis Longcore and Catherine Rich, "Ecological Light Pollution", *Frontiers in Ecology and Environment*, no. 2 (2004): 191–198.

[39] "Bird notes from Irish lighthouses", *Country Life Illustrated* (8 December 1900): 727–728.

[40] See Longcore and Rich, "Ecological Light Pollution".

[41] Walter Albion Squires and Harold E. Hanson, "The Destruction of Birds at the Lighthouses on the Coast of California", *The Condor*, no. 20 (1918): 6–10.

apparently trapped within illuminated cities, unable to find their way out.[42] More subtly, perhaps, life scientists have observed since the 1970s that street lighting (acknowledging that different kinds of street light have diverse ecological effects), in particular, has a dual ecological function, repelling and attracting organisms—often, for example, attracting insects that in turn attract foragers like some species of bats. Providing new niches such as these might sound like an opportunity, but in reality, anthropogenic light contorts the sensory dimensions of the environment, disrupting ecosystem rhythms in the process.[43]

Narratives of access, inhabitation, marginalization, and exclusion permeate the histories of nocturnal human-animal relations, and they are all fundamentally founded on the senses and diverse sensory experiences. On the one hand, these narratives are really about sensory adaptation, augmentation, and limitation, and the vulnerability of bodies to environments to which their sensory systems are not ideally suited. We hesitate to call this a '*sensory negotiation*', because the story of the twentieth century, in most contexts where light has been widely used as an accessibility tool, is one of an intense human capacity to manipulate environmental conditions to suit themselves as the dominant—perhaps even 'normative'—species.

The example of the archetypal lighthouse illuminating the skies illustrates that the sensory differences between species and the consequent shaping of the natural world to suit the senses of one organism over another has had significant material implications. Understanding of these sensory experiences is critical to a full understanding of human-animal relations across past and present contexts. Beyond these, however, sensory diversity has inspired other kinds of human-animal relationships. Not least, scientific communities have sought to understand the sensory apparatus and the sensory experiences of other beings. In so doing, imaginary gymnastics (diverse empathies) have been necessary in order to conceptualize ways of being in the world that are often radically removed from our own. Scientific interest in the senses has ranged beyond the nocturnal context—there is long-standing fascination in canine sensory systems, from highly developed olfaction to the mysterious 'sixth sense' that supposedly allows a dog to know precisely when their human companion is about to drive up the road after a day at work. The interest in 'Other' lives

[42] See, for instance, "Correspondence: animal casualties on the road", *Country Life* (3 February 1966): 245.

[43] See Longcore and Rich, "Ecological Light Pollution".

speaks volumes for the capacity of radical sensory difference to grab and hold the collective attention of scientific communities all over the world. In the context of darkness, though, this scientific fascination with sensory exotica has arguably been particularly pronounced. From the star-nosed mole to the hyper-sensitive olfactory organs of the European hedgehog, and from the tactile cilia of dark-adapted cave species to the echolocating abilities of insectivorous bats, dark-adapted senses are frequently radically different to our own diurnal sensorium, and this has produced, over the course of the past several centuries, a set of invasive, manipulative, and unbalanced human-animal relationships.[44]

Most fundamentally, nocturnal species have been awarded cultural positions—at least in the West—which construct them as otherworldly, perhaps even spectral. Take this quote from *Country Life*: 'of all the cries in the night, the sound of the badger is the most unearthly: a dying scream, the shriek of a soul in torment…'[45] Add to that the chaotic flight of the bat, the screech of the barn owl, and the rustling of the hedgehog as it ambles through Autumn's fallen leaves and you have a soundscape framed by the darkness itself—a place and time which looms large and monstrous in the western cultural imagination.[46] The cultural baggage of the night combines with mysteries of the nocturnal sensorium to render dark-dwelling animals 'freakish' in their apparent marginality in relation to life on earth with what we construe as more 'normal' sensory apparatus and ways of being.

The apparent abnormality of some of these senses—an inability to even begin to conceptualize their realities—has had important (and quite ghoulish) consequences for some nocturnal individuals—as if plucked from the lurid imaginations of Gothic writers. The echolocating capacities of some species of bat was not well understood until Donald Griffin and Robert Galambos joined dots in the early 1940s, which had lain unconnected and quite scattered for many decades.[47] In the late eighteenth century—a time when natural phenomena were increasingly catalogued and taken apart to be fully understood—several naturalists performed horrific

[44] See Catania, *Great Adaptations*.

[45] "Night cries in the wood", *Country Life* (16 September 1954): 902.

[46] See Nina Edwards, *Darkness: A Cultural History* (London: Reaktion Books, 2018).

[47] Donald R. Griffin and Robert Galambos, "The Sensory basis of obstacle avoidance by flying bats", *Journal of Experimental Zoology*, no. 86 (1941): 481–506 and Donald R. Griffin and Robert Galambos, "Obstacle avoidance by flying bats: the cries of bats", *Journal of Experimental Zoology*, no. 89 (1942): 475–490.

experiments in the name of uncovering the secrets of bat navigations. Lorenzo Spallanzani, working in the late eighteenth century, sought to get to the bottom of bats' apparent 'sixth sense' by gouging out their eyes with red hot pokers and filling their ears with starchy material. In so doing, he attempted to construct bat sensoria through a process of sensory elimination: which senses could be removed and yet still produce a bat able to navigate in the darkness?[48] At the same time, Louis Jurine was conducting his own gruesome experiments for the same purpose.[49]

Later experiments—not really initiated in any meaningful sense until the twentieth century—were admittedly less Gothically grotesque but were still founded upon the manipulation of bat bodies, though not before early twentieth-century experimenter Hahn coated bat wings with Vaseline, cutting off their ears and tragi, and covering their heads with hoods in the name of deciphering the perplexing problem of apparently 'sightless' bat navigation.[50] Griffin and Galambos, working in the 1930s and early 1940s, were more humane, eschewing bodily mutilations in favour of using 'reversible' means of covering eyes, ears and mouth before encouraging the bats to navigate among several vertical metal wires in a laboratory room.[51] Invasive, injurious, or otherwise, all of these experimental strategies were complex simply because of the sensorial chasm separating human experience from that of bats.

Such experiments may well have been initially founded on the unquenchable desire of Western science, intensifying from the Enlightenment onwards to comprehensively comprehend and catalogue natural phenomena. But, 'knowing' nocturnal more-than-human sensoria has important practical applications—rooted in this sense of empathy for the worlds of nocturnal others—in our present circumstances, where Anthropocene nights are perilous indeed. Historical attempts to understand how other animals experience the world has permitted—in several, all-too-few, instances—adaptations to anthropogenic and anthropocentric structures. At some lighthouses, whose sweeping rays of light so distorted

[48] On Spallanzani, see Sven Dijkgraaf, "Spallanzani's unpublished experiments on the sensory basis of object perception in bats", *ISIS*, no. 51 (1960): 9–20.

[49] René Sigrist, « Louis Jurine, Chirurgien et naturaliste (1751–1819)», *Le Rhinolophe*, no. 11 (1995): 1–8.

[50] Walter Louis Hahn, "Some habits and sensory adaptations of cave-inhabiting bats", *Biological Bulletin*, no. 15 (1908): 135–164.

[51] Griffin and Galambos, "The Sensory basis of obstacle avoidance by flying bats", 481–506.

the dark skies for migrating birds, as early as 1935, structures were designed to allow birds who came too close to the light to perch safely around the lighthouses' edge. Such structures were, however, criticized as fairly ineffectual. Instead, it was argued that cup-like structures ought to be designed to catch stunned birds as they fell, preventing a cataclysmic (for them) impact with the ground.[52] In the UK, the Royal Society for the Protection of Birds successfully campaigned for most lighthouses to be floodlit, allowing migrating birds to better navigate past them on their long nocturnal journeys. Birds migrating under cover of darkness remain at risk at being 'blinded' by the light emitted from skyscrapers, television aerials, and burning oil and gas at offshore platforms.[53]

Thinking with the senses and sensations of creatures of the night, then, illustrates various ways in which sensory approaches to human-animal histories might serve the purpose of substantively widening conceptualizations of—and empathy surrounding—past interspecies realities. Certainly, nocturnal engagements between people and animals were frequently violent, manipulative, and exploitative. But more than that, we find that the context of the night provides evidence of human-animal relations figured by sensory negotiation and, later, some degree of empathy; negotiations that were often in flux. Nocturnal human-animal histories are about diverse abilities, and shifting vulnerabilities when bodies, senses, and environments do not cohere. Such histories also provide evidence of an 'arm's race' of sorts, as human technological adaptation rendered the darkness evermore live-able (for people). In the process, nocturnal creatures, whose sensoria have been configured over eons, were forced to endure the de-synchronization of their bodies with environments in which they had lived for generations. Nocturnal human-animal histories show that sensory negotiations impact lives. Sensory negotiations in the midst of Anthropocene nights change the world.

[52] "Birds and Lighthouses", *Country Life* (9 February 1935): 142.

[53] "Dangers of Migration For Migratory Birds", https://magpie.rspb.org.uk/birds-and-wildlife/natures-home-magazine/birds-and-wildlife-articles/migration/dangers-of-migration/#:~:text=Dangers%20of%20migration%201%20Bad%20weather%20Bad%20weather,7%20Sticky%20death%20...%208%20Protecting%20migrants%20 [accessed October 13, 2021).

Sensory Horsetory

On a moonlit night, horses can see as well as we do at noon. Although they are not nocturnal, their ancient enemies are—the canids and felids that have long hunted them. Horses have evolved excellent night vision, their pupils able to dilate enough to capture sparse photons at night, and their retinas are generally rod dominated.[54] The reflecting *tapetum lucidum* (a beautiful phrase in Latin for "bright tapestry") in the back of their eye provides the non-absorbed photons more chance of capture by photoreceptors, improving their night vision further. One may learn this in biology textbooks or experience it—as I have—on a night ride with one's mare, through a mountain forest, on a path lit up only by lunar luminescence, in which one eventually abandons the notion of trying to steer and gives the reins to the being with superior senses for the night: the horse. This illustrates the confluence of how we understand horse senses—experimentally and experientially.

Indeed, the etymology of the idiom of 'horse sense', first recorded in the early 1800s, might allude to this kind of unexpected sensory ability. Or it may stem from the notion that rural people working with horses have a kind of practical astuteness. Or perhaps the sense is connected to the horse itself, reflecting that horses have a kind of perceptiveness that keeps them safe. Certainly, it is an acknowledgement of the connection of horses, humans, and perception. Historically, the human senses were frequently portrayed as horses. This allegorical hippomorphism is evident in, for example, a Sanskrit text from fifth century BCE, in which the body was likened to a chariot where the horses are the human senses, and the driver was the intellect. Similarly, Plato described the challenges of the charioteer of reason as he tried to rein in a pair of ill-matched horses—the ignoble and noble parts of the soul.[55] Renaissance metaphor continued to allude to a dyad of horse and human in understanding the senses: the rider acting as the mind, the horse as metaphor for the sensual.[56] This reflects the historical relationship of humans not only to other animals (perceived as

[54] Equine colour vision resembles 'red-green blindness' in humans.

[55] Bart Westerweel, "The well-tempered lady and the unruly horses", in Theo D'haen, Rainer Grübel and Helmut Lethen (eds), *Convention and Innovation in Literature* (Utrecht: John Benjamin's Publishing Company, 1989): 115.

[56] Bart Westerweel, "The well-tempered lady and the unruly horses", in Theo D'haen, Rainer Grübel and Helmut Lethen (eds), *Convention and Innovation in Literature* (Utrecht: John Benjamin's Publishing Company, 1989): 115.

irrational, only-physical bodies) but also an ambivalent understanding of their own human senses (deemed as pleasurable and useful, and yet also as the means of sinning). It also reflects the centrality of the horse in trying to grapple historically with our own human senses. This is perhaps because of the human need to understand horse sense more than that of any other creature. Indeed, historically, humans have tried harder to understand the world from the horse's sensory point of view (as opposed to the other animals). It was necessary in domesticating, training, and riding them—dangerous and intimate processes that historically have compelled humans to at least try to 'see' the world through horses' eyes—for the safety of the human.[57] Moreover, understanding the sensory realm of the horse was utilitarian, it helped keep these valuable creatures alive.[58]

A long *sensory negotiation* aptly captures at least five millennia long close relationship between horses and humans: *two life forms, products of evolution*; of adapting bodies and behaviours in response to the wider environment—including each other—as these physical responses facilitate the building of relationships both within and between species.[59] It is a heavily debated narrative, but current consensus suggests horses and humans first entered the intense intimacy of their sensory negotiation about 5500 years ago in today's Kazakhstan. New research, predicated on archaeology and DNA analysis, is challenging older paradigms, but it is possible that the Botai culture of hunters and herders first hunted and then domesticated the caballine creatures.[60] Certainly, recent discoveries suggest that it is likely that the single, linear meta-narrative of a one-off domestication event in one place by one people is untrue and that horse domestication

[57] Draws on Sandra Swart, "The World the Horses Made – A South African case study of writing animals into social history", *International Review of Social History*, no. 55 (2010); Sandra Swart, *Riding High*. These historical sensory negotiations may now be seen in equine facilitated healing encounters where, for example, veterans suffering PTSD are encouraged to engage with each of the horse's (five) sense in order to engage with their own. Brooke Lichter, "From combat to calm: equine therapy with veterans", in Megan E. Delaney (ed.) *Nature Is Nurture: Counselling and the Natural World* (Oxford: Oxford University Press, 2020): 151.

[58] Maria Vilain Rørvang, Birte L. Nielsen and Andrew N. McLean, "Sensory Abilities of Horses and Their Importance for Equitation Science", *Frontiers in Veterinary Science* (2020).

[59] Drawn from Swart, "The World the Horses Made"; Sandra Swart, "Horses in the South African War, c.1899–1902", *Society & Animals*, no. 18 (2010); Swart, *Riding High*.

[60] M.A. Levine, "Botai and the Origins of Horse Domestication", *Journal of Anthropological Archaeology*, no. 18 (1999): 29–78.

has a complex, multi-layered history, based on a series of successful sensory negotiations.

The long and close sensory negotiation was necessitated in the making of the historical horse-human dyad. It was integral to hunting, domesticating, taming, training, sacrificing, harnessing, and riding horses. Our two species have become enmeshed in an array of affiliations: a continuum from oppression to collaboration, to an admittedly uneven partnership. The slow intimacy of a relationship older than five millennia created a shared embodied conversation. Being-with horses is akin to Lorenz's practice of 'becoming-with' his jackdaws, only we expect the horse to learn our side of the dialect too. Arguably, horses have evolved to be more empathetic *to us* than most animals, including most domesticated animals, because of the close reading of our intentionality they have needed to develop since domestication—with anthropogenic selection for horses that appeared tameable and trainable—when really, they were empathetic.[61] Indeed, there is even evidence for a possible purely olfactory recognition of *human emotions* by horses. After all, equine empathy for *us* is so clear that there is a whole body of mental health practices based on their capacity to provide feedback by responding to changes in our affective states, mirrored in instant shifts in their body language.[62]

Of course, horses live in our world, but also in a world parallel to our own. We share the five classic senses, but not the same ranges. They evolved as social, grazing, prey species—but since domestication selection for other traits have shaped their sensorium. Arguably selecting for tractable horses may have dimmed their capacity to respond to stimuli as they used to. Nevertheless, horses have retained far greater sensory capacity aimed at predator detection and flight—which explains their success when escaping human control and 'rewilding' in feral herds of mustangs or brumbies, for example. Equine eyes are on the side of the head with

[61] Or in the technical words of scientists: the process of domestication may have favoured the development of the ability to appreciate heterospecific emotions. Jan Ladewig, "Body language: Its importance for communication with horses", *Journal of Veterinary Behavior*, no. 29 (2019): 108–110; Agnieszka Sabiniewicz et al., "Olfactory-based interspecific recognition of human emotions: Horses (Equus ferus caballus) can recognize fear and happiness body odour from humans (*Homo sapiens*)", *Applied Animal Behaviour Science*, no. 230 (2020): 105–172; S.M. Stone, "Human facial discrimination in horses: can they tell us apart?", *Animal Cognition*, no. 13 (2010): 51–61.

[62] K. S. Trotter, *Harnessing the Power of Equine Assisted Counselling* (Abingdon: Routledge, 2012).

monocular vision so they can see separate objects with each eye at the same time, permitting a horse with their head lowered to eat grass in relative safety relying on their almost panoptic vision to scout for predators. Their nasal sensitivity offers an extended temporal knowledge: pheromones signal past skirmishes, allies and enemies, births and death, excitement, and seasonal sexuality.[63] Members of a herd even have a shared odour. Smell is key in social recognition. Horses identify each other by body odours and from urine and dung. Their ability to hear alerts them to silently stalking predators. As my mare has taught me, the direction of their ears reveals where their attention is directed. If her ears are pricked forward, she feels she needs to focus closely that way. Emotion impacts position: if her ears are flat back, she is angry or anxious, whereas both ears softly drooping reveal her unwinding into calm. Their skin has sensory receptors that are specifically sensitive to touch, which is integral in mutual grooming to cement alliances.[64] Tactile signals—rather than visual or aural—have been deployed predominately historically.[65]

Significantly, the human impact on a horse has changed or even curtailed the equine sensorium. In addition to their changed environments and altered social herds, the various training processes humans deployed historically fundamentally ingrained horses becoming less responsive to external environmental stimuli. Horses were said to become 'dead to the leg' or 'hard mouthed', where human training methods had led to 'learned helplessness' so the horse no longer reacts even to uncomfortable stimuli.[66] Moreover, constant stabling in urban horses may lead sometimes to myopia.

Importantly, the archives show how the horse changed our sensory world, too. The rural and urban soundscapes changed because of horses, as Coates noted. The sounds of hammer on anvil, the jingle of bits, the creak of leather saddlery, the crack of whips, the thudding of hooves, the whinny of horses, and perhaps most of all, the absence of that with which they were replaced: the automobile made the horse-era world a different

[63] Some smells—like fire—trigger action in this odd-toed ungulate who evolved in flammable steppe grasslands.

[64] S.L. Crowell-Davis, K.A. Houpt and C.M. Carini, "Mutual grooming and nearest neighbour relationships among foals of Equus caballus", *Applied Animal Behaviour Science*, no. 15 (1986): 113–123.

[65] Birte Nielsen and McLean, "Sensory Abilities of Horses".

[66] C. Hall et al., "Is there evidence of learned helplessness in horses?", *Journal of Applied Animal Welfare Science*, no. 11 (2008): 249–266.

place. The human sensory experience that was altered with the introduction of the Horse Age included the human experience of speed and the meaning of distance. There is also a body of sensory knowledge about the other that must be learned afresh with each generation of horse and human—or be lost. Past humans were able to interpret the sensory world of the horse: their flared nostrils of an eager or anxious or aroused horse (knowing what *they* are 'smelling'), or the ebullient flatulence of a joyful horse (knowing what *we* are!).

The history of empathetic understanding of the horse's othered sensorium may be traced to Plutarch's tale of Alexander the Great's 'Bucephalus'. This horse was unmanageable until Alexander saw through the horse's eyes and realized his intractability was fearfulness and, instead of beating him into submission, simply turned him into the sun so he could not see his shadow that frightened him.[67] In modern times, natural horsemanship has emphasized the need for sensory empathy. Cowboy Bill Dorrance helped pioneer modern 'natural horsemanship', predicated on 'feel': the empathetic understanding of the horse's senses, especially touch. He proposed that '*feel*' requires deep 'knowing' based on years of being with horses: "Empathy for the horse is the capacity of a person to be able to feel what the horse feels, to read a situation the same way, and to have an understanding of what the horse is going to do in response to that situation".[68]

Of course, manipulating the horse's sensory experience to alter their behaviour is integral to this kind of historical relationship. Embodied empathy, in our terms, has been at the heart of many horse-training methods historically. Horses learn to understand what is meant by the slight shift in a rider's body, a shift in pressure of seat bones, the level of stiffness or relaxation in the various muscles of the rider in contact with the horse's body. At the same time, riders need to be "kinaesthetically literate" and learn to "read what [their] skin tells [them]",[69] listening and speaking to the horse through their shared frontier of skin. Historically people have had to think about and thus record—as best they could—the act of 'feel'.

[67] J. Dryden, *Plutarch's lives* (New York: Modern Library, 1942).

[68] Bill Dorrance and Leslie Desmond, *True Horsemanship through Feel* (Guilford, CT: Lyons Press, 1999, 2007) 22; see also K. Brandt, "A Language of Their Own: An Interactionist Approach to Human-Horse Communication", *Society & Animals*, no. 12 (2004): 299–316.

[69] Vicki Hearne, *Adam's task: Calling animals by name* (New York: Alfred A. Knopf, 1986): 110–111.

Is it thus possible to use 'touch', as well as our other senses, to understand the idiographic sensory world of past horse-human relationships? Both memory and perception are strongly impacted by previous sensory experience.[70] An individual horse may reveal their own history: after all, animals do tell stories—with their bodies.[71] The horse's body offers us an archive. Firstly, clearly each animal has an individual history written on their bodies. The brands or tattoos on a horse are a rich archive. A 'dead-mouthed' riding horse or the soring *scars of a horse compelled years before to perform the 'big lick'*,[72] or the saddle-sore scars of a pack horse all bear testimony to how horses have endured human needs. Moreover, their histories are revealed through bodily behaviours and appearance. The long-retired warhorse might smell or hear something that reminds them of their martial youth and take off at a gallop. The close observation of an individual horse's history is accelerated by the practice of riding. An interspecies historian learns by listening, watching, touching, and *being with* the subject. *The exchange between the bodies of the horse and human: the kinetic methodology is part of zooethnography, riding a strange horse and in a strange place, learning new languages of the body from horses in different contexts helps us glimpse their histories with humans.*[73]

'Being with horses' offers a multi-species sensory methodology.[74] We expect such bodily or affective history to help us develop fresh lines of thinking because of the sympathetic identification with the past that Collingwood saw as the prerequisite for any historical understanding.[75] Thinking through the body can engender a form of historicised understanding through somatic sensory experience—connecting "the present

[70] A. Akrami, et al. "Posterior parietal cortex represents sensory history and mediates its effects on behaviour", *Nature*, no. 554 (2018): 368–372.

[71] Aaltola, *Varieties of Empathy*, 35.

[72] Soring is the practice of using tactile sensitivity as a brutal aid of instruction: imposing targeted pain to a horse's legs or hooves to encourage them into an exaggerated high-stepping gait. Mustard oil, diesel or kerosene are sometimes smeared on their limbs.

[73] This section draws on Sandra Swart, "Kicking over the traces? Freeing the animal from the archive", in Jennifer Bonnell and Sean Kheraj (eds), *Traces of the Animal Past: Methodological Challenges in Animal History* (Calgary: University of Calgary Press, 2022), which proposes the first tentative steps towards the intersection of animal history, sensory history and oral history.

[74] Despret uses the term anthropo-zoo-genesis to signify this relationship of human-with-horse and horse-with-human. Vinciane Despret, 'The Body We Care For: Figures of Anthropo-zoo-genesis', *Body & Society*, 10, 2004, 111–134.

[75] R.G. Collingwood, *The Idea of History* (Oxford: Clarendon Press, 1946) 215.

with the past".[76] As Corbin contended, the historian's gateway into past sensory experiences have been mediated by language.[77] But what if it does not have to be? Here the historian may foreground the sense long considered the least worthy of attention: touch. We draw on Maurice Merleau-Ponty's reading of touch as the 'double sensation'—because to touch is to be touched.[78] The riding experience is in the tactile frontier—the shared skin.[79] Mutual interspecies communication happens, albeit unevenly—as Fijn and Argent suggest—where both humans and horses convey their social knowledge.[80] The process involves kinaesthetic perception, to "develop a shared multi-species culture in which bodies can be attuned to one another".[81]

This embodied knowledge of how to ride is itself embedded in cultural and historical contexts.[82] Mongolian horses do not respond to the quiet seat of my own horse-human culture. In my fieldwork, the pony I rode anticipated my shifting in the saddle; he conveyed to me in a series of subtle bodily suggestions that I should adopt a raised light seat, hovering above his back at a trot, and should mirror his movements to one side or

[76] Katherine Johnson, "Rethinking (re)doing: historical re-enactment and/ as historiography", *Rethinking History*, No. 19 (2015): 193–206. Many historians retreat in a Pavlovian fashion from very notion of 'doing' history. See Paul Pickering, "'No Witnesses. No Leads. No Problems: The Reenactment of Crime and Rebellion", in I. McCalman & P. Pickering (eds), *Historical Reenactment: From Realism to the Affective Turn* (Hampshire: Palgrave Macmillan, 2010): 109–133, 122.

[77] Alain Corbin, "Charting the Cultural History of the Senses," in David Howes (ed.), *Empire of the Senses: The Sensual Culture Reader* (London: Routledge, 2005): 128–139. Also see Nigel Rothfels, "Touching Animals: The Search for a Deeper Understanding of Animals", in Dorothee Brantz (ed.) *Beastly Natures: Animals, Humans, and the Study of History* (Charlottesville: University of Virginia Press, 2010): 38–58.

[78] Maurice Merleau-Ponty, *Phenomenology of Perception*, trans. Colin Smith (London: Routledge, 2002): 106–107.

[79] Ashley Montagu, *Touching: The Human Significance of the Skin* (London: Harper and Row, 1978): 20–46.

[80] See Natasha Fijn, *Living with Herds: Human-animal Co-existence in Mongolia* (Cambridge: Cambridge University Press, 2011).

[81] Andrea Ford, "Sport horse leisure and the phenomenology of interspecies embodiment", *Leisure Studies*, no. 38 (2019): 329–340.

[82] Charlotte Marchina, "Follow the horse: The complexities of collaboration between the lasso-pole horse (uurgach mor') and his rider among Mongolian horse herders", in Dona Davis and Anita Maurstad, eds, *The Meaning of Horses: Biosocial Encounters* (Abingdon: Routledge, 2016): 102–113; Robin Irvine, "Thinking with horses: troubles with subjects, objects and diverse entities in eastern Mongolia", *Humanimalia: A Journal of Human/ Animal Interface Studies*, no. 6 (2014): 62–94.

the other as he moved. Of course, partly this is to do with the technology historically adopted—Mongolian riders tilt to one side to avoid their jarringly rigid saddles. (My equine interlocutor reminded me to do that also—my faulty use of the technology his humans use irked him too.) Of course, sensory experience is transient, so offers a challenge to historians, and while written descriptions of sensory experiences in the past can be used, we historians can go further by trying to recreate past sensory experiences, to better understand past ways. Thus, even just riding in a 'traditional saddle' is a method of experiential history: learning from trying to use technology of the past. Mongolian saddles sit high off the horse's back—perhaps herders ride half-standing and tilted to the side to offset not only the hard saddles but the small horses' short and concomitantly choppy gait? The saddle was interesting for an historian interested in 'agency' because it permitted less (human) control over gait and speed. It seemed as though the horse was expected to choose an appropriate gait, where necessary, so that the rider could focus on the job at hand like herding. Csordas calls these "culturally elaborated ways of attending to and with one's body in surroundings that include the embodied presence of others".[83] Bodies communicate not only biology, but also culture—and culture always has a history.[84]

In writing animal history, one may wish to 'be-with' or 'live-with'—akin to methodologies from multi-species ethnographies or anthrozoology.[85] There is an important caveat, however, in the methodology of 'being-with'. Of course, one must be careful not to extrapolate directly from present to past horses. Horses have local cultures that impact their own sensorium, or their own local genetic conditions derived through historical anthropogenic selection. Thus, the sensory is not synchronic, but diachronic. It is not for *all* time—but *a* time. But nevertheless, the senses are more genetically entrenched than other traits that are more rapidly altered by shifting historical context. Thus, spending time 'being with' with currently living animals to help one understand previous generations of the same species is still useful and makes more sense as a methodology if you are writing *sensory* history—rather than other kinds of history.

[83] Thomas Csordas, *Body/Meaning/Healing* (New York: Palgrave Macmillan, 2002): 7–8.

[84] This paragraph draws on Swart, "Kicking over the traces?"

[85] See, for example, Lindsay Hamilton and Nik Taylor, *Ethnography After Humanism: Power, Politics and Method in Multi-Species Research* (Basingstoke: Palgrave Macmillan, 2017).

Sensory history reins in conventional wisdom about the ways that horses mattered. It emphasizes the abilities but also the vulnerabilities, the life but also the mortality of horses. In the shade of the big stories about horses on the level of human societies—domestication, migration, war, conquest, and colonization—exist small slices of intimate individual sensory exchange: *of trying to imagine each other's worlds.* Both offer histories of how the sensory negotiation with horses changed how humans experienced the world—but also how horses did.

Conclusion: Making Sense of the Sensory

Most historians have investigated the past through sight instead of attempting to appreciate the auditory, olfactory, gustatory, or tactile dimensions, trusting sight as the only lens, literally. A sensory history of animals, which foregrounds senses within and beyond the classical five, is a salient reminder of this snare. It is clear that we need new ways to imagine the shared animal entanglements of the past. We have offered a sample card of how to explore the sensory—affective, bodily, and intracorporeal—experiences and connections between humans and other animals across two distinct but often quite connected contexts. In the process, we have raised critical questions that tend to revolve around tensions at the heart of attempting to access *other* worlds, and the ways in which human sensory configurations and experiences create frames of reference that are difficult to overcome, and which have very material consequences for life in all its diversity. Indeed, our case studies offer ways of moving beyond the exhausted concept of 'animal agency', urging us to think more about the richness of what it meant and means to be a body living *as it is* in the world and with diverse others.[86]

By turning to the darkness, we find a space and time in which the human sensorium was—and is—really put to the test, where diverse abilities collide and where power relations manifest in particular kinds of sensory negotiations that impacted on an array of registers on the ways in which nonhuman beings have been able to live as their whole uncompromised selves. And by tuning into a "horse sense", of time and distance, of a world of richly textured shadows, smells, and sounds unsensed by humans, rather than endlessly searching for and reifying "horse agency"

[86] Joshua Specht, "Animal History After Its Triumph: Unexpected Animals, Evolutionary Approaches, and the Animal Lens", *History Compass,* no. 14 (2016): 326–336.

(for that is self-evident), our narratives themselves might transform our understanding of the past as a *lived-in place* for and with all beings. Past humans developed idiographic and diachronic bodies of sensory knowledge about horses and exhibited sensory empathy with them. They also sought means of understanding creatures whose sensoria seemed to be thoroughly alien. These attempts at empathy (sometimes manifesting in thoroughly violent manipulations) had significant material consequences that emerged from sensory entanglements. In that sense, then, we propose a *return* rather than a new sensory *turn, in the animal context*. Rather than only pioneer a new way of thinking like animals—indeed 'seeing like animals'—based on new breakthroughs in visual ecology, for example, we also suggest how to reconstruct old ones. Indeed, we suggest that historical empathy can only be animalised with a thorough sensory turn, because such a turn can help historians move towards encounters with the unseen, the unheard, and the untouched animal in the past.

These limits take one, for instance, to the sensory world of the horse-human dyad—each with their own senses, but able to be so much more—or less—by coming together in a shared sensorium. Embodied empathy emerges from closely shared experiences. The premise is 'getting into the skin' of another—a skill integral to the art of writing history—necessitating imaginative empathy. The changing horse-human world is accessible through the intimacies of sensing between human and horse. An equihistorian attuned to the visceral might be able to employ an embodied and embedded methodology—the corporeal dynamics of 'humans being with horses'—which might offer a history not only about horses but *from* them. As we observed at the beginning, these are ways of sensing—diverse ways of *perceiving and being*—far beyond our own. The corollary is that these have shaped animals' experience of history differently to us, have made history different to us, and have not remained fixed at all moments of history. Their sensorium shaped our history, and while there may be substantial methodological challenges associated with attempts to envisage a world of sensation radically divergent from our own, we nonetheless implore animal historians to embrace the challenge of sensing life.

Bibliography

Aaltola, Elisa, *Varieties of Empathy: Moral Psychology and Animal Ethics* (Lanham, MD: Rowman & Littlefield International, 2018).

Akrami, A. et al. "Posterior parietal cortex represents sensory history and mediates its effects on behaviour", *Nature*, no. 554 (2018): 368–372.

Anon, "Birds and Lighthouses", *Country Life* (9 February 1935): 142.

Anon, "Bird notes from Irish lighthouses", *Country Life Illustrated* (8 December 1900): 727–728.

Anon, "Correspondence: animal casualties on the road", *Country Life* (3 February 1966): 245.

Aristotle, *De Anima* ii 4, 415b27–415b28.

Baár, Monika, "Prosthesis for the body and for the soul: the origins of guide dog provision for blind veterans in interwar Germany", *First World War Studies*, no. 6 (2015): 81–98.

Bell, G.P. "The sensory basis of prey location by the California leaf-nosed bat *Macrotus californicus* (Chiroptera: Phyllostomidae)", *Behavioral Ecology and Sociobiology*, no. 16 (1985): 343–347.

Benner, Stephen, "Defining life", *Astrobiology*, no. 10 (2010): 1021–1030.

Brandt, K, "A Language of Their Own: An Interactionist Approach to Human-Horse Communication", *Society & Animals*, no. 12 (2004): 299–316.

Burnett, Stephen, "Perceptual Worlds and Sensory Ecology", *Nature Education Knowledge*, no. 3 (2011).

Classen, Constance, *The Color of Angels: Cosmology, Gender and the Aesthetic Imagination* (London: Routledge, 1998).

Coates, Peter A. "The Strange Stillness of the Past: Towards an Environmental History of Sound and Noise", *Environmental History*, no. 10 (2005): 636–665.

Collingwood, R. G. *The Idea of History* (Oxford: Clarendon Press, 1946).

Connor, Steven, "The Menagerie of the Senses", *Senses and Society*, no. 1 (2006): 9–26.

Corbin, Alain, *The Foul and the Fragrant: Odor and the French Social Imagination* (Leamington Spa: Berg, 1986).

Corbin, Alain, "Charting the Cultural History of the Senses," in David Howes (ed.), *Empire of the Senses: The Sensual Culture Reader* (London: Routledge, 2005): 128–139.

Cronin, Thomas W, Sönke Johnson, N. Justin Marshall and Eric J. Warrant, *Visual Ecology* (Princeton; Princeton University Press, 2014).

Crowell-Davis, C. L., K.A. Houpt and C.M. Carini, "Mutual grooming and nearest neighbour relationships among foals of Equus caballus", *Applied Animal Behaviour Science*, no. 15 (1986): 113–123.

Csordas, Thomas, *Body/Meaning/Healing* (New York: Palgrave Macmillan, 2002).

"Dangers Of Migration For Migratory Birds", https://magpie.rspb.org.uk/birds-and-wildlife/natures-home-magazine/birds-and-wildlife-articles/migration/dangers-of-migration/#:~:text=Dangers%20of%20migration%201%20Bad%20weather%20Bad%20weather,7%20Sticky%20death%20...%208%20Protecting%20migrants%20 [accessed October 13, 2021).
Dangles, O. et al., "Variability in sensory ecology: Expanding the bridge between physiology and evolutionary biology", *Quarterly Review of Biology* no. 84 (2009): 51–74.
Davis, Lennard, *The End of Normal: Identity in a Biocultural Era* (Ann Arbor: University of Michigan Press, 2013).
Despret, Vinciane 'The Body We Care For: Figures of Anthropo-zoo-genesis', *Body & Society*, 10, 2004, 111–134.
Despret, Vinciane, "Responding Bodies and Partial Affinities in Human–Animal Worlds", *Theory, Culture & Society*, no. 30 (2013): 51–76.
Dijkgraaf, Sven, "Spallanzani's unpublished experiments on the sensory basis of object perception in bats", *ISIS*, no. 51 (1960): 9–20.
Dorrance, Bill, and Leslie Desmond, *True Horsemanship through Feel* (Guilford, CT: Lyons Press, 1999, 2007).
Dryden, J, *Plutarch's lives* (New York. Modern Library, 1942).
Earth at Night (Apple, 2020).
Edwards, Nina, *Darkness: A Cultural History* (London: Reaktion Books, 2018).
Endacott, Jason L. "Negotiating the Process of Historical Empathy", *Theory & Research in Social Education*, no. 42 (2014): 4–34.
Febvre, Lucien, "Smells, Tastes, and Sounds", in *idem*, *The Problem of Unbelief in the Sixteenth Century* (Cambridge: CUP, 1982).
Feng, A. S. et al. "Ultrasonic communication in frogs", *Nature*, no. 440 (2006): 333–336.
Fijn, Natasha, *Living with Herds: Human-animal Co-existence in Mongolia* (Cambridge: Cambridge University Press, 2011).
Flack, Andrew, "Dark Trails: Animal histories beyond the light of day", *Environmental History*, no. 27 (2022a): 215–241.
Flack, Andrew, "Dark Degenerations: Life, Light and Transformation Beneath the Earth, 1840-c. 1900", *ISIS*, no. 113 (2022b): 331–351.
Ford, Andrea, "Sport horse leisure and the phenomenology of interspecies embodiment", *Leisure Studies*, no. 38 (2019): 329–340.
Fudge, Erica, "Milking other men's beasts", *History and Theory*, no. 52 (2013): 13–28.
Gangopadhyay, Nivedita, "Introduction: Embodiment and Empathy, Current Debates in Social Cognition", *Topoi*, no. 33 (2014): 117–127.
Geertz, Clifford, "'From the Native's Point of View': On the Nature of Anthropological Understanding in Culture Theory", in Richard Shweder and

Robert LeVine (eds), *Culture. Theory: Essays on Mind, Self, and Emotion* (Cambridge: CUP, 1984).

Griffin, Donald R. and Robert Galambos, "The Sensory basis of obstacle avoidance by flying bats", *Journal of Experimental Zoology*, no. 86 (1941): 481–506.

Goldman, Alvin and Frederique de Vignemont, "Is Social Cognition Embodied?", *Trends in Cognitive Sciences,* no. 13 (2009): 154–159.

Gonzalez-Liencres, Cristina, Simone G. Shamay-Tsoory and Martin Brüne, "Towards a neuroscience of empathy: Ontogeny, phylogeny, brain mechanisms, context and psychopathology", *Neuroscience & Biobehavioral Reviews,* no. 37 (2013): 1537–1548.

Griffin, Donald R. and Robert Galambos, "Obstacle avoidance by flying bats: the cries of bats", *Journal of Experimental Zoology*, no. 89 (1942): 475–490.

Griffin, Donald R. *Listening in the Dark* (New Haven: Yale University Press, 1958).

Gruen, Lori, "Navigating Difference (again): Animal Ethics and Entangled Empathy", *Strangers to Nature: Animal Lives and Human Ethics* (Lanham: Rowman and Littlefield, 2012): 213–233.

Hahn, Walter Louis, "Some habits and sensory adaptations of cave-inhabiting bats", *Biological Bulletin*, no. 15 (1908): 135–164.

Hall, C. et al., "Is there evidence of learned helplessness in horses?", *Journal of Applied Animal Welfare Science*, no. 11 (2008): 249–266.

Hamilton, Lindsay and Nik Taylor, *Ethnography after Humanism: Power, Politics and Method in Multi-Species Research* (Basingstoke: Palgrave Macmillan, 2017).

Haraway, Donna, "A Cyborg Manifesto: Science, Technology, and Socialist-Feminism in the Late Twentieth Century", in *Simians, Cyborgs and Women: The Reinvention of Nature* (New York: Routledge, 1991): 149–181.

Harvey, Elizabeth D. "The Portal of Touch", *American Historical Review*, no. 116 (2011): 385–400; and Hearne, Vicki, *Adam's task: Calling animals by name* (New York: Alfred A. Knopf, 1986).

Hollan, Douglas and C. Jason Throop, "Whatever Happened to Empathy?: Introduction", *Ethos,* no. 36 (2008): 385–401, 385.

Irvine, Robin, "Thinking with horses: troubles with subjects, objects and diverse entities in eastern Mongolia", *Humanimalia: A Journal of Human/Animal Interface Studies,* no. 6 (2014): 62–94.

Jackle, John, *City Lights: Illuminating the American Night* (Baltimore: Johns Hopkins University Press, 2001).

Jay, Martin, *Downcast Eyes: The Denigration of Vision in Twentieth Century French Thought* (Berkeley: University of California Press, 1994).

Johnson, Katherine, "Rethinking (re)doing: historical re-enactment and/ as historiography", *Rethinking History,* No. 19 (2015): 193–206.

Katana, Kenneth, *Great Adaptations: Star-nosed moles, electric eels & other tales of evolution's mysteries solved* (Princeton: Princeton University Press, 2020).

Kiechle, Melanie A. *Smell Detectives: An Olfactory History of Urban America* (Seattle: University of Washington Press, 2019).
Kyba, Christopher, Sara Pritchard et al., "Night Matters—Why the Interdisciplinary Field of 'Night Studies' Is Needed", *Multidisciplinary Scientific Journal*, no. 3 (2020): 1–6.
Ladewig, Jan, "Body language: Its importance for communication with horses", *Journal of Veterinary Behavior*, no. 29 (2019): 108–110.
Land, Michael F. and Dan-Eric Nilsson, *Animal Eyes* (Oxford: Oxford University Press, 2012).
Levine, M. A. "Botai and the Origins of Horse Domestication", *Journal of Anthropological Archaeology*, no. 18 (1999): 29–78.
Lichter, Brooke, "From combat to calm: equine therapy with veterans", in Megan E. Delaney (ed.) *Nature Is Nurture: Counselling and the Natural World* (Oxford: Oxford University Press, 2020).
Longcore, Travis and Catherine Rich, "Ecological Light Pollution", *Frontiers in Ecology and Environment*, no. 2 (2004): 191–198.
Lorenz, Konrad, *King Solomon's Ring* (London: Routledge, 1952, 2002).
Marchina, Charlotte, "Follow the horse: The complexities of collaboration between the lasso-pole horse (uurgach mor') and his rider among Mongolian horse herders", in Dona Davis and Anita Maurstad, eds, *The Meaning of Horses: Biosocial Encounters* (Abingdon: Routledge, 2016): 102–113.
Mason, R. T. and R.M. Parker, "Social behaviour and pheromonal communication in reptiles", *Journal of Comparative Physiology A*, no. 196 (2010): 729–749.
McComb, K. et al. "Long-distance communication of acoustic cues to social identity in African elephants", *Animal Behaviour*, no. 65 (2003): 317–329.
Merleau-Ponty, Maurice, *Phenomenology of Perception*, trans. Colin Smith (London: Routledge, 2002).
Montagu, Ashley, *Touching: The Human Significance of the Skin* (London: Harper and Row, 1978).
Nagel, Thomas, "What is it like to be a bat?", *Philosophical Review*, no. 83 (1974): 435–450.
Night on Earth (Netflix, 2020).
Pickering, Paul, "'No Witnesses. No Leads. No Problems: The Reenactment of Crime and Rebellion", in I. McCalman & P. Pickering (eds), *Historical Reenactment: From Realism to the Affective Turn* (Hampshire: Palgrave Macmillan, 2010): 109–133.
Plumwood, Val, "Being Prey", *Kurungabaa*, no. 4 (18 January 2011), https://kurungabaa.wordpress.com/2011/01/18/being-prey-by-val-plumwood/
Poole, J.H. et al. "The social contexts of some very low frequency calls of African elephants", *Behavioural Ecology and Sociobiology*, no. 22 (1998): 385–392.
Plumb, Christopher, "Reading menageries: using eighteenth-century print sources to historicise the sensorium of menagerie spectators and their encounters with

exotic animals", *European Review of History: Revue européenne d'histoire*, no. 17 (2010): 265–286.

Rothfels, Nigel, "Touching Animals: The Search for a Deeper Understanding of Animals", in Dorothee Brantz (ed.) *Beastly Natures: Animals, Humans, and the Study of History* (Charlottesville: University of Virginia Press, 2010): 38–58.

Sabiniewicz Agnieszka, et al., "Olfactory-based interspecific recognition of human emotions: Horses (Equus ferus caballus) can recognize fear and happiness body odour from humans (*Homo sapiens*)", *Applied Animal Behaviour Science*, no. 230 (2020): 105–172.

Saha, Jonathan, "Colonizing elephants: animal agency, undead capital and Imperial science in British Burma", *BJHS*, Themes 2 (2017): 2–17.

Schivelbusch, Wolfgang, *Disenchanted Night: The Industrialisation of Light in the Nineteenth Century* (Berkeley: University of California Press, 1995).

Shakespeare, Tom, *Disability: the basics* (London: Routledge, 2018).

Sigrist, René, « Louis Jurine, Chirurgien et naturaliste (1751–1819) », *Le Rhinolophe*, no. 11 (1995): 1–8.

Smith, Mark M. *Sensing the Past: Seeing, Hearing, Smelling, Tasting and Touching in History* (Berkeley: University of California Press, 2007).

Specht, Joshua, "Animal History after Its Triumph: Unexpected Animals, Evolutionary Approaches, and the Animal Lens", *History Compass*, no. 14 (2016): 326–336.

Squires, Walter Albion and Harold E. Hanson, "The Destruction of Birds at the Lighthouses on the Coast of California", *The Condor*, no. 20 (1918): 6–10.

Steck, Kathrin, Markus Knaden and Bill Hansson, "Do desert ants smell the scenery in stereo?" *Animal Behaviour*, no. 79 (2010): 939–945.

Stone, S.M. "Human facial discrimination in horses: can they tell us apart?", *Animal Cognition*, no. 13 (2010): 51–61.

Stoneman, Betty, "The Dichotomy of Sight vs. Touch: The Harm of the Primacy of Sight for Empathy", *Sloth: A Journal of Emerging Voices in Human-Animal Studies*, no. 2 (2016).

Swart, Sandra, "The World the Horses Made – A South African case study of writing animals into social history", *International Review of Social History*, no. 55 (2010a).

Swart, Sandra, *Riding High: horses, humans and history in South Africa* (Johannesburg: Witwatersrand University Press, 2010b).

Swart, Sandra, "Kicking over the traces? Freeing the animal from the archive", in Jennifer Bonnell and Kheraj, Sean (eds), *Traces of the Animal Past: Methodological Challenges in Animal History* (Calgary: University of Calgary Press, 2022).

Trifonov, Edward N. "Vocabulary of Definitions of Life Suggests a Definition", *Journal of Biomolecular Structure and Dynamics*, no. 29 (2011).

Trotter, K. S. *Harnessing the Power of Equine Assisted Counselling* (Abingdon: Routledge, 2012).

Van Dyke, J. U. and M.S. Grace, "The role of thermal contrast in infrared-based defensive targeting by the copperhead, *Agkistrodon contortrix*", *Animal Behaviour*, no. 79 (2010): 993–999.

Vilain Rørvang, Maria, Birte L. Nielsen and Andrew N. McLean, "Sensory Abilities of Horses and Their Importance for Equitation Science", *Frontiers in Veterinary Science* (2020).

Walker, Brett L. "Animals and the Intimacy of History", *History and Theory*, no. 52 (2013): 45–67.

Westerweel, Bart, "The well-tempered lady and the unruly horses", in Theo D'haen, Rainer Grübel and Helmut Lethen (eds), *Convention and Innovation in Literature* (Utrecht: John Benjamin's Publishing Company, 1989).

Would, Alice, "Fleshing Out the Animals of British Taxidermy in the long Nineteenth Century (1820–1914)", Doctoral dissertation, University of Bristol, 2021.

Zallen, Jeremy, *American Lucifers: The Dark History of Artificial Light, 1750–1865* (Chapel Hill: University of North Carolina Press, 2019).

CHAPTER 6

History According to Cattle

Terike Haapoja

Introduction

In the beginning there is a void. A void between us and history, between these words and our muted existence. How to cross this void? When language is by definition something we don't possess? You think that because of your writing you are the author of the world, but you're wrong. You were just an accident like the rest of us, floating in the sea of time. Everybody tries to explain the world. Even the stone, with its stony

The chapter is originally from Laura Gustafsson and Terike Haapoja's book *History According to Cattle* (2015) and is an introduction to Gustafsson & Haapoja's project *Museum of the History of Cattle*, which is a travelling ethnographic museum aimed at acknowledging that alternative modes of experience exist. Laura Gustafsson and Terike Haapoja, 'History According to Cattle', in Laura Gustafsson and Terike Haapoja *History According to Cattle* (New York: Punctum Books, 2015), 2–3.

T. Haapoja (✉)
Espoo, Finland
e-mail: mail@terikehaapoja.net

M. J. Glover, L. Mitchell (eds.), *Animals as Experiencing Entities*, The Palgrave Macmillan Animal Ethics Series,
https://doi.org/10.1007/978-3-031-46456-0_6

reasoning, finds order in its rocky little world. You are nothing special. There's an inside in everything.

But here I am, with my horns and tail and compartmented stomachs and a line of ancestors queuing and pushing behind me as if entering the spring field. Why? When history itself has rejected us and rendered us invisible, language as its weapon? The answer is simple: because we were there. We saw it all. But to break my silence, or what you take as silence, I must enter your language and domesticate you, like the cow whisperer tames a wild bull by talking to him with words he knows. So I borrow your words and carve myself into them, make a hole through them the shape of a cow. You might not see me, but you'll see my absence. This is where my story begins.

The Time Before History

The history of cattle is divided into three eras. The Time Before History emerges gradually from the cooling climates of the Pliocene epoch more than two million years ago. There, in the grasslands of India our ancestors, the great Auroch, come to life. Tall and heavy, they graze in groups of 20 or 30, changing location when in need of water and fresh edible grass. Their life is peaceful, as they do not practice war, and the power relations of the community are tested out in display fights in which both females and males take part. The young ones are born in the spring, staying at their mother's side until strong enough to join the community on its endless travel. Beasts of the era, saber toothed cats, hyenas, hominins, hunt them down when they can, but cannot conquer them, as they are far too many in number. Roaming in millions they gradually expand their presence Eastward and Westward as far as the great grasslands extend. Over a million years later, they inhabit most of the old world from Asia and northern Africa all the way to the western coasts of Europe.

The Historical Time

With the emergence of the common ape and its culture, we enter The Historical Time. At the beginning of the Holocene, more than 10,000 years ago, the great Auroch starts to live side by side with the ape, gradually forgetting its traditional customs and learning a new way of life. Why this anomaly in history, this exchange, takes place, we don't know. The last free Auroch who could have passed down this knowledge died alone in the remote forests of Jaktórow in 1627, taking the secret with her. We have only guesses and interpretations, unreliable traces of evidence. What we do know is that we, all of the 1.3 billion of us living today, are the successors of approximately 80 individuals living with the Mesopotamian common ape 8000 years ago. From that moment on our destinies have been intertwined. The great Auroch had become cattle, and the ape the human as we know it today.

If the time of the Auroch had been cyclical, determined by the subtle changes in the weather and by the signs and traces of its companions, this new era was moving forward like a bull. From our manure, milk and flesh, the history of the human rose like a wave, generating wealth and prosperity beyond imagining.

Cultures emerged, wars were fought. New lands were conquered and more and more of our kind were needed to support the lives of the human. Technology and writing and the rituals of the afterlife were invented. Richness emerged wherever there was a need: what was excess to some was luxury to others. Corn, potatoes, antilope skins, pearls, rubber, children, women, men of all sizes exchanged owners. Animals crossed oceans but so did parasites and diseases. Populations collapsed as new ones emerged. Kings were declared and then beheaded. The generation and distribution of wealth sought no balance but threw around destinies, fortune and misery in a random order. Soon hurricanes joined together with rains and flooding, walls of fire and drought. Sucked by economic vacuums and pushed by ecological masses, everything was on the move.

Meanwhile, unaware of our life as richness itself, we lived a modest life. The great tides of history arrived to us as streams so small they could hardly be noticed. Wars killed us, but so did peace. The food was the same, century after century. Barns were smaller or wider, our companions fewer or more, but the daily routine remained. Birth, grazing, feeding, sleeping. Our centuries were defined by the hot breath of our companion in the silence of the shelter, night after night, the thousand returns to the same fields and by the light absence of the weight of the plough or the sledge or the carriage when the day fell. So deep inside history were we, that we did not see it happening to us—nor did we notice, when it abandoned us altogether.

In the early ages we lived in the world knowing it was only for us on loan. Everything passed, and we accepted it. As we did not possess things, we did not have a word for owning. We were poor in words for things: our vocabulary was verb-based, built from the doings of ourselves and of our companion beings. But after joining the flow of human history, there was not much to do and the native names we knew were lost; soon we did not have a name for who we were. In replacement we got a name given to us by the human. That name rendered us objects, subjected to the other's doings. Our world had been reversed: around us, everything remained—it was we who passed on, eternally.

The Ahistorical Time

The Ahistorical Time has no linearity, nor cyclicity. Inside the factory the passing on of heritage became impossible. Calves were taken from us immediately when they were born, and family lines were scattered out of our sight. Doing was reduced to so little that all that was left of our habits died away. We did not learn from our mothers but from the machine that told our bodies how to stand and how to eat. Stuck in the industrial process we would live in collective isolation, cut off from all relations that could anchor us to time, history, culture. For how could we have culture, if culture was the transforming of things into objects? How could we have history, if history was the weaving together of times from the present moment into the past with a chain of words? We did not even have time—the only time we had was clock-time, ticking away in unison, counting. All we had was destiny, as unchangeable as the rotation of the sun in the sky, and even that we did not possess before it took us.

If we were to see outside the factory, we would have seen the humans digging and carving, moulding, melting, cutting, jointing, burning and growing everything that passed their way. Like the earthworm that eats its way through the soil, the humans worked through the world, processing everything into products of their culture. But what they could not see was that they were followed by eyes, eyes everywhere.

Conclusion: A Hole in the World, Through Which a Cow Could Enter

So here I am, claiming what is mine and my ancestor's by law: history, that which we so generously gave to you. Eager, enlightened, generations and generations push behind me, waiting for a verdict. By incorporating your tongue we, the foundation, the mute, are pulled into existence, into the spotlight of human thought. There, once acknowledged, we become real.

But as we face the threshold of history, we realize that outside language we are still nothing. You only hear the foundation when you can teach it to speak. I can only point to my absence, hoping that in this failure of properly portraying me a hole would appear in the world, through which a cow could enter. That's what this is. A trial, nothing more.

But as I leave you now, I do not evaporate into the realm of ideas and imagination. Instead, I melt, I dissolve into your body, as my bovine colleagues have dissolved into the bodies of your family and friends. I remain close, hidden between your concepts, curled up in your muscles, waiting to be noticed. And, some day, I will enter.

PART II

Animals' Experiences: Narratives and Histories

CHAPTER 7

A Historiography of Great Animal Massacres

Jonathan Saha

Introduction: After Darnton

As animal history has 'come of age' as sub-field within the wider discipline, one title has reoccurred over and again. It originated with Robert Darnton's landmark essay, 'Workers Revolt: The Great Cat Massacre of the Rue Saint-Séverin', published in his 1984 book *The Great Cat Massacre and Other Episodes in French Cultural History*. The essay is justly famous for its innovative engagement with cultural anthropology, being an early exponent of 'thick description' in history writing.[1] It is also a lively and vivid piece of historical interpretation, although, as I discuss below, a

[1] Thick description is the attempt to capture in a richly detailed descriptive narrative participating actors' subjective understandings of the meaning and context of their social actions. Clifford Geertz, 'Deep Play: Notes on the Balinese Cockfight', *Daedalus* 134, no. 4 (2005): 56–86.

J. Saha (✉)
Department of History, University of Durham, Durham, UK
e-mail: jonathan.saha@durham.ac.uk

M. J. Glover, L. Mitchell (eds.), *Animals as Experiencing Entities*, The Palgrave Macmillan Animal Ethics Series,
https://doi.org/10.1007/978-3-031-46456-0_7

narrative drafted entirely from the view of human actors.[2] No doubt in part because of its historiographic fame, animal historians have recycled his arresting title for their own studies. Writing in *The American Historical Review* in 2001, Nancy Jacobs entitled her essay about the apartheid state's violence in one of its quasi-independent 'Bantustan' territories, 'The Great Bophuthatswana Donkey Massacre'. Shortly after this appeared, in a study of the failings and unanticipated consequences of colonial sanitation systems in French Hà Nội, Michael Vann published a short article with the subtitle, 'The Great Hanoi Rat Massacre'. This contribution has gone on to have an unusual career, becoming a graphic novel with the altered title, 'The Great Hanoi Rat Hunt'; a title that still betrayed its roots in Darnton's work. This has then been followed by S. Hoon Song's edited chapter, 'The Great Pigeon Massacre in a Deindustrializing American Region', Yin Kit Chan's article, 'The Great Dog Massacre in Late Qing China', Ian Jared Miller's chapter, 'The Great Zoo Massacre', in his monograph on Ueno Imperial Zoological Gardens in Tokyo, Michael Pettit's essay, 'The Great Cat Mutilation', and Hilda Kean's book, *The Great Dog and Cat Massacre*.[3]

There are no doubt other studies that have riffed off Darnton's title that I have overlooked. However, beyond the title and the presence of the mass killing (or harming, in Pettit's essay) of animals, there is little that holds these studies together in terms of either their methodology or deeper philosophy. As a result, the reoccurrence of the title should prompt some reflection on what work the designation of an event as a 'great massacre' of animals is doing. Taking Erica Fudge's important historiographic

[2] Robert Darnton, *The Great Cat Massacre and Other Episodes in French Cultural History* (New York: Basic Books, 1984), 79–106.

[3] Nancy J. Jacobs, 'The Great Bophuthatswana Donkey Massacre: Discourse on the Ass and the Politics of Class and Grass', *The American Historical Review* 106, no. 2 (2001): 485–507; Michael G. Vann, 'Of Rats, Rice, and Race: The Great Hanoi Rat Massacre, an Episode in French Colonial History', *French Colonial History* 4, no. 1 (2003): 191–203; S Hoon Song, 'The Great Pigeon Massacre in a Deindustrializing American Region', in *Natural Enemies: People-Wildlife Conflicts in Anthropological Perspective*, ed. John Knight (London: Routledge, 2013), 212–29; Ying-kit Chan, 'The Great Dog Massacre in Late Qing China: Debates, Perceptions, and Phobia in the Shanghai International Settlement', *Frontiers of History in China* 10, no. 4 (2015): 645–67; Ian Jared Miller, *The Nature of the Beasts: Empire and Exhibition at the Tokyo Imperial Zoo* (Berkeley: University of California Press, 2013), 120–64; Michael Pettit, 'The Great Cat Mutilation: Sex, Social Movements and the Utilitarian Calculus in 1970s New York City', *BJHS Themes* 2 (2017): 57–78; Hilda Kean, *The Great Cat and Dog Massacre* (Chicago: University Of Chicago Press, 2017).

intervention in 2002 as a point of departure, the return to Darnton's essay would seem an ill-conceived choice. Fudge argued that animal historians should write from the position of losers, to choose to identify with animals, and to expose the brutality intrinsic to the history of human relations with animals.[4] Darnton's essay breaks much new ground, but it does not show much care for the cats, either their wider lives in early modern Paris or their experiences of this bloody event. Instead, Darnton uses an episode in which apprentices kill cats as 'joke' to protest their poor treatment at the hands of their masters, in order to tease out the symbolic resonances of felines as metonyms for class privilege. This was a history of humans' subjective experiences of marginalisation.[5] As Dominick LaCapra pointed out in his analysis of the immediate controversies that followed its publication, the debates that erupted did not explore the possibility of a reading that made 'room for the excluded perspective of the cat', let alone the 'species imperialism' sustained by the violence that Darnton exposed.[6] Darnton's essay then is more about the motivations and the worldviews of the perpetrators of the killings than it is about the mass death of sentient creatures. His analytical work offers the reader a view into a world in which it is funny, even cathartic, to kill a cat. It is a move that leaves human mastery over other living beings uninterrogated—reduced, in a sense, to a repertoire of symbolic resources.

The authors of the essays and books cited above go far beyond Darnton in recognising the presence, and on occasions the agency and subjective experiences, of animals in the past. It would seem that the reason for the recurring use of 'massacre', prefixed with the adjective 'great', may be the authors' desire to give a knowing rhetorical flourish to their titles, rather than stemming from any affinity with the essay's approach. Certainly, the implications of the framing of an event or, as Darnton would have it, episode in animal history as a 'great massacre' have not been rigorously considered. Indeed, in the essays that deploy this language in their titles, quite different types of violence are actually being connoted. In the essays of Chan and Vann, the killings of dogs and rats are largely bureaucratised and routinised.[7] In Jacobs, Kean and Miller's work, these are exceptional

[4] Erica Fudge, 'A Left-Handed Blow: Writing the History of Animals', in *Representing Animals*, ed. Nigel Rothfels (Bloomington: Indiana University Press, 2003), 3–18.

[5] Darnton, *The Great Cat Massacre*, 79–106.

[6] Dominick LaCapra, 'Chartier, Darnton, and the Great Symbol Massacre', *The Journal of Modern History* 60, no. 1 (1988): 95–112.

[7] Vann, 'Of Rats, Rice, and Race'; Chan, 'The Great Dog Massacre in Late Qing China'.

episodes of material and symbolic import in times of widespread human conflict.[8] Without deeper consideration, the uncritical use of 'great massacre' as a turn of phrase can have two unwanted effects. It can detract from the use of the term 'massacre' to connote excessive, collective lethal violence deployed by humans against their conspecifics.[9] And it can also deflect from the everyday, routine, and legally sanctioned mass slaughter of animals in human societies.[10] Historicising moments when the mass killing of animals was deemed by contemporary human actors as excessive and worthy of comment can enable historians to avoid these potential misreadings. Contextualising so-called great animal massacres in the wider necropolitical regimes that humans deployed against animals and alongside the use of excessive violence between humans provides us with a methodology for better comprehending when and why animals' suffering haphazardly enters into human public consciousness.[11] Finally, LaCapra's critiques still rings true. A thorough historicization of massacres must also give space to engage with the impact on the animal victims themselves, so far as is possible, through their own perspective.

The Great Crow Massacre of Colonial Rangoon

In 1920, the Rangoon Municipal Committee sanctioned the killing of crows, many of whom had built their nests in the tree-lined environs of the city's hospital. This came after several years of frustration among white British residents in the city, whose writings exposed a hatred of the birds, citing baseless assumptions concerning the nameless but myriad diseases that they were said to spread. Prior to 1920, the white members of the Municipal Committee had their plans—to launch a concerted campaign to systematically destroy crow nests, eggs, and chicks—blocked by representatives of the city's Hindu and Buddhist communities. That Asian religious sensibilities could be extended to crows was, for some white commentators, a racially coded sign of the colonised population's overly

[8] Jacobs, 'The Great Bophuthatswana Donkey Massacre'; Miller, *The Nature of the Beasts*, 120–64.

[9] Joanna Bourke, *What It Means to Be Human: Reflections from 1791 to the Present* (London: Virago, 2011), 164–201.

[10] Dinesh Joseph Wadiwel, 'The War Against Animals: Domination, Law and Sovereignty', *Griffith Law Review* 18, no. 2 (2009): 283–97.

[11] Achille Mbembe, 'Necropolitics', trans. Libby Meintjes, *Public Culture* 15, no. 1 (2003): 11–40.

sympathetic attitude to animals.[12] These concerns were neatly captured in 1914 by one G.C. Buchanan writing in the *Journal of the Royal Society of Arts*:

> One of the greatest trials, and probably one of the most lively carriers of disease, is the common crow. This abominable bird has been allowed to increase and multiply until it has become a veritable plague, and its filthy habits and raucous voice have caused the medical officers of the town to urge upon the municipality the necessity for a sustained effort in the way of destruction of eggs and young… unfortunately… an agitation headed by some Hindu and Burmese gentlemen, whose religion enforces a respect for life, ended in a complete victory for the crows…[13]

In 1920, for reasons that were not stated in the annual report for the municipality, the Committee was now minded to push ahead with the eradication of the 'crow nuisance' in spite of opposition from Hindu and Buddhist communities. The question that remained was how best to eliminate them. The Committee sought the expertise of the amateur ornithologist and colony's Conservator of Forests, C. J. Hopwood, as to the most effective method of killing them. Hopwood furnished them with a plan to deploy poison gas in the tree tops. This led the Committee to approach Lieutenant-Colonel S. J. M. Auld, who worked for the Anglo-Persian Oil Company and who had, alarmingly enough, a reputation as being an expert in poisonous gases. His reportedly detailed and lengthy response to the question led, ultimately, to the crude and blunt conclusion that destroying nests and killing the young through direct physical violence was the most cost effective and efficacious method open to the municipality. Turning, as the organs of the British colonial government so often did in Myanmar, to the lowest cost option available to them, the Municipality employed three gangs of Indian labourers, totalling 78 individuals, to carry out the destruction.[14]

There is no doubt a great deal that could be read from this brief episode through a Darnton-*esque* 'thick description'. The juxtaposition of a place

[12] Jonathan Saha, 'Among the Beasts of Burma: Animals and the Politics of Colonial Sensibilities, c.1840–1950', *Journal of Social History* 48, no. 4 (2015): 933–55.

[13] G. C. Buchanan, 'The Port and City of Rangoon', *Journal of the Royal Society of Arts* 62, no. 3207 (1914): 534.

[14] *Report on the Working of the Rangoon Municipality For the Year 1920–21* (Rangoon: British Burma Press, 1921), 18.

of human physical recuperation and the flocking of supposed avian vectors of disease calls to mind Mary Douglas' universalistic claim of the perennial importance of notions of purity and danger.[15] The 'plague' of crows in a region of the city dominated by grand red-brick buildings intended to display imperial authority, not just the hospital but also the general jail, indicates an underlying attempt to police animal presences in modern urban geographies.[16] And the clash between white British sensibilities and the religious mores of Hindu and Buddhist populations can be read against the secularist architecture and planning of the colonial city.[17] But it suffices to contrast the discussion of the killing of crows with the lack of discussion regarding the killing of other animals mentioned in the report, whose deaths were unremarkable and reduced to perfunctory sentences. For instance, under the heading 'The Destruction of Pariah Dogs', the report unceremoniously noted 'Two dog poisoners were employed throughout the year. The dead bodies of 7338 dogs in all were collected and brought to the carcase crematorium.'[18] The killing of 345,750 rats during the year was passed over even more cursorily. And the operation of licenced slaughterhouses was more focussed on revenues than death rates.[19] The religious dimension to the opposition to the campaign to kill crows is likely what made their deaths more noteworthy in the report. Such reformist political agitation was of interest to the report's authors. The presence of the All-India Cow Conference and their advocation of cow protection was blamed for the falling off in the slaughter of oxen in the city.[20] In other words, what made the killing of crows an exceptional deployment of mass lethal violence for human commentators at the time were the tensions between religious communities in the city.

The killing of the city's crows was noted in the nationalist vernacular press. *Sūriya* (The Sun) newspaper's regular gossip column, 'Town Gossip by Town Mouse', mentioned the event, albeit in a highly circuitous and

[15] Mary Douglas, *Purity and Danger: An Analysis of Concepts of Pollution and Taboo* (Psychology Press, 2003).

[16] Chris Philo, 'Animals, Geography, and the City: Notes on Inclusions and Exclusions', *Environment and Planning D: Society and Space* 13, no. 6 (1995): 655–81.

[17] Alicia Turner, 'Colonial Secularism Built in Brick: Religion in Rangoon', *Journal of Southeast Asian Studies* 52, no. 1 (2021): 26–48.

[18] *Report on Rangoon Municipality 1920–21*, 17.

[19] *Report on Rangoon Municipality 1920–21*, 3–17.

[20] *Report on Rangoon Municipality 1920–21*, 16.

mediated fashion.[21] The eponymous ‘Town Mouse’, as the pseudonym was given in the accompanying English title—it was *krvak‘cut‘krī”* meaning ‘Big Shrew’ in the Burmese script—mostly wrote observational pieces drawing from their human writers’ experiences of day-to-day urban life in the city. The small mammal penname enabled the human authors to adopt the position of a candid commentator, one embedded among the city’s denizens but also at a remove from the affairs of the humans they describe. Their article of 20 April 1920, which discussed the municipal government’s sanctioned destruction of nests, eggs, and nestlings, was an atypical one. Rather than reporting on conversations that occurred between the city’s *Homo sapien* residents, in this longer piece Town Mouse reported on a conversation between two cows (they were clearly gendered as female oxen in the Burmese, *nvā”ma*). The two cows were munching grass in the heat of the early summer sharing sorry tales of the abuses suffered by animals in the colony. They start with the sufferings of Rangoon’s crows relayed to them by a female crow (*kyī”ma*) and a female greater coucal, or crow pheasant (*bhut‘ma*). They reported how after an Indian man, referred to pejoratively as a ‘coringee kalar’ (*gorṅ‘gyī kulā”*), was defecated upon by a crow, the destruction of eggs began. They relate to the cows how gangs of ‘kalar’ were ordered to kill crow fledglings (*kyī”pok‘ca*) in their nests. The cows then go on to discuss their own fears for Burmese new year, *sakran‘*, in which Burmese children would twist the tales of oxen, singe them with torches, and eventually drive them into the hands of ‘kalars’ and their slaughterhouse (*kulā” lak‘roṅ‘”luik‘lui’| nvā”sathat‘ruṃ goṅ‘”rarhā pā ro*). The cows hatch a plan to avoid this fate by escaping with their friends, the goats, pigs, and ducks, to the safety of those who follow the eight Buddhist precepts (*upus‘choṅ‘ tai’ lūsū*) and practice meditation (*kammaṭhān‘”bhāwana lup‘*).[22]

The article was entitled ‘Save Us!’ and was accompanied by a cartoon which juxtaposed celebrating Burmese youth, with weeping pigs, ducks, sheep, and goats circling a chopping block with a large knife lodged in the top. The message was clear, even through the layers of animal ventriloquism at work in the piece; human authors using a small mammalian pseudonym reporting on a conversation between two cows recalling the experiences of a crow and a coucal. It was a plea for the implicitly Burmese

[21] I have transliterated the Burmese script according to the American Library Association—Library of Congress standards.

[22] *Sūriya*, 10 April 1920, 13–4.

Buddhist reader to behave in a more ethical way towards animals. It achieved this on the one hand by exposing the complicity of the reader with 'kalar' or Indian cruelty towards animals, and, on the other, by indicating the salvific role of devout Buddhists.[23] In the context of the inter-war years, the perceived dangers of the growing presence of non-Buddhist foreigners, particularly Indians, and the desire to preserve the teachings of the Buddha through reforming the behaviour of lay people, both fed into the emergence of Burmese nationalism and remained central concerns within it.[24] These two crucial internal elements of Burmese nationalism were clearly present in Town Mouse's article. It was one of a number of newspaper articles that wrote from the perspective of oxen pleading with human readers to treat them better and to refrain from eating them, part of a broader cow-protection movement that was emerging in the colony in the interwar years.[25] In the case of the killing of crows, the colonial municipal government's cost-cutting decision to override the religious objections of Buddhist and Hindu communities and employ Indian labourers to destroy the birds' nests as well as their young and unborn offspring played into these emerging nationalistic rifts. By the late 1920s and throughout the 1930s, violence against Indians and Muslims in Rangoon—a city in which Burmese people were a minority—was a frequent occurrence as intercommunal tensions were heightened by a febrile political climate, the gradual separation of Myanmar from British India, and a marked economic downturn. Thousands of Indian people started to leave the city, fearing for their personal safety in the colony.[26]

The mass killing of nestling crows and the destruction of crows' eggs in Rangoon during 1920 gained visibility not because of the intrinsic horror of

[23] For a more detailed analysis of this article see: Jonathan Saha, *Colonizing Animals: Interspecies Empire in Myanmar* (Cambridge: Cambridge University Press, 2021), 132–42.

[24] Alicia Turner, *Saving Buddhism: The Impermanence of Religion in Colonial Burma* (Honolulu: University of Hawai'i Press, 2014), 75–109; Chie Ikeya, *Refiguring Women, Colonialism, and Modernity in Burma* (Honolulu: University of Hawai'i Press, 2011), 120–42.

[25] Turner, *Saving Buddhism*, 91–93; Erik Braun, *The Birth of Insight: Meditation, Modern Buddhism, and the Burmese Monk Ledi Sayadaw* (University of Chicago Press, 2013).

[26] Rajashree Mazumder, '"I Do Not Envy You": Mixed Marriages and Immigration Debates in the 1920s and 1930s Rangoon, Burma', *The Indian Economic & Social History Review* 51, no. 4 (2014): 497–527; Matthew J. Bowser, 'Partners in Empire? Co-Colonialism and the Rise of Anti-Indian Nationalism in Burma, 1930–1938', *The Journal of Imperial and Commonwealth History* 49, no. 1 (2021): 118–47; Matthew J. Bowser, '"Buddhism Has Been Insulted. Take Immediate Steps": Burmese Fascism and the Origins of Burmese Islamophobia, 1936–38', *Modern Asian Studies* 55, no. 4 (2021): 1112–50.

such an episode in terms of the loss of avian life, but because of the racial and religious politics of the time. It was a small episode in the rise of anti-Indian xenophobia that resulted in pogroms in the city during the 1930s. The prominence given to the event in the annual report on municipal governance was disproportionate to the coverage given to the routine yet widespread eradication drives against stray dogs and rats, or the everyday slaughter of cattle for human consumption. In Town Mouse's article in *Sūriya*, to the contrary, the episode was an opportunity to surface up a wider set of cruel practices against animals, appealing to their readers to adopt a more Buddhist ethics towards other creatures—albeit in the process setting up Indian populations as a threatening other to this reformed Buddhist self. Attunement to these dynamics allow us to see how contemporary writings came to see this anti-corvid violence as exceptional and even, in the case of *Sūriya*, excessive; something that we might identify as a massacre. But there is still the 'excluded perspective' of the crow to attend to,[27] a perspective for which recent scientific scholarship provides a basis for reasoned conjecture.[28]

Crows thrive in the ecological niches produced through human urbanisation. The easy sources of food and favourable habitation provided in cities makes the birds perennial urban denizens across the globe.[29] Their intelligence is famed, and now firmly established in scientific literature. Their memories, tool-use and adaptation, ability to learn, and capacity for social communication have led to a complete reappraisal of the structure of birds' brains in general.[30] Most pertinent to the history uncovered in

[27] LaCapra, 'Chartier, Darnton, and the Great Symbol Massacre'.

[28] For compelling examples of the use of scientific knowledge for engaging the subjective experiences of animals in the past, see: Erica Fudge, 'Milking Other Men's Beasts', *History and Theory* 52, no. 4 (2013): 13–28; and Michael John Glover, 'Cattle and Colonialism: An Animal-Centred History of Southern Africa, 1652–1980s' (PhD Thesis: University of the Free State, 2021), 52–65; for an attempt at a similar approach, which tries to simultaneously embed a critique of science, see: Jonathan Saha, 'Do Elephants Have Souls? Animal Subjectivities and Colonial Governmentality', in *South Asian Governmentalities*, ed. Stephen Legg and Deana Heath (Cambridge: Cambridge University Press, 2018), 159–77.

[29] John M. Marzluff and Erik Neatherlin, 'Corvid Response to Human Settlements and Campgrounds: Causes, Consequences, and Challenges for Conservation', *Biological Conservation* 130, no. 2 (2006): 301–14.

[30] Alex H. Taylor, 'Corvid Cognition', *WIREs Cognitive Science* 5, no. 3 (2014): 361–72; Andreas Nieder, 'Inside the Corvid Brain—Probing the Physiology of Cognition in Crows', *Current Opinion in Behavioral Sciences*, Comparative cognition, 16 (2017): 8–14; John Marzluff and Tony Angell, *Gifts of the Crow: How Perception, Emotion, and Thought Allow Smart Birds to Behave Like Humans* (New York: Simon and Schuster, 2013), 11–41.

this chapter, crows respond to the deaths of their conspecifics in diverse but marked ways. In stark contrast to the corpses of other birds, when confronted with their own dead crows are reluctant to touch the remains. They have been observed watching over their dead in small groups. Whether this is the result of an emotional connection to the deceased individual, or whether it is an attempt to learn from the manner of their death, being confronted with the lifeless remains of a fellow corvid is likely a stressful event for them. The hormonal releases in these confrontations would trigger heightened awareness of their situation. In addition, there is some evidence that would suggest that crows experience anger at the deaths of close relatives, lashing out at other crows perceived to be at blame.[31] They are also highly responsive to the dangers posed by humans—not just as a group but even down to identifying dangerous individual humans—and adapting their behaviours accordingly.[32] All of this observed behaviour in crows today tells us that, at a minimum, this concerted campaign to destroy their homes and kill their nestlings would not have been unnoticed by the crows themselves. It would likely have been a distressing experience, leaving the birds anxious, afraid, and angry. They may have even changed how they lived in the city as a result of the trauma of the event, the effect that this orchestrated violence was trying to bring about.

Neither Sacrifice nor Scapegoat

There is a tendency, of which Darnton's essay represents the most obvious example, for the mass killing of animals to be understood as a synecdoche for some larger historical process. For Darnton, the killing of cats was *really* about the rumbling class conflict of early modern Paris. For Jacobs, the massacre of donkeys was revealing of the wider intertwined environmental, religious, class, and racial politics of apartheid. For Vann, the vain attempt to eradicate rats revealed the limits and self-defeating ideologies of French imperialism. For Miller, the sacrifice of Ueno Zoo's animals was a response to, and public display of, the escalating demands of total war.

[31] Marzluff and Angell, *Gifts of the Crow*, 137–55.

[32] Heather N. Cornell, John M. Marzluff, and Shannon Pecoraro, 'Social Learning Spreads Knowledge About Dangerous Humans Among American Crows', *Proceedings of the Royal Society B: Biological Sciences* 279, no. 1728 (2012): 499–508; Barbara Clucas and John M. Marzluff, 'Attitudes and Actions Toward Birds in Urban Areas: Human Cultural Differences Influence Bird Behavior', *The Auk* 129, no. 1 (2012): 8–16.

For Chan, the campaign to kill potentially rabid dogs was about the spatial regulation of the modern urban environment in colonial Shanghai. I could go on. In each example, the massacres are interpreted to simultaneously elucidate why they happened, and for what they can tell us about wider social relations—relations mostly, but by no means solely, between humans. These are important concerns to address. Episodes of mass lethal violence against animals evidently make illuminating lenses for examining past historical conjunctures. They also make compelling narratives for micro-historical studies that lend themselves to thick descriptions of layered cultural meanings. But as well as reading these episodes as a means to understanding a larger whole, they can be historicised as acts of violence in themselves. This entails engaging with the production of the archive.[33] Why were certain acts of human violence directed against animals deemed worthy of remark and record in a context of routine violence against animals? This question forces us to not lose sight of what Dinesh Wadiwel calls the 'war' for sovereignty over animals.[34] It pushes us to read even the documents decrying excessive violence against animals as simultaneously upholding speciesist hierarchies, documents marked by the barbarity at the heart of human dominance over other species, as Erica Fudge has put it.[35]

As well as situating violence deemed by contemporaries as excessive within its contingent historical context of human domination over animals, this process of historicization must be attuned to the use of collective violence against humans. As many of the articles that adopt the 'Great Massacre' motif in their titles show, the killing of animals was intimately bound up with the differentiation of humans on grounds of race and gender. This insight can be pushed further. As Zakiyyah Iman Jackson has persuasively argued in her recent book on animals and antiblackness, the connection between animalisation and racism is not primarily one of unjust exclusions from the category 'human'—a framing that exposes the liberal humanism still lurking in some posthumanist scholarship. Her work details how notions of animality have been mobilised by colonisers against both black(ened) bodies and nonhuman animal bodies through

[33] Michel-Rolph Trouillot, *Silencing the Past: Power and the Production of History* (Boston: Beacon Press, 1995); Ann Laura Stoler, *Along the Archival Grain: Epistemic Anxieties and Colonial Commonsense* (Princeton, NJ: Princeton University Press, 2009).

[34] Wadiwel, 'The War Against Animals: Domination, Law and Sovereignty'.

[35] Fudge, 'A Left-Handed Blow'.

intersecting but irreducibly specific historical practices. To undo the effects of these intersecting histories, she provides an unruly counternarrative through the texts of African diasporic writers whose work reveals the ongoing violence of colonialism in *making* that speciesist divide.[36] Underscoring Jackson's insights, a focus on the mass killing of animals that does not also acknowledge the intersecting history of racialisation misses how animal massacres often served to differentiate and dominate other humans.

Finally, greater attention being paid to the archive also facilitates the historian's excavation of the material presences, and even subjective experiences, of the nonhuman creatures who fell victim to human deployments of lethal collective violence. The written record contains within it the traces of the animals themselves. Their actions and behaviours shaped what humans observed about them, albeit in ways often submerged beneath layers of colonial discourse. They were necessary to the production of human-authored texts, informing its content. Acknowledging the generative presence of animals within the archive means these 'traces' can be identified, thereby creating space for historians to creatively attempt reconstructions of their subjective experiences.[37] Histories that fail to attend to the experiences of the animal victims of these massacres risk reproducing the same relations of dominance over animals intrinsic to the very violence that they study. But animals were not just sacrifices or scapegoats caught up in human affairs, or rather, not *only* sacrifices and scapegoats. They were complex, sentient, and sensate beings in their own right. Recognition of their rich lives and experiential worlds cut short in acts of blunt, indiscriminate violence transforms these episodes into crooked mirrors for the present. They can provide a view of human dominance over animals in the past that unsettles the historian and their readers in their continued place at the apex of interspecies power relations.

[36] Zakiyyah Iman Jackson, *Becoming Human: Matter and Meaning in an Antiblack World* (New York: New York University Press, 2020).

[37] Etienne Benson, 'Animal Writes: Historiography, Disciplinarity, and the Animal Trace', in *Making Animal Meaning*, ed. Linda Kalof and Georgina M. Montgomery (East Lansing: Michigan State University Press, 2011), 3–16; Brett L. Walker, 'Animals and the Intimacy of History', *History and Theory* 52, no. 4 (2013): 45–67; Angela Cassidy et al., 'Animal Roles and Traces in the History of Medicine, c.1880–1980', *BJHS Themes* 2 (2017): 1–23.

Bibliography

Benson, Etienne. 'Animal Writes: Historiography, Disciplinarity, and the Animal Trace'. In *Making Animal Meaning*, edited by Linda Kalof and Georgina M. Montgomery, 3–16. East Lansing: Michigan State University Press, 2011.

Bourke, Joanna. *What It Means to Be Human: Reflections from 1791 to the Present.* London: Virago, 2011.

Bowser, Matthew J. '"Buddhism Has Been Insulted. Take Immediate Steps": Burmese Fascism and the Origins of Burmese Islamophobia, 1936–38'. *Modern Asian Studies* 55, no. 4 (2021a): 1112–50.

———. 'Partners in Empire? Co-Colonialism and the Rise of Anti-Indian Nationalism in Burma, 1930–1938'. *The Journal of Imperial and Commonwealth History* 49, no. 1 (2021b): 118–47.

Braun, Erik. *The Birth of Insight: Meditation, Modern Buddhism, and the Burmese Monk Ledi Sayadaw.* University of Chicago Press, 2013.

Buchanan, G. C. 'The Port and City of Rangoon'. *Journal of the Royal Society of Arts* 62, no. 3207 (1914): 531–46.

Cassidy, Angela, Rachel Mason Dentinger, Kathryn Schoefert, and Abigail Woods. 'Animal Roles and Traces in the History of Medicine, c.1880–1980'. *BJHS Themes* 2 (2017): 1–23.

Chan, Ying-kit. 'The Great Dog Massacre in Late Qing China: Debates, Perceptions, and Phobia in the Shanghai International Settlement'. *Frontiers of History in China* 10, no. 4 (2015): 645–67.

Clucas, Barbara, and John M. Marzluff. 'Attitudes and Actions Toward Birds in Urban Areas: Human Cultural Differences Influence Bird Behavior'. *The Auk* 129, no. 1 (2012): 8–16.

Cornell, Heather N., John M. Marzluff, and Shannon Pecoraro. 'Social Learning Spreads Knowledge about Dangerous Humans among American Crows'. *Proceedings of the Royal Society B: Biological Sciences* 279, no. 1728 (2012): 499–508.

Darnton, Robert. *The Great Cat Massacre and Other Episodes in French Cultural History.* New York: Basic Books, 1984.

Douglas, Mary. *Purity and Danger: An Analysis of Concepts of Pollution and Taboo.* Psychology Press, 2003.

Fudge, Erica. 'A Left-Handed Blow: Writing the History of Animals'. In *Representing Animals*, edited by Nigel Rothfels, 3–18. Bloomington: Indiana University Press, 2003.

———. 'Milking Other Men's Beasts'. *History and Theory* 52, no. 4 (2013): 13–28.

Geertz, Clifford. 'Deep Play: Notes on the Balinese Cockfight'. *Daedalus* 134, no. 4 (2005): 56–86.

Hoon Song, S. 'The Great Pigeon Massacre in a Deindustrializing American Region'. In *Natural Enemies: People-Wildlife Conflicts in Anthropological Perspective*, edited by John Knight, 212–29. London: Routledge, 2013.
Ikeya, Chie. *Refiguring Women, Colonialism, and Modernity in Burma*. Honolulu: University of Hawai'i Press, 2011.
Jackson, Zakiyyah Iman. *Becoming Human: Matter and Meaning in an Antiblack World*. New York: New York University Press, 2020.
Jacobs, Nancy J. 'The Great Bophuthatswana Donkey Massacre: Discourse on the Ass and the Politics of Class and Grass'. *The American Historical Review* 106, no. 2 (2001): 485–507.
Kean, Hilda. *The Great Cat and Dog Massacre*. Chicago: University Of Chicago Press, 2017.
LaCapra, Dominick. 'Chartier, Darnton, and the Great Symbol Massacre'. *The Journal of Modern History* 60, no. 1 (1988): 95–112.
Marzluff, John, and Tony Angell. *Gifts of the Crow: How Perception, Emotion, and Thought Allow Smart Birds to Behave Like Humans*. New York: Simon and Schuster, 2013.
Marzluff, John M., and Erik Neatherlin. 'Corvid Response to Human Settlements and Campgrounds: Causes, Consequences, and Challenges for Conservation'. *Biological Conservation* 130, no. 2 (2006): 301–14.
Mazumder, Rajashree. '"I Do Not Envy You": Mixed Marriages and Immigration Debates in the 1920s and 1930s Rangoon, Burma'. *The Indian Economic & Social History Review* 51, no. 4 (2014): 497–527.
Mbembe, Achille. 'Necropolitics'. Translated by Libby Meintjes. *Public Culture* 15, no. 1 (2003): 11–40.
Miller, Ian Jared. *The Nature of the Beasts: Empire and Exhibition at the Tokyo Imperial Zoo*. Berkeley: University of California Press, 2013.
Nieder, Andreas. 'Inside the Corvid Brain—Probing the Physiology of Cognition in Crows'. *Current Opinion in Behavioral Sciences*, Comparative cognition, 16 (2017): 8–14.
Pettit, Michael. 'The Great Cat Mutilation: Sex, Social Movements and the Utilitarian Calculus in 1970 New York City'. *BJHS Themes* 2 (2017): 57–78.
Philo, Chris. 'Animals, Geography, and the City: Notes on Inclusions and Exclusions'. *Environment and Planning D: Society and Space* 13, no. 6 (1995): 655–81.
Report on the Working of the Rangoon Municipality For the Year 1920–21. Rangoon: British Burma Press, 1921.
Saha, Jonathan. 'Among the Beasts of Burma: Animals and the Politics of Colonial Sensibilities, c.1840–1950'. *Journal of Social History* 48, no. 4 (2015): 933–55.
———. Colonizing Animals: Interspecies Empire in Myanmar. Cambridge: Cambridge University Press, 2021.

———. 'Do Elephants Have Souls? Animal Subjectivities and Colonial Governmentality'. In *South Asian Governmentalities*, edited by Stephen Legg and Deana Heath, 159–77. Cambridge: Cambridge University Press, 2018.

Stoler, Ann Laura. *Along the Archival Grain: Epistemic Anxieties and Colonial Commonsense*. Princeton, NJ: Princeton University Press, 2009.

Taylor, Alex H. 'Corvid Cognition'. *WIREs Cognitive Science* 5, no. 3 (2014): 361–72.

Trouillot, Michel-Rolph. *Silencing the Past: Power and the Production of History*. Boston: Beacon Press, 1995.

Turner, Alicia. 'Colonial Secularism Built in Brick: Religion in Rangoon'. *Journal of Southeast Asian Studies* 52, no. 1 (2021): 26–48.

———. Saving Buddhism: The Impermanence of Religion in Colonial Burma. Honolulu: University of Hawai'i Press, 2014.

Vann, Michael G. 'Of Rats, Rice, and Race: The Great Hanoi Rat Massacre, an Episode in French Colonial History'. *French Colonial History* 4, no. 1 (2003): 191–203.

Wadiwel, Dinesh Joseph. 'The War Against Animals: Domination, Law and Sovereignty'. *Griffith Law Review* 18, no. 2 (2009): 283–97.

Walker, Brett L. 'Animals and the Intimacy of History'. *History and Theory* 52, no. 4 (2013): 45–67.

CHAPTER 8

From French Guinea to Florida: Chimpanzees as Multi-purpose Objects of Research (1920s–1940s)

Marion Thomas

Introduction

In the late 1990s, when Alain Corbin wrote *The Life of an Unknown: The Rediscovered World of a Clog Maker in Nineteenth-Century France* (2001), he aimed at reconstituting the existence of a person, whose memory has been abolished, and giving him a second chance to become a part of the memory of his century.[1] Corbin's challenging undertaking also matches a "history from below" approach, as he examines the tumultuous and eventful nineteenth-century France—from the Napoleonic Wars to the Third

[1] Alain Corbin, *The Life of an Unknown: The Rediscovered World of a Clog Maker in Nineteenth-Century France*, trans. Arthur Goldhammer (New York: University of Columbia Press, 2001).

M. Thomas (✉)
University of Strasbourg, Strasbourg, France
e-mail: marion.thomas@unistra.fr

M. J. Glover, L. Mitchell (eds.), *Animals as Experiencing Entities*, The Palgrave Macmillan Animal Ethics Series,
https://doi.org/10.1007/978-3-031-46456-0_8

Republic—through the everyday experiences and perspectives of an ordinary man, a forester, and clog-maker called Louis-François Pinagot (1798–1876). History is full of ordinary people, and, since E. P. Thompson's seminal 1968 work on the making of the English working class, there have been a flurry of studies on peasants, patients, women, and colonized peoples, for example, to retrieve the ways in which these invisible people have shaped history. A recent "Animal turn" within history also added animals to this already extensive list. As noted by Georgina Montgomery and Linda Kalof, recent work in the interdisciplinary scholarship in animal studies clearly shows that "animals occupy the most under-represented of all historically invisible groups." Adopting a perspective "from below" enables us to explore the history of animals "as subjects rather than objects," as well as helps us "to rethink the relationships between humans and other animals."[2] Animal histories sprout from fertile historical grounds in which animals exist in interaction with humans: the field, the zoo, the farm, and the laboratory, to cite a few. While it is certain that animals have been enrolled into multiple roles in science—including as victims of diseases, as experimental material, as models, and thereby featuring as passive objects manipulated by their human handlers—one challenge is to engage more deeply with animals' lived experiences and explore the ways in which they are intertwined with human histories, and the extent to which animals shape them.[3]

In this chapter, the possibility of an animal-centred history takes us to the African bush and primate laboratories in French Guinea and in the US. I examine the significant roles chimpanzees were made to play in modern science in the context of the Pasteur Institute in Kindia (French Guinea). In the early twentieth century, French Guinea was allegedly inhabited by 200,000 chimpanzees.[4] This made it a perfect location for the bacteriologist and Pastorian Albert Calmette (1863–1933) to

[2] Georgina M. Montgomery and Linda Kalof, "History from Below: Animals as Historical Subjects," in *Teaching the Animal: Human-Animal Studies Across the Disciplines*, ed. Margo DeMello (Brooklyn, NY: Lantern Books, 2010), 35.

[3] Angela Cassidy et al., "Animal Roles and Traces in the History of Medicine, c.1880–1980," *British Journal for the History of Science Themes* 2 (2017), p. 13; Abigail Woods et al., *Animals and the Shaping of Modern Medicine: One Health and Its Histories* (London: Palgrave Macmillan, 2017), 3 & 5.

[4] Robert Wilbert and Maurice Delorme, "'Pastoria', Centre de recherches biologiques et d'élevage de singes, Institut Pasteur de Kindia, Guinée française," *Bulletin de la Société de pathologie exotique et de ses filiales de l'Ouest africain et de Madagascar* 24 (1931): 132.

implement a chimpanzee facility which would use primates as models to study human infectious diseases. So the Pasteur Institute in Kindia, also called “Pastoria,” was established in 1922. The next idea was to develop a psychological programme, apart from a biomedical one. Calmette was aware that apes could be used as a bargaining chip to attract foreign scientists and international funding at a time when primate research was emerging in different places in the world. Robert Mearns Yerkes (1876–1956) and Henry Wieghorst Nissen (1901–1958), two important figures of twentieth-century American psychology and primatology, visited Pastoria in 1929 and 1930. Yerkes used Pastoria not only as a platform to venture into the wild to study chimpanzees but also as a supplier for his primate laboratories in the US.

My aim is to reconstruct the histories of chimpanzees who stayed in or transited Pastoria by using the traces they left on the medical and psychological historical records. I especially focus on two female chimpanzees: Rose, who was subjected to TB trials at Pastoria, and Kambi, who was shipped from Kindia to the US, where she was enrolled into psychobiological studies at the Yale Laboratories of Primate Biology. With these two cases, I aim at understanding and elucidating what their lived experiences might have been. As such, the chapter aligns itself with recent work that focuses on individual animals.[5] On the one hand, I examine their lives as individuals who were experimented upon, and explore the interactions between theories and practices of biomedical and psychological experiments on chimpanzees; the socio-political and ideological conditions during the French colonial era; and the development of bacteriology and primatology. On the other hand, I examine the ways in which their lived experiences were impacted by colonialism. I then seek to shed light on issues of power and recognition at stake in the circulation and appropriation of knowledge about chimpanzees and on how they intermesh with questions of race and colonialism. Finally, I examine how individual chimpanzees were used in accruing scientific knowledge (here bacteriology and

[5] Georgina M. Montgomery, “‘Infinite Loneliness’: The Life and Times of Miss Congo,” *Endeavour* 33, no. 3 (2009): 101–105; Tania Munz, “‘My Goose Child Martina’: The Multiple Uses of Geese in the Writings of Konrad Lorenz,” *Historical Studies in the Natural Sciences* 41, no. 4 (Fall 2011): 405–446; Amy Nelson, “The Legacy of Laika: Celebrity, Sacrifice, and the Soviet Space Dogs,” in *Beastly Natures: Animals, Humans and the Study of History*, ed. Dorothee Brantz (Charlottesville and London: University of Virginia Press, 2010), 204–224; Éric Baratay, *Animal Biographies: Toward a History of Individuals*, trans. Lindsay Turner (Athens, GA: University of Georgia Press, 2022).

primatology) and also inadvertently contributed to the scientists' public image. In a nutshell, I want to argue for the value and the interest of the lives of these forgotten actors in science, and show how making them visible in history helps to rethink the relationship between humans and animals and shed a different light on the history of French colonial science.

Rose (circa 1916–1927) and the Success of TB Trials at Pastoria

Unlike Kambi whose lifetime spans across two continents, Africa and America, Rose remained all of her short life in Guinea. Her fate, however, was largely determined by French scientific matters. But before going to Guinea, we first have to stop in Lille, in the north of France. Rose's story starts there, in connection with the history of the BCG (Bacillus Calmette-Guérin) vaccine, which Albert Calmette fashioned with the veterinarian Camille Guérin (1872–1961) in the early twentieth century. The attempt to develop a vaccine against tuberculosis would bring Calmette worldwide fame and lasting recognition in the history of medicine. Calmette got involved in the war on tuberculosis in the late 1890s, while working at the Pasteur Institute in Lille. Unlike social hygienists, who grounded tuberculosis infections in environmental conditions, Calmette focused exclusively on contagion, and viewed vaccination as the only way to eradicate the microbe.[6]

From 1900 onwards, alongside Guérin, Calmette became involved in meticulous and long-term studies on how animals infected with an attenuated form of the tuberculosis bacterium could develop resistance (or immunity) to the disease. The disease organism (*Mycobacterium bovis*) was grown in a sterile artificial environment composed of potatoes mixed with ox's bile. Repeated culture and subculture in this medium reduced the virulence of the organism and once attenuated it was passed through a set of animals, including guinea pigs, rabbits, bovines, and horses, to evaluate its safety and efficacy. On 5 January 1921, after the 230th passage through an animal host, Calmette and Guérin proclaimed that they had fashioned

[6] Aro Velmet, "The Making of a Pastorian Empire: Tuberculosis and Bacteriological Technopolitics in French Colonialism and International Science, 1890–1940," *Journal of Global History* 14, no. 2 (2019): 205; Aro Velmet, *Pasteur's Empire: Bacteriology and Politics in France, Its Colonies, and the World* (Oxford and New York: Oxford University Press, 2020), 115–141.

a live bacillus tuberculosis vaccine completely devoid of virulence, even at high doses, for all animal species. They then proceeded to perform a series of tests, which consisted in administrating the vaccine by way of injection or oral ingestion to induce immunity in animals. The vaccinated animals were then exposed to animals artificially infected with normal TB as well as other animals who were used as controls. The latter, control animals, died after a few weeks of exposure, while the vaccinated animals showed no signs of the disease.[7]

The last stage, before moving on to using the vaccine on humans, involved an evaluation of the immunizing effects of the BCG vaccine by undertaking long-term, controlled, large-scale experiments on primates.[8] Such experiments in the mainland, however, were almost impossible since, as Calmette reckoned, animals caught tuberculosis on boats bound for France, either from the cook or from their caretakers, making them unsuitable for TB trials. Since the apes could not be kept in laboratories in temperate climates, Calmette decided to relocate his tuberculosis work directly to the primates' native environment.[9]

Calmette chose to institute a chimpanzee facility in French Guinea for two main reasons. First, with steamers departing regularly from Bordeaux or Marseille, Conakry, Guinea's harbour capital, could be reached within two weeks. Second, chimps were abundant and could be purchased at a low cost.[10] This, however, amounted to a very high cost from a chimpanzee's perspective: kidnapping, social isolation, incarceration in cages, invasive biomedical experiments, illness, and in most cases death. The primate station was officially launched in November 1922 in Kindia, a city located ninety miles away from Conakry and accessible by train.[11] Although

[7] Christian Bonah, *L'Expérimentation humaine. Discours et pratiques en France 1900–1940* (Paris: Les Belles Lettres, 2007), 216–218.

[8] Ibid., 224.

[9] Albert Calmette to Sir Walter Fletcher, 23 October 1928, Medical Research Council FD 1/803, PRO. See also Delphine Berdah, *Abatttre ou vacciner. La France et le Royaume-Uni en lutte contre la tuberculose et la fièvre aphteuse (1900–1960)* (Paris: Éditions de l'École des Hautes Études en Sciences Sociales, 2018), 120–123.

[10] The price of a chimpanzee in mainland France was between 6,000 and 10,000 francs, while in Guinea it was possible to buy a chimpanzee for less than 250 francs. See Fernand Honoré, "Les singeries de l'Institut Pasteur à Kindia et à Paris," *L'Illustration* no. 4390, (Avril 23, 1927): 408.

[11] "Convention entre le Gouverneur Général de l'A.O.F. et le Directeur de l'Institut Pasteur de Paris (I.P. de Kindia), 20 Novembre 1922," folder "Direction des IPOM. Conventions. IP de Kindia," IPO.Conv.3, AIP.

Calmette was attracted to far-off lands from very early in his career—let's remember that, in 1891, he had been entrusted by Louis Pasteur (1822–1895) with the founding of the first overseas Pasteur laboratory in Saigon, Indochina—he never set foot in French Guinea. Instead, he sent to Kindia two military veterinarians, Robert Wilbert (1877–1945) and later Maurice Delorme (1890–1943), to carry out experiments on chimpanzees and monkeys. Sure the Kindia station embodied not only Calmette's long-term endeavour with the BCG vaccine and his enduring involvement in the Pasteur Institute's overseas work, it also chimed with his conviction of the essential role of the Pasteur Institute in "sending throughout the world [...] a large number of missionaries disciplined by [Pasteur's] methods and impregnated with an ardent scientific faith, to support France's civilizing expansion and effectively promote the *mise en valeur* of its colonies."[12]

Wilbert and Delorme proved very diligent and dedicated Pastorian emissaries and were successful with the primate BCG trials. In two years, Wilbert, first Director of Pastoria, gathered evidence which helped actualize Calmette's plan to conclusively demonstrate the safety and efficacy of TB prevention through the BCG vaccine. Among the fifteen chimpanzees and fifty-eight monkeys Wilbert used in his experiments, one was singled out: the female chimpanzee Rose (*Troglodytes niger*).

What do we know about Rose? Her name comes up in the station's reports, animal register, laboratory notebook, and scientific publications. The ape register especially helps to yield information about her. Based on the age indicated on her file (8 years old) and her date of arrival in Pastoria, in March 1924, Rose was likely born in 1916. Her date of death appears on the animal register and the laboratory notebook, but with a one-year difference, while a station report appears to confirm the date of 24 September 1927.[13] Upon her arrival at Pastoria, Rose was given a name and an ID number, 107—the 107th ape to be brought to Pastoria. Given her date of admission and place of purchase, Mamou, a city that sat along the Conakry-Kankan railway, ninety-five miles from Kindia, it is likely that Rose was acquired through a man by the name of "Victor Godeffrois, who was appointed as second-class deputy in the Civil Services of French West

[12] Albert Calmette, "Les missions scientifiques de l'Institut Pasteur et l'expansion coloniale de la France," *Revue scientifique* 50, no. 5 (February 3, 1912): 133.

[13] Laboratory notebook "Animaux d'expérience (Singes)," tome I, Index card "Rose," AIP-Kindia.

Africa" and "assigned for a six-month mission to capture anthropoids for the Pasteur Institute,"[14] in Mamou, in October 1923.[15]

In September 1923, a change in the decree regulating hunting in French West Africa authorized the capture and trade of live chimpanzees, but only on specific grounds, namely for use in and by scientific and medical institutions. A scientific capture permit was required and European hunters should be escorted by indigenous people acting as scouts. Only European nets and traps would be allowed.[16] For instance, Godeffrois was provided with equipment, including nets and weapons, and assigned with local guards—Morlaye Yattara, Yamoussa Kamara, and Karamoko Diallo—to help him to requisition men from a local village to "pick up chimps." Once the target chimps were located, the men surrounded the tree where he had taken refuge and cut it down, causing the animals to fall.[17] Obviously, this decree made chimpanzee hunting a colonial enterprise where Guinean people were enrolled for the purpose of colonial science and compelled to use European methods to capture chimpanzees alive.

Wilbert viewed Godeffrois as a "daring and passionate hunter" and was grateful to him for "largely supply[ing] [Pastoria] with chimps and monkeys, in very good condition and without wounds." In contrast, Wilbert pointed an accusatory finger at the Guineans, who, he claimed, abused the animals, causing them to die quickly: a "massacre" as he called it.[18] In his eyes, the only way to stop it was to hire "an indigenous labour force

[14] Jules Carde, "Décision du Gouverneur Général de l'Afrique Occidentale Française, 29 August 1923," and Victor Godeffrois to the Governor of French Guinea, 17 October 1923, 1D34, folder "Mission Voronoff (Godeffrois), 1924," ANG.

[15] I wish to thank Vincent Leblan for pointing me to the existence of documents on the colonial hunter Victor Godeffrois in the National Archives of Guinea. I am also very indebted to him for mentioning archives kept at the library of the IRBAG, formerly Pastoria, at a time I was wondering about their existence.

[16] Jules Carde, "Actes du Gouvernement Général. Arrêté du Gouverneur Général fixant les conditions dans lesquelles peuvent s'opérer la capture et l'exportation des chimpanzés," *Journal Officiel de la Guinée française* 23, no. 539 (October 1, 1923): 315.

[17] The Cercle Commander of Mamou to the Governor of French Guinea, 30 January 1925, 1D34, folder "Mission Voronoff (Godeffrois), 1924," ANG.

[18] Robert Wilbert, "Rapport sur le fonctionnement du Centre de Recherches biologiques sur les singes anthropoïdes de l'Institut Pasteur de Kindia en Guinée française (Institut Pasteur de Kindia) en 1926," 10 February 1927, p. 11–12, IPO.RA.23, AIP.

guided, controlled, even in repressive ways, by Europeans specialized in the capture of animals."[19]

The term "massacre," which recurs in colonial administrative writings, also appears in the popular press, in reference to the animal deaths that colonists largely blamed on African hunters.[20] In contrast, the difficulties and dangers inherent in the capture of live animals and the chance for men to sustain serious wounds during the hunts were rarely mentioned. An acknowledgement of them, however, came from Jean Louis Georges Poiret (1872–1932), the Governor-General of French Guinea at that time. In 1925, he wrote to Wilbert that "evidently [...] whatever natives are commandeered to catch apes, they would rather cudgel them than to get themselves devoured or mutilated by these [...] animals." It was then "necessary to supply these hunters with trained dogs and special equipment paid for from the local budget."[21] Likewise, the Cercle Commander of Forécariah reported to Wilbert about the death of a man, killed by a falling tree, as he was trying to capture a chimpanzee. Consequently, he asked for grants to pay his people, so that they do not get discouraged.[22] There were also serious disagreements between advocates on the use of lines, lassoes, or nets—the so-called "white way", and those of the "black way," which consisted in waiting for the animals to tire themselves out.[23] A colonial administrator also reported about the unsuccessful attempts to make local people use European equipment. [24] This is another evidence of the difficulty of hybridizing colonial and local practices.

Rose was certainly captured in the "white way," as Godeffrois was in charge of the hunt. She was probably acquired for twenty francs, an amount that Godeffrois claimed to have paid to the village chief for female chimps. She was then reportedly transported to the Mamou ape house in a locally woven liana basket.[25] Having survived the trauma of her capture

[19] Robert Wilbert, "'Pastoria'," in Comptes rendus des séances/communications, tome XII 1928-1929, ed. Académie des sciences coloniales (Paris: Société d'Éditions géographiques, maritimes et coloniales, 1932), 486.

[20] Jean Perrigault, *Bêtes et gens de brousse* (Paris: Éditions de la Nouvelle Revue Critique, 1931), 162.

[21] Georges Poiret to Robert Wilbert, 17 June 1925, AIP-Kindia.

[22] H. Martin-Chartrie to Robert Wilbert, 5 November 1925, AIP-Kindia.

[23] Perrigault, *Bêtes et gens de brousse*, 72–3.

[24] The Cercle Commander of Mamou to the Governor of French Guinea, 30 January 1925, 1D34, folder "Mission Voronof (Godeffrois), 1924," ANG.

[25] Ibid.

and making it through the first three months of her captivity without developing any lethal pulmonary or intestinal disease, she certainly was judged as a good subject for the bacteriological research conducted in Pastoria.

During her life in captivity, Rose was locked in cages and was never taken out for a walk, because of her rather tall size.[26] Indeed, only young (and small) chimpanzees were allowed "to go out once a day under the supervision of a Guinean caretaker to the immediate surroundings of mango trees in which to climb and have fun."[27] This emphasis on semi-freedom was something special to Pastoria. Very early on, Calmette got the idea of setting up a chimpanzee facility which would "allow the development and freedom of apes, while favouring their progressive taming by daily contact with the caretakers in charge of distributing a determined amount of food to them, in chosen places."[28] Similarly, Wilbert and Delorme were convinced that "the education in semi-liberty [would] give the best results," that apes would finally "get accustom to a man's voice, their company, and get the habit of obeying."[29] This was especially true for young chimpanzees whom Wilbert compared to "schoolchildren frolicking in a schoolyard. Their cries of joy [...] strangely assimilat[ing] them to humans."[30] This education also helped to better handle the chimps within the laboratory's walls. As a French veterinarian who visited Pastoria in 1931 reckoned: "[A] permanent contact with civilized man gives [the chimpanzees] the essentials of an education for their use as experimental animals."[31] Another visitor of Pastoria, the French zoologist Paul Rode (1901–1948) envisaged this outdoor time as a way for the chimpanzees "to live a few hours with the illusion of a freedom they greatly appreciate,"

[26] Robert Wilbert to Albert Calmette, 18 October 1927, AIP-Kindia.

[27] Robert Wilbert, "Rapport sur le fonctionnement du Centre de recherches biologiques sur les singes anthropoïdes de l'Institut Pasteur de Kindia en Guinée française (Institut Pasteur de Kindia) en 1926," 10 February1927, p. 43, IPO.RA.23, AIP.

[28] Robert Wilbert, "Rapport sur le fonctionnement de l'Institut Pasteur de Kindia en 1926," [For the year] 1927, pp. 7–8, AIP-Kindia.

[29] Wilbert and Delorme, "'Pastoria', 140.

[30] Wilbert, 'Rapport sur le fonctionnement du Centre de recherches biologiques sur les singes anthropoïdes de l'Institut Pasteur de Kindia en Guinée française (Institut Pasteur de Kindia) en 1926', p. 43, IPO.RA.23, AIP.

[31] A. Donatien, "'Pastoria', Centre de recherches biologiques et d'élevage de singes (Institut Pasteur de Kindia, Guinée française)," *Revue vétérinaire et Journal de médicine vétérinaire et de zootechnie réunis* LXXXIII (September 1931): 528.

and "deterred them from escaping to the close bush, all the more as abundant food was waiting for them, when brought back to their cages."[32]

Clearly, letting out the chimpanzees secured several advantages. It provided an opportunity to train chimpanzees, to ease their life as captive animals, and eventually set the stage for better use of them in the lab. In this process, the work of the Guineans was instrumental, and indeed vital, but it was acknowledged only in passing in station reports or correspondence. So were other contributions of the Guineans. For instance, they helped Wilbert and Delorme with choosing plants germane to the chimpanzee's diet or plants with therapeutic virtues against intestinal infections, a major scourge for animals kept in cages.[33] The Guineans were also entrusted with caring for chimpanzees who had been experimented on and local women, the so-called nurses, were hired in Pastoria to look after and feed baby chimps. Then, local knowledge and skills played an important role in the functioning of the chimpanzee facility.

As French and Guineans interacted in the care of the chimpanzees, this resonated with colonial discourses. The colonial logic of paternalistic domination permeated scientific practices, including Wilbert and Delorme's relationships with Guinean staff and animals intended for experiments. Wilbert and Delorme were entrusted by Calmette with the mission of training the Guineans towards specific jobs, such as laboratory assistants and animal carers. Feeding the animals was a "delicate task": Calmette recommended that "all the food [except raw fruits which provide vitamins] would be sterilized in an autoclave, or given very fresh to animals without too much handling by the animal-carers."[34] This implied that animal caretakers would be "required to take severe disinfection measures before and after each contact with the animals."[35] Wilbert easily embraced these recommendations. He was convinced that "the natives provide effective care when under the constant and direct supervision of a European," and bemoaned not being able to take "immediate and required control" of his indigenous staff, as he was snowed under with multiple

[32] Paul Rode, "'Pastoria', centre de recherches sur les singes en Guinée française," *La Terre et la vie* 7, no. 4 (July-August 1937): 114.

[33] Robert Wilbert, "Rapport sur le fonctionnement de l'Institut Pasteur de Kindia en 1925," 10 February 1926, p. 10, IPO.RA.23, AIP.

[34] Albert Calmette to Robert Wilbert, 6 February 1925, AIP-Kindia.

[35] Maurice Delorme, "Rapport sur 'Pastoria'—Institut Pasteur de Kindia, 1936," 11 June 1937, p. 16, IPO.RA.23, AIP.

tasks.[36] The rules regarding the caring of animals demonstrate the extent to which the training of Guinean people and the educating of chimpanzees could be intertwined. This supports Donna Haraway's statement that "nonhuman primates were a fundamental part of the apparatus of colonial medicine" and Vincent Leblan's argument that, in French Guinea, not only local people but also chimpanzees were enrolled in the French civilizing mission.[37]

Rose was vaccinated orally and then exposed to cohabitation with artificially infected and therefore contagious chimps and with non-infected chimps serving as controls. This cohabitation in the same cage allowed for the simulation of what happened in a familial environment.[38] This reflected Wilbert's effort to adapt the ape model to the human model and to replicate the most favourable conditions for the onset of the disease in the domestic realm. Indeed, Calmette had advised him "to have the chimpanzees receive their vaccines immediately upon being introduced to the contaminated environment [in order to] realize as accurately as possible what happens in everyday life among human beings and what would happen if some of them were vaccinated."[39]

In July 1924, after being administrated with a 250 mg dose of BCG vaccine (split into five doses of 50 mg each), Rose was placed in a cage with Butor and Gertrude, two other vaccinated chimps, five artificially infected chimps, and seven controls. In December 1924, Rose got a second series of five doses. Rose, Gertrude, and Butor emerged unharmed from several months of cohabitation, while all the infected chimps and most of the controls died of tuberculosis.[40] In December 1926, Rose was vaccinated for the third time.[41] From 1924 to 1927, Wilbert and Delorme

[36] Robert Wilbert to Albert Calmette, 10 October 1924, AIP-Kindia.

[37] Donna Haraway, *Primate Visions: Gender, Race, and Nature in the World of Modern Science* (New York and London: Routledge, 1989), 19; Vincent Leblan, "Une frontière coloniale homme/animal: captures de chimpanzés et fabrique des identités raciales à l'Institut Pasteur de Kindia, Guinée française, 1920–1930," in *Le Spécimen et le collecteur. Savoirs naturalistes, pouvoirs et altérités (XVIIIe–XXe siècle)*, ed. Dominique Juhé-Beaulaton and Vincent Leblan (Paris: Publications scientifiques du Muséum national d'histoire naturelle, 2018), 402–409.

[38] Robert Wilbert, "Expériences de vaccination des singes contre la tuberculose par le BCG," *Annales de l'Institut Pasteur* 39, no. 8 (August 1925): 643–644.

[39] Albert Calmette to Robert Wilbert, 12 December 1923, AIP-Kindia.

[40] Wilbert, "Expériences de vaccination des singes," 644–45.

[41] Laboratory notebook "Animaux d'expérience (Singes)," tome I, Index card "Rose," AIP-Kindia.

reported succinctly but triumphantly to Calmette that Rose was "in excellent health."[42] Rose became the poster child for TB vaccination, her name appearing in the monthly reports of the station, but interestingly also in scientific publications by Wilbert and Calmette. Rose eventually died on 24 September 1927. Wilbert reported that she "very slowly passed away from starvation; she no longer wanted to eat or drink, and was extremely dejected."[43]

Whatever her woeful living conditions during five years of captivity and her miserable end, the most important thing for Wilbert was that Rose did not die of tuberculosis (she indeed died of jaundice).[44] With the success of primate TB trials in Kindia, Calmette could legitimately move on to the human model. "We are no longer allowed," he claimed in 1927, "not to apply the same method to immunize children born to TB-infected mothers, or exposed to family contagion."[45] Thanks to the bodies of Rose and her TB companions, Pastorian science had accumulated evidence to show that it was working for the benefit of humanity, a refrain that was used to justify the presence of Pastorians in the tropics and served as a rationalization of colonization.

Transition: Monkey Business in the Tropics—Yerkes, Nissen, and the Pastorian Connection

Calmette's ambition to use apes as an animal model seemed unlimited. Next to the development of a biomedical programme, Calmette harboured hopes that psychological experiments would be carried out at Pastoria.[46] By widening the scientific scope of the primate station, Calmette also aimed at hosting foreign scientists and attracting international funding. For instance, he approached the Yale professor of psychology, Robert Yerkes, who enjoyed special relations with the Carnegie Institution and

[42] Robert Wilbert to Albert Calmette, 18 October 1927, AIP-Kindia.

[43] Ibid.

[44] Ibid. and Laboratory notebook "Animaux d'expérience (Singes)," tome I, Index card "Rose," AIP-Kindia.

[45] Albert Calmette, "L'état actuel de nos connaissances sur la vaccination antituberculeuse," *Revue vétérinaire et Journal de médicine vétérinaire et de zootechnie réunis* LXXIX (March 1927): 149.

[46] Albert Calmette, "Sur l'utilisation des singes en médecine expérimentale: le laboratoire Pasteur de Kindia (Guinée française)," *Bulletin de la Société de pathologie exotique* et de sa filiale de l'Ouest africain 17 (1924): 14.

the Rockefeller Foundation.[47] This proved successful: Yerkes visited the Kindia facility in August 1929.

At that time, he was at the peak of his academic career. A professor of comparative psychology at Yale University, he had recently received a huge grant from the Rockefeller Foundation which would allow him to fulfil his dream of founding a second, but much bigger, primate laboratory in Florida.[48] In the late 1920s, Yerkes was also eager to see that the fledgling science of primatology could develop beyond the walls of the laboratory. He set up a programme for what he called the "naturalistic" study of primates and which resulted in three field expeditions. One of them took place in the Guinean bush.[49] In October 1929, back to the US from his short visit in Kindia, Yerkes prompted Henry Nissen, a new recruit at his Institute for Human Relations at Yale, to go to Guinea to observe chimpanzees in the wild.

Born in Chicago in 1901 and educated in the Lutheran faith, Nissen received an MA degree and a PhD from Columbia University in New York. In late 1928, during a presentation at the American Psychological Association, Nissen, then a doctoral student, impressed Yerkes, who eventually hired him at Yale. During his career at the Yale-then Yerkes Laboratories of Primate Biology, Nissen "was generally acknowledged to know more about chimpanzees than anyone else and to have an unusual rapport in working with them."[50] No doubt, Nissen, like Yerkes, shared a passion for apes to which he devoted all his life. Unlike Yerkes, however, there are only scant mentions of his name in the historiography, maybe owing to his shy, patient, careful, and modest personality.[51]

[47] Albert Calmette to Robert Yerkes, 4 March 1924, group 569, series II, box 69, folder 1314, RMYP. See also Marion Thomas, "Between Biomedical and Psychological Experiments: The Unexpected Connections between the Pasteur Institutes and the Study of Animal Mind in the Second Quarter of the Twentieth Century France." *Studies in History and Philosophy of Biological and Biomedical Sciences* 55, no. 1 (2016): 31 & p. 38.

[48] Haraway, *Primate Visions*, 59–83; Donald A. Dewsbury, *Monkey Farm: A History of the Yerkes Laboratories of Primate Biology, Orange Park, Florida 1930–1965* (Lewisburg, PA: Bucknell University Press, 2006), 58–64; Georgina M. Montgomery, *Primates in the Real World: Escaping Primate Folklore and Creating Primate Science* (Charlottesville and London: University of Virginia Press, 2015), 29–33.

[49] Montgomery, *Primates in the Real World*, 46–51.

[50] Dewsbury, *Monkey Farm*, 208.

[51] Leonard Carmichael, "Henry Wieghorst Nissen February 5, 1901–April 27, 1958," *Biographical Memoirs, National Academy of Sciences* 38 (1965): 214.

Fitting into Yerkes' "naturalistic" goals the Guinean expedition aimed primarily at "testing the feasibility of field studies and in making a start in discovering a workable methodology and technology for naturalistic observation of the chimpanzee."[52] Yerkes also envisaged it as "a plan of cooperation between Pastoria and Yale University which [he] had every reason to believe might prove possible of realization and valuable in its results."[53] Last, but not least, Yerkes entrusted Nissen with "the collection of a group of chimpanzees suitable for use in the anthropoid research laboratories of Yale University."[54] Concretely it meant that Pastoria would be a platform for Nissen's expeditions in the bush. Delorme, Deputy-Director of Pastoria at that time, spared no pains to provide Nissen with all the essential assistance, such as "securing for [him] three native assistants" to venture into the wild and "enlisting the cooperation of colonial officials."[55]

The African assignment was certainly a challenge for a young man of twenty-eight, freshly out of university, and who had never left his country, let alone trod on the African soil. It would also be a memorable experience. As he was departing from Africa, Nissen wrote to his dearest friend Elmer: "Really hate to leave Africa; it's a great country. We must come here together some day, what say?"[56] Beyond the fatigue and the demanding job, Nissen had good reasons to rejoice about his sojourn in Guinea. In his luggage were three field notebooks, which would be the base of his 1931 article on the chimpanzee in French Guinea. Nissen envisaged his article as just "the start of the story of the chimpanzee in his native habitat."[57] Nonetheless, it would earn him good reviews, as well as an invitation to the Academy of Sciences in Washington. In the reminder of the chapter, I am not going to develop Nissen's observations in the field,

[52] Henry W. Nissen, "A Field Study of the Chimpanzee: Observations of Chimpanzee Behavior and Environment in Western French Guinea," *Comparative Psychology Monographs* 8, no. 1, serial no. 36 (December 1931): 2.

[53] Robert M. Yerkes, foreword to Henry W. Nissen, "A Field Study of the Chimpanzee: Observations of Chimpanzee Behavior and Environment in Western French Guinea," *Comparative Psychology Monographs* 8, no. 1, serial no. 36 (December 1931): iii.

[54] Nissen, "A Field Study of the Chimpanzee," 2.

[55] Ibid., 5.

[56] Henry Nissen to Elmer, 18 May 1930, box 78, folder "African Correspondance-1930," EU14, YNPRCR.

[57] Nissen, "A Field Study of the Chimpanzee," 104.

but I will examine the trajectory of a baby chimpanzee, Kambi, who Nissen brought back to the US.

From the Jungle to Florida: The Multiple Uses of the Chimpanzee Kambi (circa 1929–1946)

Out of the eighteen chimpanzees that Nissen finally assembled with the help of Delorme to ship over to the US, the female baby chimp Kambi would play a specific role in Nissen's life and career. Nissen dedicated an entire notebook to the infant chimpanzee, whom he himself captured alive, near the village of Kambi (after which she was named), Nissen's second base during his field journey in Guinea.[58] Unlike Rose, who was exclusively used for the purpose of TB trials, I will show how Kambi benefitted from preferential care after her capture, and, once in the US, was fashioned into a multiple-purpose "experimental chimpanzee." Ideally, this means for Yerkes that the animal would be healthy, robust, gentle, quickly tamed, and overall "actively cooperative."[59] In reality, Yerkes knew "there [were] individual patterns and types," and that "in any group of experimental chimpanzees, [...] one [animal] may quickly grasp the requirements of a situation, cooperate perfectly [...], whereas another may be slow, unpredictable, relatively uncooperative or even refractory."[60] Then, for Yerkes, the experimenter had to be cognizant of individualized personality of his subjects, so his work could be facilitated. The combination of an extreme care for the animals and the adjustment of an individual animal to a specific research topic evidence not only for Yerkes' goal to optimize the "precious booty" Nissen brought back from Guinea but also for the different ways in which the French and Americans took advantage of the availability of chimpanzees in Guinea.

Unlike the sixteen chimpanzees who were "captured for [Nissen] by natives at the instigation of officials of the French Government and of the Pasteur Institute," and Bimba who "was purchased from a native who had shot the mother," Kambi, the second baby chimp Nissen brought back to

[58] Henry Nissen, Notebook Kambi #1, box 77, EU14, YNPRCR.

[59] Robert M. Yerkes and S. D. Shirley Spragg, "La mesure du comportement adapté chez les chimpanzés," *Journal de psychologie normale et pathologique* XXXIV (1937): 452.

[60] Robert M. Yerkes, "The Life History and Personality of the Chimpanzee," *The American Naturalist* 73, no. 745 (March-April 1939): 101.

the US, was captured during field observations in late April 1930.[61] Nissen's guides had led him where the chimpanzees had established their nests for the night. It was dawn, and the guides marched directly under the trees, instead of forming a circle around them. No doubt this scared off the animals, who started coming down the trees. Hearing them descending, Nissen "decided that [their] only chance to keep what animals remained up was by starting noise and burning grass immediately. [He] therefore fired the Winchester, which was the signal for the hubbub to begin." And then something unexpected happened: "[T]here was crying of a baby chimpanzee in an isolated tree." As Nissen commented: "The natives were standing around with clubs, ready to knock the animal on the head when it dropped."

Adrenaline inspired him to the following feat: he climbed up the tree and "grabbed the infant by the back of the neck, [...] then wrapped it in the robe of one of the blacks."[62] In contrast with common narratives on Guinean hunts, as seen above, there were no casualties, neither on the hunters' nor on the chimpanzees' side. Quite the contrary, a baby chimp had been rescued. Nissen wrote proudly to Yerkes about a "capture by myself [...] of a five and one-half pound female chimpanzee without injury to captor, captee or, remarkable to say, to the mother of the infant."[63] Nissen probably insisted on the fact that the mother was unscathed, as it was common that when a baby chimp was captured the hunters kill the mother to get the baby. Sure, Kambi's mother wasn't shot, unlike the mother of Bimba, as mentioned above. But, as Nissen inferred, after having a careful look at the nest after he had picked up the young female chimpanzee: "[T]he adult with whom the infant had slept, [...] had descended rapidly, being in great fear, and the little one had lost its hold."[64] From Kambi's mother's perspective, having her presumably crying baby left in the tree, amidst frightening gunshots, would certainly be a traumatic experience, and presumably also for Kambi.

When Nissen mentioned his guides, porters, and scouts in the field, he underscored their laziness, lack of intelligence, and sometimes their dishonesty, thereby reproducing the colonial stereotypes of his day.[65] Nissen,

[61] Nissen, "A Field Study of the Chimpanzee," 10.

[62] Ibid., 38.

[63] Henry Nissen to Robert Yerkes, 27 April 1930, group 569, series II, box 36, folder 689, RMYP.

[64] Nissen, "A Field Study of the Chimpanzee," 49.

[65] Ibid., 13.

however, singled out Moussa Camara and Ben Reffell, whose services Delorme secured for him.[66] Moussa Camara, a former infantry man, served as cook and supervisor of the Guinean team.[67] In the field, not only did Moussa care about the logistics, but he also provided Nissen with a great deal of relevant information with regard to the chimpanzees' food, nesting, and social behaviour. Nissen regarded him as "an exceptionally intelligent and able native [whose] services were worth those of a dozen others." So too did Nissen view as "honest and loyal" Ben Reffell, a Susu who hailed from Sierra Leone, and who "could read and write, as well as speak English," and who served as his interpreter and personal assistant.[68] I found pictures which feature Nissen's favourite assistants taking good care of Kambi. Those photographs show Moussa making little Kambi stretch, while others feature Kambi being kindly fed with a cup of milk and helped to walk by Ben, or being held gently in Ben's arms. Those pictures conveyed simple pleasure and joy of caring for a small baby chimp. Nissen, too, embraced this role of chimp carer, but certainly with more strain, as showed by the ways in which his journey back home with his "precious" animals proceeded.

From the time the convoy left Kindia by train, stopped over at the docks in Conakry, and finally crossed to the US, Nissen fussed over the health and well-being of his chimpanzee captives. At Conakry's harbour, the quiet Nissen shouted at Jules Burki, the steamship agent, to have the cages out of the freight car so he could put a tarpaulin over them, to protect the chimpanzees from inclement weather. Ensuring the right amount and diversity of food stores for travel was also one of Nissen's greatest worries before boarding.[69] Once at sea, Nissen, albeit suffering from bouts of malaria, continued to be obsessed with the welfare of his animals. He succeeded in getting the cages placed on a shelter deck and made sure the animals got fresh air and exercise.[70] Nissen also carefully checked that the apes were eating properly and was watchful in singling out the "grabber" in each cage trying to hog all the food.[71] Nissen had heeded Delorme's

[66] Ibid., 5.

[67] Henry Nissen, Certificate for Moussa Kamara, 23 May1930, box 78, folder "African Correspondence—1930," EU14, YNPRCR, and Nissen, 1931, p. 13.

[68] Henry Nissen, Certificate for Ben Reffell, Pasteur Institute at Kindia, French Guinea, 26 May 1930, folder "African Correspondence—1930," box 78, EU14 and Nissen, 1931, p. 13.

[69] Nissen, Notebook #4, p. 16 & p. 20, p. 23–24, box 77, EU14, YNPRCR.

[70] Ibid., p. 44, p. 28 & p. 73.

[71] Ibid., p. 23.

advice to place two animals in a cage since "the social stimulation thus provided tends to make the chimpanzees eat and thus to keep them healthy."[72] However, not only good, but also bad feelings could be passed on from one animal to another. As Nissen noted: "[T]he little man in cage 2, Boy, is so scared that 1) he loses out on food, and 2) he infects his two companions with his fear."[73] Nissen also tried to tame the animals, but received a small bite on his left thumb and a hitch on the leg and abdomen.[74] At some point, unable to cope as a unique caretaker, he "made arrangements with an ordinary seaman, [...] a nice fellow—willing intelligent and handy, [...] to help him 4 hrs a day, at 3 frcs a day."[75] Among the chimps, Kambi and Bimba were especially Nissen's focal point. Nissen was very much bothered about Kambi's health, as she could have high chest temperatures and difficulty breathing. He gave her oil rub and Vaporub, wrapping her in flannel, feeding her only with orange juice, even keeping her in his room, as "her entire body was unbelievably hot." Despite Nissen's diligent pampering of his animals, two of them died of pneumonia.[76] No doubt, Kambi survived thanks to Nissen's special care.

Obviously, Nissen knew that chimpanzees were delicate organisms, with a high susceptibility to get sick or injured. He was also aware that "to keep the youngster alive and bring it safely to America will be no easy task."[77] One must not forget, however, that the Guinean chimpanzees were intended to "serve as subjects in various psychobiological investigations" in the US.[78] Then, it would probably be a mistake to see Nissen as being exceptionally concerned about "his" animals for themselves. Because chimpanzees were expensive and essential for Yerkes' laboratories, they had to be treated well, at least from the point of view of physical well-being. When they landed in New York on 17 June 1930, Kambi and Bimba were transferred with four other chimps to Yerkes' laboratory at

[72] Henry Nissen to Robert M. Yerkes, 5 April 1930, group 569, series II, box 36, folder 689, RMYP.

[73] Henry Nissen, Notebook #4, p. 56, box 77, EU14, YNPRCR.

[74] Ibid., p. 29.

[75] Ibid., p. 33.

[76] Ibid., p. 58, p. 61, p. 39 & p. 41.

[77] Henry Nissen to Robert Yerkes, 27 April 1930, group 569, series II, box 36, folder 689, RMYP.

[78] Nissen, "A Field Study of the Chimpanzee," 10.

Yale.[79] Among them, Lorma died from a hook-worm infection shortly after her arrival.[80] Boyd (formerly called Boy) and Kindia were transferred to John F. Fulton's physiology laboratory at Yale, where they lost their lives in lobotomy experiments.[81] Moussa (also called Moss) "served for seven years as experimental subject before being transferred to Yerkes' laboratory in Florida, where he died from dysentery."[82] In the meantime, Kambi and Bimba were enrolled in studies on cooperation and problem-solving conducted by Nissen and his colleague Meredith P. Crawford (1910–2002).

Like Nissen, Crawford earned a PhD in Comparative Psychology from Columbia University, and had been hired by Yerkes right after his PhD to work in New Haven, and then later at Orange Park, Florida.[83] Crawford collaborated with Nissen on a study of food-sharing behaviour, which included three chimpanzees from the Pastoria convoy: Moos, Bimba, and Kambi, aged five-and-one-half-years at that time. Unlike Bimba and Moss, Kambi was "not employed as subject," but "included for a complete record of social environment." Nissen and Crawford concluded that "food-sharing [was] influenced strongly by previously established social relationships." They voiced the idea that one could "attribute to chimpanzee what is called altruism in man," albeit with reservations as "the psychological nature of 'altruism' [was] at present time very imperfectly understood."[84]

Cooperative behaviour was also at stake in other experiments that Crawford conducted, this time alone. Drawing on Kimball Young's *An Introductory Sociology* (1934), Crawford defined "cooperation" as "sympathy, identification of one person with another, and teamwork, the

[79] Henry Nissen to Robert Yerkes, 9 July1930, group 659, series II, box 36, folder 689, RMYP. See also Nissen, Notebook #5, p. 31 (recto), box 77, EU14, YNPRCR.

[80] Henry Nissen to Robert Yerkes, 15 July 1930, group 659, series II, box 36, folder 689, RMYP.

[81] Wesleyan University, "Kindia," The First 100 Chimpanzees, accessed June 15, 2023, http://first100chimps.wesleyan.edu/kindia.html.

[82] Wesleyan University, "Moos," The First 100 Chimpanzees, accessed June 15, 2023, http://first100chimps.wesleyan.edu/moos.html; Robert Yerkes to Henry Nissen, 9 May 1933, group 569, series I, box 36, folder 692, RMYP.

[83] Dewsbury, *Monkey Farm*, 208.

[84] Henry W. Nissen and Meredith P. Crawford, "A Preliminary Study of Food-Sharing Behavior in Young Chimpanzees," *Journal of Comparative Psychology* 22, no. 3 (1936): 389 & 415.

coordination and integration of action and not merely additive effort."[85] One group of Crawford's experimental animals included Kambi and Bimba. A good follower of Yerkes, Crawford was interested in animals' life histories, with particular emphasis on social relationships, to help in the decision to pick them for a specific experiment. For instance, he noticed that, "brought from African and being cared for together for more than a year, [Kambi and Bimba] were strongly attached to each other and mutually dependent in a manner characteristic of infant chimpanzees."[86] An anecdote illustrates well the two infants' attachment. In late August 1930, as Nissen was returning to New Haven from holiday, he discovered that Kambi and Bimba were full of lice. After treating them, he put them in separate boxes and they "seemed pretty unhappy about it." Kambi especially screamed a lot.[87] However attached the two young chimps may have been, they did not have the same level of intelligence, according to Crawford. "Bimba has been rated one of the most intelligent animals in the colony," while Kambi was of "phlegmatic temperament" and rated "of average intelligence."[88]

The general plan of the experiment room consisted of a restraining cage and an observation booth, together with the apparatus, which comprised a wooden track on which a heavy box was drawn by two ropes, and an electrical mechanism delivering food to the top of the box. One of the experimental problems required "the pulling in of a weighted box, too heavy for a single animal to move, with two ropes." It's important to note that "the observer, retired to the booth [...] so that he was completely concealed from the animals."[89] In such a situation, he may observe both solicitational and directive gestures. For example, Bimba "reached directly for Bula's neck or shoulders, pushed her lightly toward the front of the cage and sometimes patted her on the back or shoulder when she did not immediately begin to pull."[90] As Crawford concluded: "[T]he intensity and duration of solicitation probably depended on factors such as

[85] Meredith P. Crawford, "The Cooperative Solving of Problems by Young Chimpanzees," *Comparative Psychology Monographs* 14, no. 2, serial no. 68 (June 1937): 2.

[86] Crawford, "The Cooperative Solving of Problems," 13.

[87] Henry Nissen, Notebook #5, p. 76 (verso) & p. 75 (verso), box 77, EU14, YNPRCR.

[88] Crawford, "The Cooperative Solving of Problems," 14.

[89] Ibid., 16–18 & 6–7.

[90] Meredith P. Crawford, "The Cooperative Solving by Chimpanzees of Problems Requiring Serial Responses to Color Cues," *The Journal of Social Psychology* 13, no. 2 (1941): 271.

responsiveness of the partner, degree of friendship between partners, strength of individual motivation toward food, dominance relationship, and intelligence."[91]

Crawford also shot some of his experiments. Not only did the filmed scenes play a role as scientific proof testifying to cooperative behaviour among chimpanzees, but they also were staged to endear these apes to a general public, and thereby promote the work done at the Yale Laboratories of Primate Biology. The footage of Crawford's experiments even crossed the Atlantic. For instance, it was used in a six-minute documentary titled "À quoi pensent les animaux?" ("What Animals Think Of?") released in July 1964 by the French National Institute of Audiovisual Media.[92] After showing chimpanzees able to adjust sticks of different lengths to reach a food placed out of their cages, the final scene featured another form of intelligent behaviour: cooperative behaviour. Dubbed with a swing-band music, Kambi, Bimba, Alpha, and Bula cheerfully made the demonstration that "there is strength in numbers," making Crawford's experiments the highlight of the documentary. No doubt, next to the smart dancing bees (another example of animal behaviour featured in the documentary), the Guinean chimpanzees helped convince the watcher that symbolic language and intelligence were not only human prerogatives.

When they arrived in Orange Park, in the mid-1930s, Kambi and Bimba would be more callously manipulated by their experimenters than they had been in New Haven. Bimba was also put in a breeding programme, and she eventually mothered Banka whose father was Frank.[93] As far as Kambi was concerned, she was enrolled into a toxicology study under the supervision of Sidney Durward Shirley Spragg (1909–1995). Spragg completed his BA and MA in Psychology at the University of Washington, entered a PhD in Experimental Psychology at Yale University, and then carried on with a post-doctoral research at Orange Park. Between 1935 and 1937, he conducted a study on morphine addiction in chimpanzees to "determin[e] whether or not an infrahuman organism can develop a genuine addiction to morphine." Spragg was convinced that chimpanzees were the right experimental animals to understand certain basic problems of

[91] Crawford, "The Cooperative Solving of Problems," 86.

[92] Lumni Enseignement, "L'intelligence animale," Video "Visa pour l'Avenir. À quoi pensent les animaux?," accessed June 15, 2023, https://enseignants.lumni.fr/videos/liste?search=intelligence%20animale&fiche-media=00000000628.

[93] Wesleyan University, "Bimba," The First 100 Chimpanzees, accessed June 15, 2023, http://first100chimps.wesleyan.edu/bimba.html.

drug addiction, because of their specific physiological affinity with humans. Four chimpanzees, three males and one female, served as subjects for his chronic morphinism experiments. The female was Kambi, whom Spragg described as "[a]n adolescent, sexually immature animal [at that time she was about eight-year-old] quiet and rather stolid; in good health."[94]

It is interesting to note that "[p]rior to receiving morphine injections, all subjects were trained to cooperate voluntarily in receiving hypodermic injections of physiological saline solution." For instance, the animals were "trained to lean across the experimenter's leg, when he placed his foot up on the bench," or to "tolerate being stabbed on the end of the finger with a small lancet."[95] Ironically then, the cooperative studies Kambi had been through in New Haven somehow helped her to develop a behaviour which eased the morphine addiction experimental protocol.

Spragg was satisfied with Kambi's reaction to the experimental protocol: she showed the prominent effects of the early daily doses of morphine: a great deal of scratching, picking at hair, loose bits of skin and scabs, and then loss of appetite.[96] One of the most marking symptoms was a withdrawal from social interests and contact. As Spragg's laboratory notes disclosed:

> April 9, 1937. [...] she used to become quite excited when led close to the cages of other animals, would grab at them, and frequently throw handfuls of pine-needles at them. Since she has been getting the drug this has entirely disappeared, and she pays very little, if any, attention to other animals. The drug seems to have made her more 'introverted', if one may use that term for a chimpanzee![97]

Finally, on 28 April 1937, the first signs of dependence appeared. Kambi "seem[ed] to be restless. She sat near the experimenter, and licked the soles of her feet, then the wire mesh, [...] she [was] also yawning frequently."[98]

Spragg demonstrated not only the possibility of addiction in non-human species, but also resemblance of symptoms (observed in

[94] S. D. Shirley Spragg, "Morphine Addiction in Chimpanzees," *Comparative Psychology Monographs* 15, no. 7, serial no. 79 (April 1940): 17–18 & 22–23.

[95] Spragg, "Morphine Addiction," 40.

[96] Ibid., 28 & 32.

[97] Ibid., 53.

[98] Ibid., 57.

chimpanzees) in many ways with those reported for human addicts, behaviourally as well as physiologically. Against a prevalent theory of the time regarding the importance of societal factors in the development of addiction, Spragg concluded that morphine addiction was fundamentally determined by physiological factors.[99] Clearly then, Kambi's sentient and captive body proved a good model to help him demonstrate the biological nature of morphine addiction. Spragg's studies with Kambi also chimed with Yerkes' intent "to convert the [chimpanzee] into as nearly ideal subject for biological research," and Yerkes' hope that "the eventual success [of such research] might serve as an effective demonstration of the possibility of re-creating man himself in the image of a generally acceptable ideal."[100] These wishes eventually fitted into Yerkes' larger belief that eugenics and human engineering could provide answers to the disorder of modern society.[101]

In 1941, Kambi (and also Bimba) was used in experiments on chimpanzee laterality. The study was conducted by Glen Finch, a researcher at Orange Park, on a cohort of thirty chimpanzees. After 800 trials, in four test situations, Finch concluded that "detected right- and left-handedness were almost equally distributed in the group of animals."[102] This contrasted with the dominance of right-handedness so manifest in humans. Commenting on Finch's results Yerkes, who had been forced to learn to write right-handed, provocatively wrote: "[unlike chimpanzees] humans [...] [were] slaves to convention, having surrendered [their] boasted freedom of individual choice, decision, and direction of effort." He concluded that "the free natural development observable in the chimpanzee [was] preferable to the custom-imposed handedness development of civilized man."[103]

A few years before she died, Kambi was enrolled into studies bearing on neuroses under the supervision of the Canadian psychologist Donald Hebb (1904–1985). Hebb presented a wide-ranging discussion of spontaneous neuroses in chimpanzees and humans and developed a theory for them. Looking at Kambi, Hebb identified a pattern of mood swings,

[99] Ibid., 128.

[100] Robert M. Yerkes, *Chimpanzees: A Laboratory Colony* (New Haven, CT: Yale University Press, 1943).

[101] Haraway, *Primate Visions*, 59–64.

[102] Glen Finch, "Chimpanzee Handedness," *Science* 94, no. 2431 (August 1, 1941): 118.

[103] Yerkes, *Chimpanzees*, 114–115.

between periods of euphoria and profound depression.[104] Nissen had been aware of Kambi's changing emotional state, and of her shift, when adolescence came, from introversion to emotional instability.[105] In the post-adolescence period, Hebb noticed that her "troughs of depression were deeper, and the crests of euphoria less high than before adolescence." Kambi could be extremely unresponsive to the environmental events and show an extreme lack of spontaneous activity. For Hebb, this echoed "in many ways the objective picture of human depression." Let's note also that Kambi got spayed at the age of 12 to check if the removal of the ovaries would impinge on her mood. As Hebb justified, "[S]he was chosen for the experimental operation [...] since she was a useless breeder." No apparent effect was observed after operation; she was as "crazy" before as after the operation, in the words of the caretaking staff.[106] This last description certainly testifies to the damaging effects of the invasive experiments, such as the morphine addiction study, had on the already psychologically fragile Kambi. This must surely leave us to question how Nissen, who supposedly cared so much for Kambi, bringing her safe and sound to the US, allowed her to be used in experiments which would devastate her physically and mentally. Kambi, like Bimba two years earlier, died from dysentery in January 1946.[107]

Conclusion: What Knowledge Do We Gain from Making Animals Visible in History?

Rose and Kambi are two examples chosen from the thousands of chimps who passed through the Pasteur Institute in Kindia during the colonial period. It's no easy task to bring the lives of those animals into the narrative as experiencing beings. Rose, Kambi, and Bimba were snatched from the forest, even stolen from their family as babies to be caged; they lost their agency and freedom and were experimented upon, and so much more. How can we capture that? The direct evidence is scant, but knowing

[104] Dewsbury, *Monkey Farm*, 202.

[105] Nissen cited by Donald O. Hebb, "Spontaneous Neurosis in Chimpanzees: Theoretical Relations with Clinical and Experimental Phenomena," *Psychosomatic Medicine* 9, no. 1 (January-February 1947): 5.

[106] Hebb, "Spontaneous Neurosis in Chimpanzees," 5–7.

[107] Wesleyan University, "Kambi," The First 100 Chimpanzees, accessed June 15, 2023, http://first100chimps.wesleyan.edu/kambi.html.

about these beings, I have tried to infer and tell their stories "from beneath."

Exploring the biographies of Rose and Kambi helps us to understand how chimpanzees were co-constitutive of the establishment of bacteriology and primatology in the early twentieth century. These biographies indeed illustrate the multiple uses made of the chimpanzees in the laboratory—Kambi's being a paradigmatic case as she was subjected to both psychological and biomedical experiments—and the ways in which the chimpanzees could play a role in the scientists' recognition and public image. Rose embodied the success of the TB trials at Pastoria, while Kambi was associated with Nissen's successful field expedition—which acted as a trigger for Nissen's career—as well, and she participated in the promotion of the Yale Laboratories of Primate Biology in the 1930s–1940s. In the laboratory, Rose and Kambi were forced to give the most of themselves until their bodies gave up. While the average chimpanzee life expectancy in captivity is thirty to forty years, Rose died at the age of eleven and Kambi at the age of seventeen. By bringing attention to the kidnapped primates who were observed, manipulated, and abused in the Kindia's laboratory and the Yale Laboratory of Primate Biology, this chapter aims also to pay tribute to animals who died in the name of science.

The examination of Rose and Kambi's lived experiences provides insight into the ways in which ape knowledge circulated and got internationalized between France and Guinea and Guinea and the US, as well as evidenced for different research and political agendas. Calmette envisaged apes as commodities that could yield knowledge about human diseases, while Yerkes was dedicated to the study of chimpanzees themselves and aimed at fashioning the chimpanzee into an "ideal experimental subject" which would yield knowledge that would ultimately serve that of man. For Calmette chimpanzees were also an asset for the French Empire: they could be used as a bargaining chip to attract foreign scientists and funding. For instance, Calmette expected "once Mr Nissen had emptied the Kindia chimpanzee reserve [...] to get the Rockefeller Foundation or Yale University to come to our rescue a little."[108] This indeed happened: in August 1930, Yerkes sent a cheque of 25,000 francs on the behalf of Yale

[108] Albert Calmette to Robert Wilbert, 9 July 1930, AIP-Kindia.

University "for support of work in the Kindia laboratories," and the donation of the animals to supply his American laboratories.[109]

The study of the life trajectories of Rose, Kambi, and Bimba also enables us to make visible another actor who has long been underrated, if not altogether ignored by scientific literature—the local field guides, laboratory assistants, and caretakers. I have tried to integrate the experiences of colonized people in the narratives of these chimpanzees and demonstrate, however imperfectly, the importance of the Guinean knowledge and skills in the caring and handling of chimpanzees in the laboratory. By retrieving the role played by the Guineans in the production of Western scientific knowledge on apes, I have challenged a historiography, which undervalues, or simply overlooks, it by relying almost exclusively on colonial scientific discourses.

Finally, exploring the stories of Rose and Kambi helps to demonstrate that the colonial process often consisted in blurring the frontier between colonized humans and animals. For instance, at Pastoria, Rose and her fellow companions were "educated through a permanent contact with civilized man," in much the same way the indigenous population was controlled by the colonists. Another example testifies for the control, if not exploitation, of both animals and colonized humans by the colonials: chimpanzee hunting.[110] Rose was captured by Guinean villagers supervised by a colonial hunter for the purpose of Western science, while Kambi was captured by accident by Nissen with the help of his field team, itself assigned to him by French colonials. Then, using both animals and colonized humans as a lens to explore colonial science helps to retrieve the ways in which scientific discourses and practices often intertwined with racist discourse and to identify the particular experiences of domination of both animals and colonized people, and the extent to which they could overlap.

[109] Robert Yerkes to H. J. Ostrander, 8 August 1930, folder "French Guinea Expedition," box 77, EU14, YNPRCR.

[110] On that point, see Leblan, "Une frontière coloniale homme/animal," 412–425.

Bibliography

Unpublished Sources

Archives de l'Institut Pasteur, Paris (AIP).
Archives de l'Institut Pasteur de Kindia, Guinea (AIP-Kindia).
Archives Nationales de Guinée, Conakry, Guinea (ANG).
Robert Mearns Yerkes Papers, Sterling Memorial Library, Yale University, New Haven, CT (RMYP).
The National Archives. Public Record Office, Kew, Richmond, Medical Research Council (PRO).
Yerkes National Primate Research Center Records, Stuart A. Rose Manuscript Archives and Rare Book Library, Emory University, Atlanta, GA (YNPRCR).

Published Sources

Baratay, Éric. *Animal Biographies: Toward a History of Individuals.* Translated by Lindsay Turner. Athens, GA: University of Georgia Press, 2022.
Berdah, Delphine. *Abatttre ou vacciner. La France et le Royaume-Uni en lutte contre la tuberculose et la fièvre aphteuse (1900–1960).* Paris: Éditions de l'École des Hautes Études en Sciences Sociales, 2018.
Bonah, Christian. *L'Expérimentation humaine. Discours et pratiques en France 1900–1940.* Paris: Les Belles Lettres, 2007.
Calmette, Albert. "Les missions scientifiques de l'Institut Pasteur et l'expansion coloniale de la France." *Revue scientifique* 50, no. 5 (February 3, 1912): 129–133.
Calmette, Albert. "Sur l'utilisation des singes en médecine expérimentale: le laboratoire Pasteur de Kindia (Guinée française)." *Bulletin de la Société de pathologie exotique et de sa* filiale de l'Ouest africain 17 (1924): 10–19.
Calmette, Albert. "L'état actuel de nos connaissances sur la vaccination antituberculeuse." *Revue vétérinaire et Journal de médicine vétérinaire et de zootechnie réunis* LXXIX (March 1927): 142–152.
Carde, Jules. "Actes du Gouvernement Général. Arrêté du Gouverneur Général fixant les conditions dans lesquelles peuvent s'opérer la capture et l'exportation des chimpanzés." *Journal Officiel de la Guinée française* 23, no. 539 (October 1, 1923): 315.
Carmichael, Leonard. "Henry Wieghorst Nissen February 5, 1901–April 27, 1958." *Biographical Memoirs, National Academy of Sciences* 38 (1965): 205–222.
Cassidy, Angela, Rachel Mason Dentinger, Kathryn Schoeffert, and Abigail Woods. "Animal Roles and Traces in the History of Medicine, c.1880–1980." *British Journal for the History of Science Themes* 2 (2017): 11–33.

Corbin, Alain. *The Life of an Unknown: The Rediscovered World of a Clog Maker in Nineteenth-Century France.* Translated by Arthur Goldhammer. New York: University of Columbia Press, 2001.

Crawford, Meredith P. "The Cooperative Solving of Problems by Young Chimpanzees." *Comparative Psychology Monographs* 14, no. 2, serial no. 68 (June 1937): 1–88.

Crawford, Meredith P. "The Cooperative Solving by Chimpanzees of Problems Requiring Serial Responses to Color Cues." *The Journal of Social Psychology* 13, no. 2 (1941): 259–280.

Dewsbury, Donald A. *Monkey Farm: A History of the Yerkes Laboratories of Primate Biology, Orange Park, Florida 1930–1965.* Lewisburg, PA: Bucknell University Press, 2006.

Donatien, A. "'Pastoria', Centre de recherches biologiques et d'élevage de singes (Institut Pasteur de Kindia, Guinée française)." *Revue vétérinaire et Journal de médicine vétérinaire et de zootechnie réunis* LXXXIII (September 1931): 528–531.

Finch, Glen. "Chimpanzee Handedness." *Science* 94, no. 2431 (August 1, 1941): 117–118.

Haraway, Donna. *Primate Visions: Gender, Race, and Nature in the World of Modern Science.* New York and London: Routledge, 1989.

Hebb, Donald O. "Spontaneous Neurosis in Chimpanzees: Theoretical Relations with Clinical and Experimental Phenomena." *Psychosomatic Medicine* 9, no. 1 (January-February 1947): 3–19.

Honoré, Fernand. "Les singeries de l'Institut Pasteur à Kindia et à Paris." *L'Illustration* no. 4390, (Avril 23, 1927): 407–409.

Leblan, Vincent. "Une frontière coloniale homme/animal: captures de chimpanzés et fabrique des identités raciales à l'Institut Pasteur de Kindia, Guinée française, 1920–1930." In *Le Spécimen et le collecteur. Savoirs naturalistes, pouvoirs et altérités (XVIIIe–XXe siècle)*, edited by Dominique Juhé-Beaulaton and Vincent Leblan, 385–427. Paris: Publications scientifiques du Muséum national d'histoire naturelle, 2018.

Lumni Enseignement. "L'intelligence animale." Video "Visa pour l'Avenir. À quoi pensent les animaux?" https://enseignants.lumni.fr/videos/liste?search=intelligence%20animale&fiche-media=00000000628. Accessed June 15, 2023.

Montgomery, Georgina M. "'Infinite Loneliness': The Life and Times of Miss Congo." *Endeavour* 33, no. 3 (2009): 101–105

Montgomery, Georgina M. *Primates in the Real World: Escaping Primate Folklore and Creating Primate Science.* Charlottesville and London: University of Virginia Press, 2015.

Montgomery, Georgina M. and Linda Kalof. "History from Below: Animals as Historical Subjects." In *Teaching the Animal: Human-Animal Studies Across the Disciplines*, edited by Margo DeMello, 35–47. Brooklyn, NY: Lantern Books, 2010.

Munz, Tania. "'My Goose Child Martina': The Multiple Uses of Geese in the Writings of Konrad Lorenz." *Historical Studies in the Natural Sciences* 41, no. 4 (Fall 2011): 405–446.

Nelson, Amy. "The Legacy of Laika: Celebrity, Sacrifice, and the Soviet Space Dogs." In *Beastly Natures: Animals, Humans and the Study of History*, edited by Dorothee Brantz, 204–224. Charlottesville and London: University of Virginia Press, 2010.

Nissen, Henry W. "A Field Study of the Chimpanzee: Observations of Chimpanzee Behavior and Environment in Western French Guinea," *Comparative Psychology Monographs* 8, no. 1, serial no. 36 (December 1931): 1–122.

Nissen, Henry W. and Meredith P. Crawford. "A Preliminary Study of Food-Sharing Behavior in Young Chimpanzees." *Journal of Comparative Psychology* 22, no. 3 (1936): 383–419.

Perrigault, Jean. *Bêtes et gens de brousse*. Paris: Éditions de la Nouvelle Revue Critique, 1931.

Rode, Paul. "'Pastoria', centre de recherches sur les singes en Guinée française." *La Terre et la vie* 7, no. 4 (July-August 1937): 109–116.

Spragg, S., D. Shirley. "Morphine Addiction in Chimpanzees." *Comparative Psychology Monographs* 15, no. 7, serial no. 79 (April 1940): 1–132.

Thomas, Marion. "Between Biomedical and Psychological Experiments: The Unexpected Connections between the Pasteur Institutes and the Study of Animal Mind in the Second Quarter of the Twentieth Century France." *Studies in History and Philosophy of Biological and Biomedical Sciences* 55, no. 1 (2016): 29–40.

Velmet, Aro. "The Making of a Pastorian Empire: Tuberculosis and Bacteriological Technopolitics in French Colonialism and International Science, 1890–1940." *Journal of Global History* 14, no. 2 (2019): 199–217.

Velmet, Aro. *Pasteur's Empire: Bacteriology and Politics in France, Its Colonies, and the World*. Oxford and New York: Oxford University Press, 2020.

Wesleyan University. "Bimba." The First 100 Chimpanzees. http://first100chimps.wesleyan.edu/bimba.html. Accessed June 15, 2023a.

Wesleyan University. "Kambi." The First 100 Chimpanzees. http://first100chimps.wesleyan.edu/kambi.html. Accessed June 15, 2023b.

Wesleyan University. "Kindia." The First 100 Chimpanzees. http://first100chimps.wesleyan.edu/kindia.html. Accessed June 15, 2023c.

Wesleyan University. "Moos." The First 100 Chimpanzees. http://first100chimps.wesleyan.edu/moos.html. Accessed June 15, 2023d.

Wilbert, Robert. "Expériences de vaccination des singes contre la tuberculose par le BCG." *Annales de l'Institut Pasteur* 39, no. 8 (August 1925): 641–651.

Wilbert, Robert. "'Pastoria'." In *Comptes rendus des séances/communications, tome XII 1928-1929*, edited by Académie des sciences coloniales, 481-488. Paris: Société d'Éditions géographiques, maritimes et coloniales, 1932.

Wilbert, Robert and Maurice Delorme. "'Pastoria', Centre de recherches biologiques et d'élevage de singes, Institut Pasteur de Kindia, Guinée française." *Bulletin de la Société de pathologie exotique et de ses filiales de l'Ouest africain et de Madagascar* 24 (1931): 131–149.

Woods, Abigail, Michael Bresalier, Angela Cassidy, and Rachel Mason Dentinger. *Animals and the Shaping of Modern Medicine: One Health and Its Histories.* London: Palgrave Macmillan, 2017.

Yerkes, Robert M. Foreword to Henry W. Nissen, "A Field Study of the Chimpanzee: Observations of Chimpanzee Behavior and Environment in Western French Guinea." *Comparative Psychology Monographs* 8, no. 1, serial no. 36 (December 1931): i–vi.

Yerkes, Robert M. "The Life History and Personality of the Chimpanzee." *The American Naturalist* 73, no. 745 (March-April 1939): 97–112.

Yerkes, Robert M. *Chimpanzees: A Laboratory Colony.* New Haven, CT: Yale University Press, 1943.

Yerkes, Robert M. and S. D. Shirley Spragg. "La mesure du comportement adapté chez les chimpanzés." *Journal de psychologie normale et pathologique* XXXIV (1937): 449–475.

CHAPTER 9

Animals and Colonial Indian Archives: Locating Nonhuman Agency and Subjectivities

Heeral Chhabra

Introduction

This chapter engages the question: why are animals ubiquitous in the history of colonial India but do not feature in historians' accounts as beings with subjectivity and agency? I raise this question with regard to colonial Indian archives. Despite having pervasive presence in the official colonial archives of India, historians seem reluctant to attend to animals as

H. Chhabra (✉)
Department of History, University of Liverpool, Liverpool, UK

M. J. Glover, L. Mitchell (eds.), *Animals as Experiencing Entities*, The Palgrave Macmillan Animal Ethics Series,
https://doi.org/10.1007/978-3-031-46456-0_9

historically participative beings endowed with agency and subjectivity.[1] Drawing on Amanda Rees' work, agency here refers to the ability to impact historical processes rather than self-conscious knowledge of the ability to impact history.[2] By subjectivity I specifically mean an individual animal's unique subjective aspects such as their individual temperament and intelligence.

The reasons for this assumed inability of animals to possess agency and subjectivity in historians' accounts of colonial India, I assert, can be located in two intersecting domains. One is the *historian-centred* domain—located in the assumptions and understanding about the concept of agency with which the historians approach the archives. The second is the *archive-centred* domain—located essentially in the nature of the archives themselves. A careful analysis of both these domains is essential to delineating the reasons for animals' exclusion as subjective beings with agential capacities, and also as a way of motivating for a conceptual basis of including animals in colonial histories via official colonial sources.

Far from being absent or scarce, a quick perusal of Indian historical sources (ranging from literary creations, iconographic representations, official records, oral sources and so on) across numerous periods brings out the widespread presence of animals within those sources.[3] When spe-

[1] There are some partial exceptions, such as Mahesh Rangarajan, "Animals with rich histories: The case of the lions of Gir Forest, Gujarat, India," *History and Theory* 52, no. 4 (2013): 109–127; Deepak Bhat Dundi, Istvan Praet, and Garry Marvin, "Good, Quarrelsome, Bad: Animal Agency and Human-Elephant Interactions in the Western Ghats, India," *Frontiers in Conservation Science* 4 (2023), https://doi.org/10.3389/fcosc.2023.1142333; Nicholas H.A. Evans, "Blaming the rat? Accounting for plague in colonial Indian medicine," *Medicine Anthropology Theory* 5, no. 3 (2018): 16–42; Niharika Dinkar, "Seeing the Elephant: Animal Spectatorship and the Imperial Gaze in Colonial India," *South Asian Studies* 37, no. 2 (2021): 117–136; Pratik Chakrabarti, "Beasts of Burden: Animals and laboratory research in colonial India," *History of Science* 48, no. 2 (2010): 125–151; Heeral Chhabra, "Animal labourers and the law in colonial India," *South Asia Research* 39, no. 2 (2019): 166–183.

[2] Amanda Rees, "Animal Agents: Can animals shape their own lives? Or the course of history?" *Aeon Essays*, February 26, 2018, https://aeon.co/essays/can-animals-shape-their-own-lives-or-the-course-of-history.

[3] Shibani Bose in the introduction of her book *Mega Mammals in Ancient India: Rhinos, Tigers, and Elephants* samples out significance of animals' presence as faunal remains (sources themselves!) crucial to understand not only the history of specific animal species but also their linkages with the surrounding ecosystems, representation in seals of Harappan civilisation and latter-day iconography, in corpus of Sanskrit, Pali, Prakrit, classical Western accounts and so on. For more details refer to Shibani Bose, *Mega Mammals in Ancient India: Rhinos, Tigers, and Elephants* (Delhi: Oxford University Press, 2020).

cifically focussed on official colonial sources, animals are found in abundance and in most departmental records including Home, Military, Legislative, Municipal, Princely State Records and so on.[4] Furthermore, the role of animals in the creation, propagation and maintenance of colonial rule has been immense.[5] Despite such an expansive role in making of the colonial realities, animals' treatment by historians has been very limited; that is to say, even though animals have featured in historian's narratives, the agential and subjective realities of animals in making of the past have been grossly overlooked. This isn't to say that animals haven't been noticed. The presence of nonhuman animals has usually been invoked as corroborative facts to strengthen larger arguments but very few historical works have weaved narratives around animals themselves. Even when animals are of central historical importance, they are rarely viewed as significant historical beings with the potential to impact historical changes. In order to understand this neglect of animals' potential as historically significant actors, the nature of historian's focus needs to be probed further.

[4] For instance, detailed references to animals are found in files pertaining to animal breeding, animal grazing, trade in animals, slaughtering of animals, animal sacrifices, cases of snake bites and poisoning, destruction of wild life, entire branches on fisheries—availability of existing varieties of fishes and introduction of new ones—details about camels and horses for military purposes, tackle presence of 'stray' dogs and so forth. Some examples include Prevention of Cruelty to Animals Act discussed extensively in Legislative Department, No. 607–639, Part A, April 1890 and Destruction of Street Dogs discussion in Home Department, Police Branch, File No. 95-105, Part A, June 1902—both available at National Archives of India (NAI).

[5] For more details refer to works of Harriet Ritvo, *The Animal Estate: The English and Other Creatures in the Victorian Age* (Cambridge, Massachusetts: Harvard University Press, 1987); Alfred Crosby, *Ecological Imperialism: The Biological Expansion of Europe, 900–1900* (New York: Cambridge University, 1986); Aaron Herald Skabelund, "Animals and Imperialism: Recent Historiographical Trends," *History Compass* 11, no. 10 (2013): 801–807 and *Empire of Dogs: Canine, Japan, and the Making of the Modern Imperial World* (Ithaca and London: Cornell University Press, 2011); Rohan Deb Roy, "Nonhuman Empires," *Comparative Studies of South Asia, Africa and the Middle East* 35, no. 1 (2015): 66–75, and *Malarial Subjects: Empire, Medicine and Nonhumans in British India, 1820–1909* (Cambridge: Cambridge University Press, 2017); Hevia, *Animal Labour and Colonial Warfare* (Chicago: University of Chicago Press, 2018); Jonathan Saha, *Colonizing Animals* (Cambridge: Cambridge University Press, 2021) and "Colonizing Elephants: Animal Agency, Undead Capital and Imperial Science in British Burma", *BJHS* Themes, 2 (2017): 169–189; Sandra Swart, *Riding High: Horses, Humans and History in South Africa* (Johannesburg: Wits University Press, 2010); H. Chhabra, 'Animal Labourers and the Law in Colonial India', *South Asia Research* 39, no. 2 (2019): 166–183.

The ability of the historian to attend to the 'animal turn' within history is much determined by the questions that are asked of the sources, and these questions in turn are much influenced by historians' conceptual frameworks and how sources are examined. It thus becomes crucial to critically investigate some of these frameworks and categories, such as agency, with which historians approach and explore the historical presence of animals and consequently determine animals' positions in history.

In order to do so, this chapter will be divided into three broad sections. The first section discusses varied understandings of the term agency and its conceptualisation, which most often precludes nonhumans as significant agential beings. The second section highlights the issues that have been central in attending to the colonial archives as 'repositories of the past', and their links with power, knowledge construction and marginality. It strives to locate the nonhumans within this matrix. The third section looks ahead to illustrate ways through which colonial archives could be analysed via animal-centric research questions. The third section focusses on correspondence of the Commissariat (Transport) Division of the Military officials in 1886 in British India about rules for purchasing of transport animals, which resulted in a set of instructions titled *Purchase of Transport* that delineates extensive details about animals who were to be used for transport. This set of instructions, written by R.C. Low (Colonel, Commissary General for Transport), was meant for officials responsible for the purchase of animals. These documents are currently held in the National Archives of India (NAI).[6] These remarkable instructions detail the types of animals and animal specificities that were important when selecting and purchasing animals for use in colonial military transport. It provides insight into how animals were selected and used and also into how colonists conceptualised animals and contributed to the formation of animals' identities.

All three sections seek to attend to the following questions in various ways. How did archives of colonial India record nonhuman animals? What status was accorded to animals in Indian colonial archives? Is the absence of animal subjectivity/agency assumed in these records, or are animals' specificities such as species-characteristic capacities, sensory faculties,

[6] This isn't an isolated attempt at creating instructions or manuals for recruiting transport animals by the military department. Detailed manuals such as Indian Army Service Corp Training, Part II, Animal Transport (1927) also provide extensive details about animal transport. This is available at the British Library India Office Records IOR/L/MIL/17/5/2186.

cognitive and experiential capabilities recognised? These questions are explored by focussing on detailed instructions, written by Low, pertaining to purchase of transport animals for the purposes of military transportation. The aim is to uncover colonial understandings of working animals and the process of according them property-like status (through their segregation and use) which excluded a recognition of their sentience and agency.

Question of Agency

The growing scope of the 'animal turn' saw rise of the 'animal question' in various disciplines, and history did not remain untouched. With nonhumans in some ways gaining long overdue importance, one of the questions often raised in this regard is: do animals have agency? This concern is imperative because 'agency', as a concept, has been crucial in determining the worth or importance of an historical being or actor. David Shaw points out that for many "possessing agency brings to the fore the very purpose and point of history. For some, to deny agency seems almost to deny historical significance."[7] In other words, to acknowledge agency was to enable historicity. So, those beings who had agency could be the subjects of history.

The notion of agency, however, has shifted over time. Susan Crane describes how the term agency came up in the 1970s and encapsulated a critique of structuralism.[8] Amanda Rees details the way in which the concept of agency was shaped and gained momentum, emphasising that "the concept of agency is itself problematic, of course, since it emerged as part of the 'universalising' humanism of the Enlightenment, when to be human meant to be white, male and financially independent".[9] According to

[7] David Gary Shaw, "A Way with Animals," *History and Theory* 52, no. 4 (2013): 7.

[8] Structuralist analyses of culture did not recognise the consciously counter-hegemonic actions of individuals and structuralism did not account for the creative and subversive interventions that people can make even as their societies 'make' them. From V. Despret, "From Secret Agents to Interagency," *History and Theory* 52, no. 4 (2013): 30.

[9] In another article Rees opines that "historically it has been far too tightly tied to the notion of liberal selfhood. It is one that emerged within a specific and contingent historic moment, but has, through its location within the 'universalizing' humanism of the Enlightenment, retained its purchase on the intellectual imagination, particularly in relation to groups that benefit disproportionately from the unequal distribution of economic, social and political power". Amanda Rees, "Animal agents? Historiography, theory and the history of science in the Anthropocene," *The British Journal for History of Science* 2, (2017): 9.

Vinciane Despret, "[T]he very notion of agency still conveys its classic understanding as intentional, rational, and premeditated, and is still embedded in humanist and Christian conceptions of human exceptionalism".[10]

Such an understanding of agency had very evident drawbacks, i.e., apart from leaving out the marginalised human groups from historical significance, it also did not attend to nonhumans,[11] and in order to be more inclusive in its reach the notion had to be expanded in its meaning and content. But when it comes to including nonhuman animals into the domain of agency there are many hurdles to be crossed conceptually and in praxis. According to Chris Pearson, "[H]istorians treat agency as a uniquely human attribute, arguing that animals lack the cognitive abilities, self-awareness, and intentionality to be agents".[12] And it is precisely here that the problem arises when we talk about animals having historical agency. The reasons can be located in the assumed human exclusivity and keeping human-centric yardsticks as the determining measures of what constitutes agency. Rees expands on the reasons: the issue is not merely about agency or an ability to influence the world, but the capacity for self-conscious agency, i.e., the knowledge of doing so deliberately and reflectively. Rees writes:

> Humans have the capacity to act as agents, because they are considered to know what they are doing and why they are doing it. But even though non-humans possess individuality and consciousness, too, the absence of self-consciousness among them has generally been taken to preclude agency.[13]

However, these notions have been increasingly questioned, resulting in the scope of agency being expanded. Scholars are not only challenging and questioning the existing basis of acknowledging agency but suggesting new nomenclatures to attend to nonhuman agency. At the same time

[10] Despret, "From Secret Agents to Interagency", 29.

[11] Rees notes that "we need to be aware that this attention to animal agency was itself produced at a specific historical juncture—in the early twenty-first century, in the aftermath of a succession of liberation movements that have progressively democratized public and private space in the West, paving the way for 'histories from below', whether conceptualized in terms of class, race, gender, sex, species or terrain". Rees, "Animal agents? Historiography", 9.

[12] Chris Pearson, "Dogs, History, And Agency," *History and Theory* 52, no. 4 (2013): 128.

[13] Rees, "Animal Agents".

these attempts are not restricting themselves to any specific discipline but are expanding their input and gaze beyond disciplinary boundaries and fostering an interdisciplinary approach.

Attempts at such a revision can be analysed through a two-pronged approach. One of the approaches is to look more closely at the lives of animals themselves, to argue for their capacities which bring forth their distinctness. As highlighted by Rees, works of scholars such as Niko Tinbergen and David Lack brought out the need to know the animals individually.[14] For them if one wished to study or adopt an animal perspective and understand the lived experience of animals then their individual identification in the environment in which they had evolved was the first step.[15] Rees also cites Sandra Swart's work on horses/horstory (horse-stories) which stands out as one crucial example because it attempts to bring in the animal perspective through a combination of interdisciplinary methods.[16] Pearson, in his essay highlighting the role of dogs during World War II, argues that human levels of intentionality are not a precondition of agency.[17] He drew on research into canine psychology to propose that dogs display some degree of intentionality and self-directed action.

The other approach is to provide terminologies which are better equipped to attend to the agency of the nonhuman animals. In order to tackle this issue conceptually, Despret uses the notion of 'agencement' which highlights the *interdependence of agency* as being an inherent feature of agency, and arguing that "[t]here is no agency that is not interagency".[18] The term agencement was also used by Gilles Deleuze but translated by his translator Brian Massoumi as 'assemblages'.[19] Despret, however, retains the term agencement. There is another approach which has made attempts to attend to 'materialities' as another form of nonhumans who have

[14] Both of them were evolutionary biologists. Lack wrote *The Life of the Robin*, dating to 1943, and Tinbergen wrote *Curious Naturalists*, published in 1958. The latter studied wasps and used coloured dots to identity each wasp individually; he then realized that such an exercise had transformed these wasps "into personal acquaintances, whose lives from that very moment became affairs of the most personal interest and concern to me". From "Animal Agents" by Rees. For further details see Niko Tinbergen, *Curious Naturalists* (New York: Doubleday Anchor Books, 1968) and David Lack, *The Life of the Robin* (London: Penguin, 1953).

[15] Rees, "Animal Agents".

[16] Rees, "Animal agents? Historiography", 6.

[17] As argued in Pearson, "Dogs, History, and Agency".

[18] Despret, "From Secret Agents to Interagency", 44.

[19] Despret, "From Secret Agents to Interagency", 38.

agency. Maan Barua points out that 'matter' can seem like an overarching term which can imply both living substances and non-living materials, but strict divisions between them is often eschewed.[20] There are other scholars who have focussed on the potential of materials, associations, relations and networks, suggesting they have an immense influence on agents. Prominent in this regard has been the Actor-Network Theory.[21] Linda Nash argues that we need to recast our definitions of agency to be less about the "self-contained individual confronting the external world" and more about social milieus, such as Bruno Latour's Actor-Network Theory or Tim Ingold's "organism-in-its-environment" perspective.[22] When historians expand definitions of "agency" to include what Donald Worster described as those "independent energies that do not derive from the drives and intentions of any culture", nonhuman animals, from mega-fauna to microbes, become key agents in the unfolding of history.[23]

While substantial headway has been made regarding the issue of agency and nonhuman animals, its applicability and acceptance by historians (other than animal historians) remains underutilised. A significant reason for this is a reluctance to alter the understanding of 'agency', especially the aspect of self-realisation of action and intention. The link agency has with human-centric formulations or the anthropocentric nature of conceptual frameworks (which have already been mentioned) plays a key role in this regard. I would like to extend this further to highlight that the human-centric nature of conceptual frameworks is an engrained feature which often determines two crucial aspects of historical research. First, human-centric conceptual frameworks influence the questions asked of the sources and, second, they influence the way sources are analysed, including sources' vocabularies and concomitant understandings with which

[20] For details look up Maan Barua, "Volatile ecologies: Towards a material politics of human-animal relations," *Environment and Planning A* 46, no. 6 (2013): 1462–1478.

[21] The Actor-Network Theory (Latour and Woolgar 1986, Latour 1987) is "a form of constructivism that rejects the idea of a social determination of scientific knowledge. … The leading thought is that scientific knowledge is an effect of established relations between objects, animals, and humans engaged in scientific practices. … Establishing a scientific belief, theory, or facts comes down, from the point of view of the actor-network theory, to placing these actors in a stable network. In this sense, scientific beliefs, knowledge, theories, and facts are taken to be constructed by translations taking place in established networks." From: W. Detel, "Social Constructivism" in *International Encyclopedia of the Social & Behavioral Sciences*, ed. Neil J. Smelser, Paul B. Baltes (Pergamon, 2001), 14264–14267.

[22] Walker, "Animals and the Intimacy of History," *History and Theory* 52, no. 4, (2013): 51.

[23] Walker, "Animals and the Intimacy of History," 50.

historical narratives are written.[24] Human-centric concepts delimit our purview and gaze (often unknowingly). This limits the expansiveness of the historian's analytical reach—making nonhumans not an obvious category of historical actors capable of agency and subjectivity. What, then, is the solution to this problem? A step forward could be a definite attempt to revisit conceptual frameworks, but more importantly to revisit the archives with broader frameworks which aren't human-exclusive and ask different questions of the sources.

Archives and Animals

The official colonial archives, in the case of modern Indian history—the colonial archives of the British rule in India—form one of the main repositories on which much historical writing about the period is based. These archives are frequently used by researchers. The significance of these sources makes them invaluable for our enquiry about animals.

Far from being absent, nonhuman animals feature abundantly in various departments of these archives (such as Municipal, Judicial, Legislative, Military and other departments). Such presence is also expected given the extent of animals' roles in making and maintaining the empire. Many scholars have researched the role that animals played in making colonial control a thriving reality. Interestingly, the ***animal turn*** in history has also encouraged researchers to probe the presence of those who have hitherto been overlooked. The question here is why the exclusion of animals as agents has occurred despite animals' abundant presence in the sources. A part of the reason can be explained and located in the nature of colonial archives themselves.

Recently scholarship, such as that of Fiona Paisley and Kirsty Reid, has examined the complexity of colonial archives through its creation, process of accumulation, sifting, collation of content matter, accessibility, reproduction and so on.[25] Investigations of many colonial archives by scholars from disciplines such as history, anthropology and archive science have

[24] I would like to add to it that this anthropocentric understanding is further bolstered by the human-centric nature of history as a discipline as well—in its formulation and intention, i.e., to say that history writing is a human endeavour (by the humans and for the humans); however, in the making of history and impact of history the nonhuman animals are by no ways marginal.

[25] Fiona Paisley and Kirsty Reid (ed.), *Sources and Methods in History of Colonialism: Approaching the Imperial Archives* (London: Routledge, 2017), Introduction.

also exposed archives' interwoven relationship between knowledge, power, empire, absences, silences, spaces and identity. The significance of archives as a site has been viewed in various ways, with some authors arguing that archives are a non-passive space.[26] These archives are far from mere repositories of facts, observations and places for collating historical evidences.[27] They are, in fact, sites of forgetting, remembering, history making and power creation.[28]

There are at least two ways in which archival content is used. One is to view and understand the past and another is to use archives for asserting or proving contemporary claims through the understandings and representation of the past.[29] With regard to the colonial state, scholars have highlighted the centrality of information collection and consumption as its integral feature, and archives as a representative of that impulse, to the extent that colonial states are regarded as information states.[30] In many ways, therefore, archives do reflect the nature of colonialism and are reflected by it. And this is very significant to note: the collection of archival material does reflect the contemporaneous trends of the British colonial state and their belief systems. Thus, the logic and intentionality of record collection and organisation are important and have the potential to reveal a lot about the ideas and conceptual frameworks pertaining to animals. The archives generate information and fashion bodies out of European knowledge systems, often neglecting, overlooking or denying recognition of non-European representations.[31] In relation to animals, this means the centrality of the animal-human divide as a philosophical concept and as a dogma (which has been highlighted by many scholars) and its role in determining the status of animals in the archives. In addition, this also facilitates convenient neglect (and negation) of local animal-centred frameworks of colonised societies.[32]

The other use of colonial archives occurs in our contemporary world, where historical knowledge and ideas about the past are utilised to bolster

[26] Reid and Paisley, *Sources and Methods*, 2.

[27] Reid and Paisley, *Sources and Methods*, 1.

[28] Reid and Paisley, *Sources and Methods*, 1.

[29] Reid and Paisley, *Sources and Methods*, 2 and 4.

[30] Reid and Paisley, *Sources and Methods*, Introduction.

[31] Reid and Paisley, *Sources and Methods*, 3.

[32] For a more specific discussion on animals and the archives, refer to Jennifer Bonnell and Sean Kharaj (ed.) *Traces of the Animal Past: Methodological Challenges in Animal History* (Calgary: University of Calgary Press, 2022).

present world sensibilities, i.e., the use of colonial archives for the understanding of colonial past for post-colonial societies.[33] This becomes even more crucial when historical events or episodes are used to bolster identity claims. Thus, the questions asked of the sources are very often determined by the present concerns of present historians. Here the role of retrospection is crucial as this facilitates or gives shape to the colonial state in its contemporaneous presence and also in hindsight. Scholars have focused on aspects regarding the relevance of past to the present and have highlighted links in various domains. With regard to polity, it is often emphasised that archival work is political in creation and consumption.[34] It entails representation of social, economic and other milieus. But the absences and silences are equally important to note for they also have the potential to speak. And, in order to fathom these nuances, there is a definite need to read the archives against the grain, i.e., to question how animals are represented, what their representation says about colonial views on animals, what animals' silencing or absence means.[35] Thus, a critical reading of the archives, that is not limited by being human-centric, is essential.

Given these insights, what potential do colonial archives hold for analysing the nonhuman animals historically? The absence of animals is not the issue within the colonial archives; their neglect is not due to the absence. As Jonathan Saha correctly points out, it is the convenient overlooking of this presence which has caused this neglect.[36] I wish to take this further, to ask what makes this oversight possible or rather convenient? Answers to this may be located at many sites; however, I argue that the nature of colonial archives has made this possible to a certain extent. The way animals are presented within the archives is reflective of the notions of those who recorded and created the records. No wonder therefore that animals feature very much as subsidiary beings mostly regarded as property, objects or commodities—which is very much reflective of the Eurocentric colonial frameworks which essentially believed in a human-animal divide but was also exclusively human-centric in gaze, concern and analysis. These human-centric ideas percolate even into documents (such as military booklets for animal management) which were explicitly con-

[33] Reid and Paisley, *Sources and Methods*, 2

[34] Reid and Paisley, *Sources and Methods*, 2.

[35] A.L. Stoler, *Along the Archival Grain: Epistemic Anxieties and Colonial Common Sense* (Princeton: Princeton University Press, 2009).

[36] This is very well argued in Jonathan Saha, *Colonising Animals* (Cambridge: Cambridge University Press, 2021).

cerned with animals. As other scholars have shown, in order to legitimise colonisers' viewpoints these archives also asserted specific epistemic frameworks which negated and delegitimised the variety of relationships between animals and humans in the colonised societies. One of the ways to achieve this was to term certain animals as dangerous, unwanted, and/or a nuisance.[37] While the making of the archives was essentially a human-centric concern, the nonhumans have been at the periphery due to neglect. The marginality that colonial archives gives to the nonhuman animals needs to be located within the logic of their creation and collation by investigating the epistemic constructs which guided the colonial actors involved in making of the archives.

So, how does one attend to the nonhuman animals in this context? Here I would like to employ the ideas about the imperial archives as suggested by Antoinette Burton, specifically emphasising the connected nature of the archives. For Burton, connections between the coloniser and the colonised are explorable in myriad ways. Two such ways applicable to this research include, first, the exploration of colonial subjectivities where one is 'subject to' and 'subject of colonialism', and, second, viewing archives as dynamic and living spaces where construction of identities and knowledge formation takes place.[38] Considering these explorations in the context of nonhuman animals who are overtly assumed to be agency-less beings, the following questions become relevant. How did archives of colonial India record nonhuman animals? Do these records and subsequent formulations essentially consider animals without their subjectivities? Are animals' specificities such as species-characteristic capacities, sensory faculties, affective and cognitive abilities, and experiential capabilities attended to? Were nonhumans colonial subjects and what status was accorded to them in these records? These questions are engaged below.

[37] In the Indian context, this was often done for the native breed of dogs who lived on the streets in abundance. In case of southern Africa it was done for the native hunting dogs who accompanied native groups into the forest. For details refer to Heeral Chhabra, "Making violent killings 'humane': State sponsored elimination of 'stray' dogs in colonial India", in *Violence Against Animals: Argos – Historical and Archaeological Animal Studies, Vol. 1*, ed. Daniel Lau (Germany: Animot Verlag, 2020), 135–156 and Lance Van Sittert and Sandra Swart, *Canis Africanis: A Dog History of Southern Africa* (Leiden: Brill, 2008).

[38] From Pailey and Reid, *Sources and Methods*, Intro, footnote 15; Antoinette Burton, 'Memory Becomes Her' and 'Archive Fever and the Panopticon of History', in *Dwelling in the Archive: Women, Writing House, Home, and History in Late Colonial India* (Oxford: Oxford University Press, 2003), 4–30, 138–144.

Nonhumans in Colonial Indian Archives: Exploring Animals in Military Transport

This section deals directly with animals and their representation in colonial archival records. The animals under investigation consist of 'transport' animals such as camels, mules, donkeys, bulls etc. who were hired, brought or recruited for the purpose of military transportation especially in Punjab in the late nineteenth century and into the twentieth century (but elsewhere too).[39] The archival sources include military correspondence aimed at the selection of 'suitable' transport animals.[40] The link between animals and transport is an old one and is marked by the use of animals to facilitate movement of humans and commodities across space. The animals used for this purpose were commonly referred to as 'beasts of burden', 'draft animals' or 'pack animals' and those who formed its rank and file included donkeys, mules, ponies, horses, camels, elephants, bullocks and others.[41] In the absence of railways and efficient roadways, and amidst tough terrains, these labouring animals formed the backbone of military transportation in the region of Punjab and at other locations as well. However, it wasn't the first time that non-combatant animals were crucial for the ruling elite. For instance, the Mughals in India (from the sixteenth to the eighteenth century) used these animals not only in military and non-military functions but also for their administrative organisation, such as the Mansabdari system, which entailed ensuring a constant supply of animals for various purposes. The dependence of ruling elites and military professionals on animals was immense, whether they were cavalry 'war-horses', 'transport camels', mules or bullocks. James Hevia provides us with much more details in this regard.[42] Evidently, the detailed management of transport animals had definite precolonial relevance, and as Hevia clearly brings out there existed a network of animal-human co-dependent

[39] The colonial Punjab of the nineteenth and twentieth centuries formed an important source of human, animal and agricultural resources for the empire but also bordered the British Indian territory, making it militarily extremely significant.

[40] NAI, Military Department, Transport, August 1886, Part A, No. 1978–89.

[41] The language used to define animals' work, status and features is more than descriptive terms. They not only delimit an animal's status but also sanction animal abuse and violence both in language and in action. For very helpful insights into these aspects, refer to Les Mitchell, *Reading the Animal Texts in the Landscape of the Dammed* (South Africa: NISC, 2019).

[42] For more details refer to James L. Hevia, *Animal Labor and Colonial Warfare* (University of Chicago Press, London and Chicago, 2018).

groups which were well integrated into the royal requirements in Mughal times. So, clearly there were well-developed networks to facilitate the use of 'transport animals' (owned by the various social groups such as tribes, pastoralists, rural inhabitants and other) for the ruling elite in times of peace and war.

The importance of these transport systems remained; what changed in the British colonial regime was the recruitment pattern and usability of these animals. This section examines detailed instructions written by Low in 1886 as part of military correspondence regarding the selection of animals for military transportation. These instructions stand out as they explicitly pertained to animals, i.e., with details about each group of transport animals. They also have a definite purpose, namely to shortlist animals for various roles to serve the military, which include travelling long distances, carrying heavy weights, tackling varied terrains and so on. The purpose of these instructions was essentially human-centric; primarily geared towards the use of animal labour for human purposes. However, my purpose here is to analyse the ways in which animals are described and understood in and through these instructions. By doing so the attempt is to bring out the status accorded to animals and whether their subjectivities were a consideration in their management.

Rules for Purchase: Elephants, Camels, Mules, Ponies, Donkeys and Bullocks[43]

As mentioned before, the use and purchase of animals for military purposes was not new. The change in the colonial context seems to be in the *method of* selecting and collecting animals. As with many other spheres, 'formal' procedures including written records and enumerations were officially adopted in the colonial period. The creation of such specific instructions was one such example, and these included detailed descriptions for each animal species. Importantly, by delineating the most desirable features of these animals from a human perspective and elaborating on undesirable traits as well, such descriptions made transport animals like a product, with specific features to be checked to make sure the most suitable and best animals were selected for human use. Also, officers who were aware of the selection procedures could be used for this work, instead of

[43]All details from NAI, Military Department, Transport, August 1886, Part A, No. 1978-89.

involving the natives or recruiting these animals directly from the market. Knowledge of the animals was crucial in making selection procedures effective for colonists. The animals used for transportation included elephants, mules, camels, ponies, donkeys and bullocks.

Significantly, a careful look at the description of each animal reveals extensive interaction with the animal. From different breeds of the same animal to nuances of each breed, both specificities and characteristic features were dealt with. Traits and behaviour patterns of each kind of animal were noted and elaborated upon with an eye to determining their suitability for military transport. The aspects of general appearance were crucial because they helped colonists identify developmental milestones and any health abnormalities. Similarly, the age-related characteristic features and signs further signified continued observation and interaction with the animals. Apart from this, the language with which animals were described and military instructions provided not only represented a very detailed and careful examination of the animals but also reflected observations about their features which went beyond the physical markers and included personality traits, signifiers of intelligence, temperamental features and so on. The important question is: in describing all these criteria, were these animals being treated as sentient, intelligent beings or just objects and resources? Can we locate the living identity of transport animals being recognised and appreciated? To answer all of these, let us first look at the details available for each animal in the military transport booklet.

Elephants

Three elephant breeds were commonly used for transport.[44] First was 'khumeria', which had the following features: "straight back sloping downwards from over the shoulders towards the tail; large barrel and powerfully constructed body; short front legs and shorter hind legs; very well-developed bump, with short and thick neck; soft skin; *the temper good and even*, and very staunch". The second was 'dausale' which in fact was a mix between the 'khumeria' and 'mirga'. Considered a very useful baggage animal, it was said to have great strength which he derived from 'khumeria', and speed which they got from 'mirga'. The last was the 'mirga'—"a long legged lanky elephant without pluck, *inclined to be timid* and difficult to get into condition when once out of it ... his pace is quick compared to

[44] All details on Elephants from 'Purchase of Transport', NAI, Military Department, Transport, August 1886, Part A, No. 1978-89, p. 2.

other elephants, and is most useful when speed is necessary". Regarding ageing elephants, the following features were emphasised, "construction and colour of the ears along with the *appearance of the eye*, skin, bones, the temple and general look of the head" (emphasis added).

The booklet noted that

> the ears of 10–15 years old are straight and unbent, and they begin gradually to bend over at the top as the age advances, and in the old age they become much more bent over; they also grow whiter and become jagged and frayed, and look like a torn frill at the ends. *The eye of a young elephant is clear and sparkling; it looks sharp and intelligent;* as the age comes on *it gradually grows dimmer and loses its lustre and prominence. The skin, as an elephant ages, loses its smoothness and purple bloom and becomes puckered.* The bones protrude and the temples grow hollow and are unfailing *sign of age.* The head becomes thin and loses its rounded and erect appearance. The trunk under the bumps becomes jiggered and rough; it also gets lighter in colour in many cases. (emphasis added)

The information in these instructions revolved around the use of the animal for the purpose of transport. Clearly, the use-value of the animal was the focus. However, the way the animal was described and that personality aspects were focussed upon is notable. For instance, the temperament of the elephant was important for their use and exploitation—hence the observation that "*the temper good and even*"; "*inclined to be timid and difficult to get into condition when once out of it*". Similarly, "*the eye of a young elephant is clear and sparkling; it looks sharp and intelligent*" (emphasis added). Such observations clearly invest the elephants with a sense of intelligence, awareness and temperament. The most unavoidable aspect of the living body—ageing—was also attested to. The features described are reflective of living beings who have personality traits, who change over time, who age and have individual variations. Though elephants were serving human transport purposes, they were described and recognised as intelligent, sentient beings.

Camels

With regard to camels, Low pointed out that they were generally found in India and Afghanistan but there were also Arabian or one-humped camels.[45] The Afghan camels were said to be "strong and thick-set, with short legs, and *could stand cold; but not heat*". "The Indian camel", Low continued, "especially the Sind, Bicknair, and the Rajputana camels, *could stand heat but not cold*".

Further, one could know the age *of the camel by their teeth* and there were certain breeds who were fit for transport purposes while others were not. It was further suggested that the camels purchased for transport should be over five and under eight years of age. "The height of the camel varies as camels coming from Punjab were the largest, and hill camels the smallest. The *height of the camel was not a test of their power*" (emphasis added).

It is crucial to point out that the living body of the camel seemed capable enough of telling stories of its own.[46] The various marks, patches, wounds etc. not only reflect their living conditions but also show their extensive use and work patterns. Camels have been used extensively for transport purposes; however, only careful interaction with and knowledge of camels could warn against erroneous assumptions. For instance, the observation that the height of the camel was not reflective of its power was crucial, for anyone who was unaware of camels' physiques and nature was likely to mistakenly believe that taller camels were stronger. Similarly, detailed knowledge about camel diseases, bodily ailments, along with techniques to hide them (by the seller) and ways to circumvent them, all attested to human handlers' very close working relations with the labouring animals and knowledge of camels' bodily capacities. More crucially, such details illustrate that each animal had its own peculiarities and working relations—without attending to them it was not possible to either make a good selection or gain effective work results. The insight here is that close and experiential knowledge of camels and individual camel's specific traits was required for colonists to make camels work effectively for them.

[45] All details on Camels from 'Purchase of Transport', NAI, Military Department, Transport, August 1886, Part A, No. 1978-89, pp. 3–4.

[46] I borrow this observation from Sandra Swart who has argued the same for working horses. For details see Sandra Swart, The World the Horses Made: A South African Case Study of Writing Animals into Social History", *International Review of Social History* 55, no. 2 (2010): 241–263.

Mules and Ponies

Mules were very important for transport and much attention was paid by the colonial authorities to their breeding and maintenance.[47] Extensive criteria were suggested for purchasing mules. Some were as follows:

> The back should be straight from wither to croup. An arched back is better than a hollow back. High wither [ridge between shoulder blades] is a fault, but not so bad as a high croup [rump or hindquarters]. Sharp prominent wither and back-bone is a fault. Many excellent mules are cow-hocked. The action should be such as to not oscillate the load. A long swinging walk is not good action, as it unsettles the load. Mules that cross their legs or brush before or behind should not be purchased. The hock should be most frequent seat of disease. No mules with a discharge from the nose or with ulcers on the body or legs or with mange is to be purchased.

As in the case of camels, some evident features which may be taken as defects or problems were actually not so. For instance, cow hocked mules were regarded as excellent. Also, the walking pattern of the mule was reflective of its use as for transport, so a long swinging walk was not desirable because the load would move or swing with the mule's walking movement. With regard to prices of the mules, a direct correlation between height/quality and prices cannot be made as a lot of variety existed in each region. One point to note is the existence of variety and unevenness of criteria—a single pattern could not be determined either for animals as a single category or for animals of the same category.

The case with ponies was similar to that of mules. Low, in his instructions, notes:

> The best transport ponies were short in the back, short in the cannon bone with low wither and broad chest. The action, as with the mule, should be such as to not oscillate the load. ... The stamp of pony varied in different districts, and *the prices of ponies vary so largely in different provinces that a scale for value according to height having general application could not be given.* ... As a general rule it was suggested that they should not be under 5 years or over 10 years old. The purchase of mares was prohibited, and when stallions were bought, they were to be castrated with discretion and

[47] All details about Mules and Ponies from 'Purchase of Transport' instructions, available in NAI, Military Department, Transport, August 1886, Part A, No. 1978-89, pp. 4–6.

> not indiscriminately. No pony with a discharge from the nose or with ulcers on the body or legs or with mange was to be purchased. (emphasis added)

Donkeys

According to the *Purchase of Transport* instructions, donkeys were the most valuable transport for supplies.[48] This was because "they are driven in droves and not led; are easily managed and *can pick up a living where animals will die*" (emphasis mine). It was noted that

> there are a few large-sized donkeys both in Punjab and Sind, but the majority were small sized animals varying from 9 hands 2 inches [97 centimetres] to 10 hands 2 inches [107 centimetres]. The Sind donkey, though small, is exceedingly strong and hardy. The points of selection for a donkey are very similar to those of a mule. ... Donkeys should not be less than 3 years or above 8 years old. No donkey with a discharge from the nose or with ulcers on the body or legs or with mange is to be purchased.

The behavioural patterns of donkeys were well observed and noted. Only if you knew the behavioural nature, temperament and characteristic of that species would you be able to enumerate its features with such certainty. Donkeys were said to be different from other transport animals as they were driven in droves and not led. This feature proved useful during treks/journeys when transporting goods. In case any transport animal collapsed the donkeys could replace the dead animal quickly. This added to their flexible use. This process of gaining knowledge was definitely through observation and interaction which appreciated the living nature of these beings.

Bullocks

These instructions inform us that the bullocks used were

> of two kinds, *viz.*, 'siege train bullocks' and 'army transport' bullocks. Siege train bullocks obtained in Bengal from the Hissar Cattle Farm; in Madras from Cattle Farm at Hunsoor; and in Bombay purchases were made departmentally. Bullocks fit for Army Transport, both draught and pack, were obtainable in large numbers at a short notice. The best field for purchase was the North-Western Provinces, especially at Allahabad, Cawnpore and

[48] All details on Donkeys from 'Purchase of Transport', NAI, Military Department, Transport, August 1886, Part A, No. 1978-89, p. 6.

> adjoining districts; Sind and Mysore and adjoining districts in Madras, also in Berar Districts.[49]

This clearly attests to a variety of breeds within areas and diverse availability of bullocks. It was further advised that "the Army Transport bullock, both for draught and pack, should not be too big, the small sized compact animal is hardier than the large bullock of the siege train class; it requires less food and is better suited for general work and transmission by the rail or sea to the base of operations." Additionally, the most important marker of its age were its teeth—the incisors (cutting teeth) and the molars (grinding teeth).

The instructions note:

> The teeth appeared in the mouth with tolerable regularity and in certain definite order, so that they afford a fairly reliable indication of the age of stock up to the time dentition is completed. ... The dentition of the bullock being completed above his fifth year, other marks must be looked for in determining the age beyond this period, and amongst others may be noticed that—the horns at 3 years' old were uniformly smooth from root to the tip every year, after 3 it has a notch on it, so counting notches and adding 3, age was approximately arrived at. The angle of the teeth, the quantity of matter ground off the top of the teeth, and the distance between the teeth themselves all assisted in approximately determining the age of animals who had become aged.

What appears from the descriptions of each animal breed is that even though the primary concern was transport, the distinctness of each animal was carefully noted and enumerated by Low. The status in terms of its use-value was that of property or a resource but, interestingly, while defining them, aspects of their individual uniqueness were collated and enumerated with diligence. The animals not only had distinct qualities and features but their bodies told stories of their occupation. Close observation enabled the officers to explicitly comment on and predict the animals' temperament and behavioural patterns. There were mentions of their intelligence and of actions which were likely to cause discomfort or pain. This implies a colonial recognition that the animals were feeling beings who were

[49] All details on Bullocks from 'Purchase of Transport', NAI, Military Department, Transport, August 1886, Part A, No. 1978-89, pp. 6–7.

capable of experiences and suffering. Thus, the animals were definitely regarded as sentient beings by the colonial authorities.

Basically, the legal identity of animals as property or owned resource (which still continues to be intact in laws of most countries) is a constructed one and goes against the animals' basic nature and identity. This identity does not seem to be constructed in the process of observation and interaction alone but most probably can be located in the realm of 'use' of these animals or in tapping into their bodily potential. This especially holds true in case of transport where the 'use' is determined by the ability to carry load, thus making bodily fitness and strength essential requirements which are portrayed as parameters to be met—making animals like a product. The essential point to highlight is the fact that attention to their specificities was very much part of the selection criteria itself. This also indicates that the use of these animals entailed accommodation of their species-specific subjective needs.[50] In fact, non-attention to their specificities often undermined their 'use-value' and led to massive loss of animal life (and as a corollary, a failure of the purpose for which they were subjugated).

Based on the above, the argument I make is as follows. As the case of military transport animals illustrates, the subjectivities (like intelligence and temperament) of the working animals as well as their individual specificities were an integral part of their use-value and also a means of understanding their agency historically. The subjectivities and specificities of each transport animal were distinct and were recognised in the process of their recruitment; these subjectivities gave them agential power, i.e., the power to impact the historical outcome in dual way. First, their specificities and subjective nature determined their use-value and also the behaviours of human recruits who were to accompany these transport animals. Second, depending upon the attention paid to the specificities, it could impact the outcome of the task they were recruited for. In other words, due cognisance of the animals' subjectivities and specificities could lead to success/failure of the military expedition. In the case of military transport animals, there seems to be a direct link between the subjectivity and agency of the nonhuman labouring animals. Based on Rees' approach, agency here is regarded as the ability to impact events rather than self-conscious

[50] For instance, a camel as a transport animal is useless without a driver (human handler) who is familiar to the camel; a donkey, on the other hand, does not require a handler as he or she can be easily led in droves.

knowledge of the ability to impact events. In that specific animals' specific temperaments or levels of intelligence meant they were suitable or unsuitable for transport labour, or that such animals had to be treated or handled differently, the animals' subjectivities were impacting colonial labour. In this sense, their subjectivity became a form of (non-self-conscious) agency. Additionally, the selection and recruitment of working animals based on such instructions correspond closely to Burton's idea of connected nature of the archives which pushed us to explore the following—what happens when one is 'subject to' and 'subject of colonialism', and how archives contribute towards constructions of identities and knowledge formation. Not only did these working animals become the 'subjects' of the empire, they were subjected to epistemic dominance of Eurocentric frameworks which determined their status and subsequent use (and misuse). The idea of animals as (a mere) property or a resource went a long way in overlooking subjectivity and agency of nonhumans and this idea was a Western construct of animal-human relationships. This nature of this relationship was strongly asserted in such booklets and in fact facilitated the use of working animals for the colonial state.

Conclusion

This chapter has attempted to understand why the ubiquity of animals in historical sources like the Indian colonial archives has not led to their presence and centrality in historical narratives in the historiography of modern Indian history. The first half of the answer can be located in the conceptual framework of 'agency' which is much human-centric and also restricts inclusion of nonhumans in historical narratives. Many trends in history writing and social sciences have attended to this drawback by either reworking the idea of agency and expanding its reach or providing alternate vocabularies to facilitate an animal-inclusive conceptual grounding. This chapter builds on the insights about agency as highlighted by Rees and other scholars, i.e., self-consciousness should not be the limiting factor to determine agential capacity. Rather the ability to impact and determine outcomes must be taken as a valid marker of nonhuman agency.

The other reason for the apparent overlooking of the nonhumans lay in the nature of colonial archives themselves which is reflective of Eurocentric notions and epistemic frameworks. Not only are relations of humans and nonhumans viewed in terms of 'animal-human divide', but the variety inherent within animal-human networks of the colonised societies was often overlooked, trivialised, penalised and negated. One way out, therefore, is to read the archives against the grain and approach them with animal-centric questions and frameworks. This chapter comprised an against the grain reading of colonial military transport correspondence through an animal-centric lens. It brought together the varied understandings of agency in an effort to analyse Low's military transport purchase instructions in an animal-centric manner. The analysis argues for a link between animal subjectivity and agency. Various labouring animals are discussed within these instructions with details about their breed, temperaments, intelligence, bodily features and other specificities. The use of animals (in whatever capacity) does necessitate in-depth understanding not only of their physical features but also of their behavioural patterns. Significantly, the requirements, features, age-related characteristics and behaviours were specific to each animal. No one method of identification, determination and observation was paramount or overarching. This in fact meant that each animal had to be treated and dealt with distinctly *as an individual*. These subjectivities and specificities form the basis on which labouring animals were selected/rejected and made to work. Only with an in-depth understanding of these variations among animals, realisations of their body's needs and features and successful catering to them could lead to completion of tasks they were recruited for. However, the absence of effective attendance to these features led to massive debacles for the labouring animals and also for the military expeditions which were dependent on their labouring capacities. The very subjectivities and specificities of these animals provided them with the agency to impact labour tasks and in turn determine the outcome of the assigned task.

However, what does this reveal about the nature of colonial archives and the status of labouring animals in the archives? While the assumed status of labouring animals was clearly that of a resource to be owned or put to work or transferred or compensated against, the process of determining and reasserting that identity may present a different picture. It is in the process of defining and determining nonhuman identities that we need to question or probe their status, role and agency in the colonial archives. The attention to the ubiquity of nonhuman presences has to be approached differently. Rather than focus on the dominant identity being reasserted within these archives, the effort could be geared towards unravelling the process through which dominant identities get formed. In the case of animals selected for transport labour, their identities were formed in relation to how suitable they were for transport labour as individuals, with individual intelligences and temperaments and bodily features.

In historians' narratives of India, animal agency and subjectivity has often been negated or neglected to the extent that nonhumans' status is regarded as a static singular identity. A critical approach towards the process of how animals' identities were formed revealed the aspects which brought forth and highlighted nonhuman agency. These included temperament, intelligence and individual bodily specificities. Adopting such an approach is likely to enable the negotiation through which the identity of property is asserted for nonhuman labouring animals but also bring forth the diversity of animal-human relationships within colonised societies. These insights can then be used further to challenge Eurocentric notions which were successful enough to shroud the ubiquity of nonhuman animals' use (abuse) and to also unmask epistemic constructs which veiled their agency, subjectivity and individual specificities.

Bibliography

Barua, M. "Volatile ecologies: Towards a material politics of human-animal relations," *Environment and Planning A* 46, no. 6, (2013): 1462–1478.

Bhat Dundi, Deepak, Istvan Praet, and Garry Marvin. "Good, Quarrelsome, Bad: Animal Agency and Human-Elephant Interactions in the Western Ghats, India." *Frontiers in Conservation Science* 4 (2023). https://doi.org/10.3389/fcosc.2023.1142333.

Bonnell, Jennifer, and Sean Kheraj. *Traces of the animal past: Methodological challenges in animal history.* Calgary: University of Calgary Press, 2022.

Bose, Shibani. *Mega Mammals in Ancient India: Rhino, Tigers, and Elephants.* Delhi: Oxford University Press, 2020.

Burton, Antoinette. *Dwelling in the Archive: Women, Writing House, Home, and History in Late Colonial India.* Oxford: Oxford University Press, 2003.

Chakrabarti, Pratik. "Beasts of burden: animals and laboratory research in colonial India." *History of Science* 48, no. 2 (2010): 125–151.

Chhabra, Heeral. "Animal labourers and the law in colonial India." *South Asia Research* 39, no. 2 (2019): 166–183

Chhabra, Heeral. "Making violent killings 'humane': State sponsored elimination of 'stray' dogs in colonial India". In *Violence Against Animals: Argos – Historical and Archaeological Animal Studies, Vol. 1*, edited by Daniel Lau, 135–156. Germany: Animot Verlag, 2020.

Crosby, A.F. *Ecological Imperialism: The Biological Expansion of Europe, 900–1900.* New York: Cambridge University, 1986.

Despret, Vinciane. "From Secret Agents To Interagency." *History And Theory* 52, no. 4 (2013): 29–44.

Detel, W. "Social Constructivism." In *International Encyclopedia of the Social & Behavioral Sciences*, edited by Neil J. Smelser, Paul B. Baltes, 14264–14267. Pergamon, 2001.

Dinkar, Niharika. "Seeing the Elephant: Animal Spectatorship and the Imperial Gaze in Colonial India." *South Asian Studies* 37, no. 2 (2021): 117–136.

Evans, Nicholas HA. "Blaming the rat? Accounting for plague in colonial Indian medicine." *Medicine Anthropology Theory* 5, 3 (2018): 16–42.

Havia, J. L. *Animal Labor and Colonial Warfare*. Chicago/London: The University of Chicago Press, 2018.

Lack, David. *The life of the robin*. London: Penguin, 1953.

Mitchell, Les. *Reading the Animal Texts in the landscape of the dammed*, South Africa: NISC, 2019.

Paisley, Fiona and Kirsty Reid. *Sources and Methods in Histories of Colonialism: Approaching the Imperial Archives*. London: Routledge, 2017.

Pearson, Chris. "Dogs, History, And Agency." *History And Theory* 52, no. 4, (2013): 128–145.

Rangarajan, Mahesh. "Animals with rich histories: the case of the lions of Gir Forest, Gujarat, India." *History and Theory* 52, no. 4 (2013): 109–127.

Rees, Amanda. "Animal agents? Historiography, theory and the history of science in the Anthropocene." *The British Journal for History of Science* 2, (2017): 1–10.

Rees, Amanda. "Animal Agents: Can animals shape their own lives? Or the course of history?" *Aeon Essays*, February 26, 2018. https://aeon.co/essays/can-animals-shape-their-own-lives-or-the-course-of-history. Accessed 11 June 2023.

Ritvo, Harriet. *The Animal Estate: The English and Other Creatures in the Victorian Age*. Cambridge, Massachusetts: Harvard University Press, 1987.

Rothfels, N. *Representing Animals*. Bloomington: Indiana University Press, 2002.

Roy, Rohan Deb. "Nonhuman Empires." *Comparative Studies of South Asia, Africa and the Middle East* 35, no. 1 (2015): 66–75.

Roy, Rohan Deb. *Malarial Subjects: Empire, Medicine and Nonhumans in British India, 1820–1909*. Cambridge: Cambridge University Press, 2017.

Saha, Jonathan. "Colonizing Elephants: Animal Agency, Undead Capital and Imperial Science in British Burma", *BJHS* 2, (2017): 169–189.

Saha, Jonathan. *Colonizing Animals*. Cambridge: Cambridge University Press, 2021.

Sittert, L.V., and Sandra Swart. *Canis Africanis: A Dog History of Southern Africa*. Leiden: Brill, 2008.

Shaw, David Gary. "A Way With Animals." *History And Theory* 52, no. 4 (2013): 1–12.

Skabelund, Aaron. "Animals and Imperialism: Recent Historiographical Trends." *History Compass* 11, no. 10 (2013): 801–807.
Skabelund, Aaron. *Empire of Dogs: Canine, Japan, and the making of the Modern Imperial World*. Ithaca and London: Cornell University Press, 2011.
Stoler, A.L., *Along the Archival Grain: Epistemic Anxieties and Colonial Common Sense*. Princeton: Princeton University Press, 2009.
Swart, Sandra. ""The World the Horses Made": A South African Case Study of Writing Animals into Social History." *International Review of Social History* 55, no. 2 (2010a): 241–263.
Swart, Sandra. *Riding High: Horses, Humans and History in South Africa*. Johannesburg: Wits University Press, 2010b.
Thomas, Keith. *Man and the Natural World: A History of the Modern Sensibility*. New York, 1983.
Tinbergen, Niko. *Curious naturalists*. New York: Doubleday, 1968.
Walker, Brett L. "Animals and the Intimacy of History." *History and Theory* 52, no. 4 (2013): 45–67.

Archival Records

British Library India Office Records IOR/L/MIL/17/5/2186.
National Archives of India (NAI), Legislative Department, April 1890, No. 607–639, Part A.
National Archives of India (NAI), Home Department, Police Branch, June 1902, File No. 95-105, Part A.
National Archives of India (NAI), Military Department, Transport, August 1886, No. 1978-89, Part A.

CHAPTER 10

Law Through the Eyes of Animals

Challie Facemire, Clayton Kinsey, Sierra Apillanes, Gwendolyn Bell, Samantha Brown, Isaac Kort-Meade, Pierce Libbey, Samuel Wu, Caitlin Doak, and Karen Bradshaw

Introduction

Critical animal studies is a scholarly approach that re-envisions various fields of studies from the perspective of animals. In this chapter, we apply this approach to law. Specifically, we re-examine several major legal cases to explicitly consider the perspective of the animals involved in the cases. To be clear, this is an exercise in creative reimagining—we do not pretend

C. Facemire
Central Arizona Conservation Alliance, Desert Botanical Garden, Phoenix, AZ, USA

Center for Biology and Society, Arizona State University, Tempe, AZ, USA

C. Kinsey • S. Apillanes • G. Bell • S. Brown • I. Kort-Meade • P. Libbey • S. Wu
Sandra Day O'Connor College of Law, Arizona State University, Phoenix, AZ, USA

M. J. Glover, L. Mitchell (eds.), *Animals as Experiencing Entities*, The Palgrave Macmillan Animal Ethics Series,
https://doi.org/10.1007/978-3-031-46456-0_10

that we can know what the specific animals in question, or species of non-human animals more generally, thought. But, engaging in the attempt to consider the perspectives—even imperfectly, perhaps wrongly—is an exercise in interspecies empathy. In attempts to avoid the purportedly cardinal sin of anthropomorphism, scholars in many academic fields have created a remove between humans and other animal species to the point that we ignore animal interests. This paves the way to creating systems and institutions that are wholly and unapologetically anthropocentric—focused wholly on human wants, needs, perspectives, and ideas. However, while there are problems with anthropomorphism in some situations, it can also be a useful tool to engender empathy in humans. One such example of this is Richard Adams' *Watership Down*[1]; Adams was an animal rights advocate who cared deeply for non-human living beings. Another example is the movie *EO*, which tells the story of a donkey and the humans that surround him over the course of his adventures.[2]

We resist wholly anthropocentric conceptions of the law. Doing so requires new, innovative, and experimental approaches. That is what this chapter attempts: envisioning new possibilities of a legal structure that considers the more-than-human world. In this chapter we will present animal perspectives with the goal of avoiding anthropomorphism, in favor of scientifically informed interpretation of events using known animal behavior and instinct to inform our portrayal of their perception—akin to

[1] Richard Adams, *Watership Down* (London: Rex Collings Ltd., 1972).
[2] Jerzy Skolimowski, dir, *EO* (2022; Poland: Skopia Film).

C. Doak
Sandra Day O'Connor College of Law, Arizona State University, Phoenix, AZ, USA

United States District Court for the District of Nevada, Las Vegas, NV, USA

K. Bradshaw (✉)
Sandra Day O'Connor College of Law, Arizona State University, Phoenix, AZ, USA

Global Institute of Sustainability, Arizona State University, Tempe, AZ, USA

Faculty Affiliate Scholar, Classical Liberal Institute, New York University School of Law, New York, NY, USA
e-mail: KBradshaw@asu.edu

Murray's "The Cows on Killing Day."[3] In blending the creative with the scientific, we present a perspective that is absent from current case law, and a tactic that might aid in future decision making and case presentations.

Radical, Activist Lawyering

We are engaged in the process of jointly radically reimagining the human relationship with nature.[4] We seek to envision a new and better future, then to lay a legal pathway between where we are and this new vision using the tools of our profession. As we see it, there are three steps to co-creating new solutions:

Step 1: Assessing the status quo and problems incumbent in it,
Step 2: Envisioning a new and better future, and
Step 3: Leveraging the tools of our profession to build legal pathways between where we are and where we want to go.

Step 1

As legal thinkers with a keen interest in Environmental Law, Animal Law, and Biodiversity Law, we are not satisfied with the state of the field. With respect to Environmental Law, it has become abundantly clear that agencies cannot wield a collection of statutes enacted in the 1970s to effectively address the problems faced by our generation: climate change, biodiversity loss, environmental justice, interspecies, and ongoing social inequities. Many of these laws are anthropocentric in nature—focusing on human health and wellbeing to the exclusion of the (often unstated) needs of non-human animals.

A talented group of environmental law and animal law scholars have taken stock of where we are today on issues ranging from the mistreatment of animals to the radical changes necessitated by climate change. It is grim.

[3] Les Murray, "The Cows on Killing Day," in *Dog Fox Field: Poems* (New York: Farrar Straus & Giroux, 1993), 27–28.

[4] "Radically reimagining the human relationship with nature" is drawn from an initiative led by Karen Bradshaw, which includes an interdisciplinary community arts event in Spring 2022 sponsored by the Desert Institute for Humanities and a series of academic works. See Karen Bradshaw, "Radically Reimagining the Human Relationship with Nature: Identifying Contemporary Rights of Nature in the United States," *Southern California Law Review* 96 (forthcoming).

Many of us know young people who are depressed, even suicidal, because of the environmental state of the world. "Research has shown that half of people between 16 and 25 years old believe the Earth may be doomed, while three-quarters feel anxiety when they think or hear about climate change."[5]

Step 2

Now, it is time to envision something different—a better future. It is not enough to criticize the past or critique the present, although both are important work. Young activists are often the leading voices on issues of climate.[6] As future lawyers, vanguards of the role in government in mediating human relationships with one another and with the extra-human world, we collectively take on the task of presenting an alternative vision for something better. We need to radically reimagine the human relationship with the natural world, exemplified here through animals. As lawyers, we can wield different legal tools to create stepping-stones between where we are today and where we hope to go.

Step 3

Some may suggest law itself is the problem; that by using traditional tools of government and governance we exacerbate the societal problems and inequities that we seek to address. We respond by noting that virtually all societies—including animal societies—rely on law and government, the idea of collective understandings ruling individual behavior.[7] Governance is omnipresent throughout societies of living things. We seek to make it more inclusive, folding in voices of traditionally excluded groups.

The authors of this project are all affiliated with Arizona State University, which has a Charter that measures its success "not by whom it excludes, but by whom it includes and how they succeed."[8] We are a new generation of lawyer-thinkers being trained by faculty who value inclusion and diversity, and who believe that arming students from historically excluded groups

[5] Oliver Milman, "Suicides Indicate Wave of 'Doomerism' Over Escalating Climate Crisis," *Guardian*, May 19, 2022, https://www.theguardian.com/environment/2022/may/19/climate-suicides-despair-global-heating.

[6] Milman, "Suicides Indicate Wave."

[7] Karen Bradshaw, *Wildlife as Property Owners: A New Conception of Animal Rights* (Chicago: University of Chicago Press, 2020), 143.

[8] "ASU Charter," Arizona State University, accessed September 9, 2022, https://president.asu.edu/asu-mission-and-goals#:~:text=ASU%20Charter,of%20the%20communities%20it%20serves.

with the tools of law is a crucial (although not sufficient) step to expanding their rights.

Our role in this unbroken lineage of legal activists is to use the tools of our profession to strive toward different and better results for our society, current and future. Doing so requires much more than parsing the minutia of section 301 of the Clean Air Act, for example—a skill that is necessary for the practice of environmental law, but not sufficient for the kind of change we hope to bring about.

Radical lawyering or activist lawyering relies upon using legal tools to achieve objectives outside the present norms. Activist lawyers act within the existing legal structures to achieve large-scale social change—they represent a cause, not a single client. Radical and activist lawyer is content-neutral: it can occur at any point in the political spectrum. More recent examples of activist lawyering include civil rights litigation, women's rights, and animal rights cases.

Critical Animal Studies

Radically reimagining a different human relationship with nature is an iterative and collective process. We are inspired by leaders in the field, including Professor Maneesha Deckha, whose work focuses on de-centering human beings by reconceptualizing colonial legal systems that reduced animals to property and instead prioritizing indigenous legal systems that treat animals as relations in need of care.[9] Professor Martha Nussbaum[10] has made tremendous inroads into imagining what justice for animals might consist of, by advancing a capabilities framework. Professor Angela Fernandez[11] has engaged in questions of animal rights as well, including a book-length analysis of the case *Pierson v. Post*, a canonical property case discussing which of two hunters owns a dead fox carcass. In this chapter, we add to the field by presenting well-known legal cases through the perspective of non-human animals. This is a legal application

[9] Maneesha Deckha, "Animalizing Law, Humanizing Animals: The Diverse and Feminist-Informed Debates in Animal Law About How to Resist Legal Anthropocentrism," *Dalhousie Law Journal* 46, no. 1 (2023, forthcoming).

[10] Martha C. Nussbaum, *Justice for Animals: Our Collective Responsibility* (New York: Simon & Schuster, 2022).

[11] Angela Fernandez, *Pierson v. Post, The Hunt for the Fox: Law and Professionalization in American Legal Culture* (Cambridge: Cambridge University Press, 2018).

of critical animal studies—seeking to reconceptualize a traditional field of academic study to be inclusive of the perspectives of non-human animals.[12]

Methodological Overview

Critical animal studies is a nascent but growing academic field. There exist broad, open methodological questions about how to envision animal histories, and philosophical questions as to whether it is even possible for humans to totally imagine being animal. There are also criticisms of attempts to attribute histories and emotions to animals in an overlay of anthropomorphism that may not serve the animals or their greater ecological community. In this endeavor we are trying to strike a balance of advocacy and empathetic imagination to give these animals a voice without overstepping.

It feels risky to be an early entrant into an emerging field. As techniques develop and improve over time, our approach might appear rudimentary or even inappropriate. Nevertheless, we are engaging in the process—acknowledging the potential to look foolish in the future—because the only way the ball moves forward is if someone is willing to push it.

This section explicitly identifies our methodological choices so that others can evaluate, critique, or replicate them. In this class, we have engaged in a variety of techniques: learning about the animals in history through legal research, reading ethnographical studies of animals (scientific studies of animal behavior, typically from peer-reviewed scientific journals), watching documentaries about the animals to engage visually with their habitats, physically observing animals in their natural habitats, observing animals in zoo, and engaging in meditation on what it is to be a particular animal, to suspend the human perspective and attempt to engage the animal perspective.[13]

[12] Karen Bradshaw, "Radically Reimagining the Human Relationship with Nature: The Current Status of Rights of Nature in the United States," *USC Law Review*, (forthcoming 2023).

[13] The idea of connecting with another through mindful contemplation is a practice throughout the world. For example, the Scottish term kything "is an almost lost Scottish word which speaks to the act of consciously communicating with an object" or the "spiritual process of focusing on one's shared connection with a loved one." See "kything," *Wiktionary: The Free Dictionary*, last edited August 15, 2022, https://en.wiktionary.org/wiki/kything#:~:text=kything%20(uncountable),connection%20with%20a%20loved%20one.

Importantly, each case study follows a consistent format. It begins with an overview of the case as it is typically understood and taught, including the historical and social context of the case. Next, it provides a first-person perspective from the animal perspective. This a blend of art, law, and science—students are drawing upon academic articles and Traditional Ecological Knowledge about animals, but also making individual choices about the creatures being reflected. It then discusses how viewing the case from the animal perspective informs current understanding. How would a professor who engaged critical animal studies teach the case differently? What might the animal perspective add to an understanding of the case, which was not present in the anthropocentric litigation?

Case Studies

This study reimagines five canonical cases from Property, Contracts, Constitutional Law, Public Lands Law, and Environmental Law from the perspective of animals. We construct case studies of how the case might be decided from an animal perspective for the following cases: *Sherwood v. Walker* (1887), *Kleppe v. New Mexico* (1976), *Gibbs v. Babbitt* (2000), *Cetacean Community v. Bush* (2004) and *Winter v. Natural Resources Defense Council* (2008), and *Naruto v. Slater* (2018).

Sherwood v. Walker

A. Title of Case, Procedural History, How it is Typically Understood and Taught

Sherwood v. Walker[14] considers the purchase of a cow called Rose who was "owned" by Hiram Walker and purchased by Theodore Sherwood for a price of 5.5 cents per pound ($80). Sherwood requested the agreed upon purchase price be laid out in writing, and Walker complied by furnishing Sherwood with a letter and naming a third party to deliver the cow. It was assumed by both parties that the cow was infertile, but Walker discovered that she was pregnant and refused to deliver the cow and honor the contract. He argued that since Rose was capable of breeding, she was worth more.

The courts decided in favor of the Plaintiff (Walker), but on appeal the Michigan Supreme Court reversed the decision. They didn't make a

[14] Sherwood v. Walker, 33 N.W. 919 (Mich. 1887).

definitive decision in favor of the defendant but concluded that if a jury could find mutual mistake, then the contract could be rescinded.[15] This case is taught in first-year contracts as an example of the doctrine of mutual mistake, where when there is a shared misunderstanding about an essential fact from both parties, a contract may be rescinded.[16]

B. First-Person Animal Perspective

I am a cow, called, by the humans, Rose 2d of Aberlone but that is not who I am. I am a member of my cattle community, a consumer of grass, and—most importantly—a soon-to-be mother. I have been moved to different places, which I didn't like, as it was scary and noisy and then different, but I got used to the new places. Perhaps this stress is why I didn't have a baby last year. Now that I am pregnant, I am restless, and uncomfortable.[17] The humans, though I am used to some of them, also keep coming around me and I want to be alone to give birth.[18] The humans I have known recently are acting differently and there have been new people.[19] Here different is not a good thing, other cows disappear and sometimes the humans and their machines smell like death. I have tried to communicate my distress, my fellows understand and have tried to comfort me, and when I gave birth, I hid in the corner of a barn. Then one human kept all the rest away so I could be with my baby and protect it. Despite the human keeping people away the new man came back, and I was forced away from my baby and onto a hard cold surface. It was loud and I tried to walk away from it but they made me stay in place for too long, then finally let me return to my baby. The humans seemed agitated for a while after that, and then the new man came back and took me away.

[15] Roberta M. Gubbins, "Mutual Mistake, the True Story of Rose of Aberlone – Otto Stockmeyer Reveals the Truth of Sherwood v Walker," *Detroit Legal News*, November 4, 2010, http://www.legalnews.com/ingham/771228.

[16] Heather J. E. Simmons, "By the Law Inspired," *Michigan Bar Journal* 92, no. 8 (August 2013): 58, http://www.michbar.org/file/journal/pdf/pdf4article2241.pdf.

[17] Marina A.G. von Keyserlingk and Daniel M. Weary, "Maternal Behavior in Cattle," *Hormones and Behavior* 52, no. 1 (June 2007): 107, https://doi.org/10.1016/j.yhbeh.2007.03.015.

[18] Maria Vilain Rørvang et al., "Prepartum Maternal Behavior of Domesticated Cattle: A Comparison with Managed, Feral, and Wild Ungulates," *Frontiers in Veterinary Science* (March 12, 2018), https://doi.org/10.3389/fvets.2018.00045.

[19] Gary M. Landsberg and Sagi Denenberg, "Social Behavior of Cattle," *Merck Veterinary Manual*, last modified June 2016, https://www.merckvetmanual.com/behavior/normal-social-behavior-and-behavioral-problems-of-domestic-animals/social-behavior-of-cattle.

Now I am again in a new place with a new community, maybe I'll get to have another baby, but I never get to keep them.

C. What Would the Result be in the Animal Kingdom?

In natural, free-roaming contexts, the rules would dictate that Rose would be kept with her calf and the calf would be raised with her and her herd. Since this is a domesticated breed (by which we mean a population that was genetically manipulated by humans for human gain), it is difficult to say what would be a natural occurrence. However, studies have shown that the removal of a calf from its mother later than was standard (i.e., the earliest age a farmer can separate a calf from its mother, in the interests of maximizing profits[20]) is better for the health of the cow and the calf. The bond between mother and calf grows over time, and groups are bonded through behaviors that take place early on, like licking.[21]

D. How Does Viewing this Case from the Animal Perspective Inform Current Understandings?

The human perspective in this case hinges on how much Rose's fertility is worth in terms of later meat or dairy output. Aside from that she is only seen as worth her weight in flesh. Even when her fertility is considered, she is seen as little more than a producer of more meat.[22] The animal's perspective can show that they are more than meat, but also can cast a light on their conditions. If she didn't produce a calf in her time with Walker before the sale issue, then it might have been due to stress and her conditions. Her body and life are dependent on a system in which she has no rights because she is outside the protection of the law.[23] The animal perspective can also be used to inform rulings that are in favor of animal welfare with the goal of improving production through improving quality

[20] Kathrin Wagner, Daniel Seitner, Kerstin Barth, Rupert Palme, Andreas Futschik, and Susanne Waiblinger, "Effects of Mother Versus Artificial Rearing During the First 12 Weeks of Life on Challenge Responses of Dairy Cows," *Applied Animal Behaviour Science* 164(2015): 1–11. DOI: 10.1016/j.applanim.2014.12.010.

[21] Compassion in World Farming, "Dairy Cow-Calf Separation and Natural Weaning," *Compassion in World Farming*, February 2013, https://www.compassioninfoodbusiness.com/media/5822658/cow-calf-separation-and-natural-weaning.pdf.

[22] Les Mitchell, "Farming: Animals or Machines?" *Southern African Linguistics and Applied Language Studies* 31, no. 3 (2013): 299–309; Les Mitchell, "Nonhumans and the Ideology of Purpose," *Anthrozoös* 25, no. 4 (2015): 491–502.

[23] Ron Broglio, *Animal Revolution* (Minneapolis: University of Minnesota Press, 2022).

of life.[24] While this still isn't ideal for the cow (who we assume wants to just live her life with her calves and herd), it would be an improvement over the living conditions at many factory farms if they are able to stay in bonded groups and with calves for as long as possible.[25]

Kleppe v. New Mexico

A. Title of Case, Procedural History, How it is Typically Understood and Taught

In *Kleppe v. New Mexico*, the United States Supreme Court determined the extent of Congress' constitutional power under the Property Clause through a state's challenge to the Wild Free-Roaming Horses and Burros Act (WFRHBA).[26] The WFRHBA provides protection from harassment, branding, capture, or death, for free-roaming wild horses and burros on federal land.[27] In this case a rancher (Kelley Stephenson) had a grazing permit on federal land. Several unbranded burros roamed into a watering area and Stephenson complained to the New Mexico Livestock Board that the burros were interfering with his livestock. The Board removed the burros and sold them at public auction, but the Bureau of Land Management found out the burros were removed and demanded they be returned pursuant to the WFRHBA. This resulted in the Board, the state of New Mexico, and an individual who purchased three of the burros suing the Secretary of the Interior. The District Court for the District of New Mexico found in favor of the State's side. However, the Secretary of the Interior appealed, and the Supreme Court reversed the decision, determining that Congress has complete power over public lands, including wildlife regulation on such lands. This case is usually taught in

[24] Peter Singer and Jim Mason, *The Ethics of What We Eat: Why Our Food Choices Matter* (Emmaus: Rodale Books, 2007).

[25] This highlights the tension between welfare and abolition, which is a large source of debate. The form of animal farming at issue in this case exists in order to kill calves, adult animals, and also their mothers, when they become no longer "efficient producers" on a continual basis—exchanging money for dead animal bodies. One framing of Rose's desire might be a desire not to be farmed, not to be property, and not to have her children farmed or property. Other, alternative framings might include freedom, being in a less abusive system, or remaining within the same system. We are indebted to an Animal Ethics Scholar for these helpful observations in a private communication.

[26] Kleppe v. New Mexico, 426 U.S. 529, 531–35 (1976).

[27] The Wild Free-Roaming Horses and Burros Act of 1971, Pub. L. No. 92–195, 85 Stat. 649 (1971).

Constitutional Law courses, as it is about balancing state and federal interests. It could also be taught as a part of Natural Resource Law courses.

B. First-Person Animal Perspective

I munch on grass as we wander between the scrub on the New Mexico plains, headed for a river. One of the older mares watches as we march up to the water's edge and start drinking.[28] Across the river cattle lift their heads to cast us a glance, and then return to drinking and socializing. We don't mind the cattle; we make friends easily and they are very nice and mostly calm. They also seem to like us since we know how to fight mountain lions and other predators.[29] The only problem with the cattle is that the humans come around them. When they do we try to stay away because we have seen other bands have members stolen by the humans. The watchful mare dips her head to the water, drinking before ushering everyone along for more foraging. All of a sudden, I feel something around my neck, and members of my family are screaming. Some of my family pulled the things around their neck, fighting as hard as they could, and escaped the humans on the other end. I could see them running toward the gorge. I couldn't get out of mine, though. I am not the biggest of our group. My heart feels like it will explode and my neck hurts as I am dragged into a small dark box.[30] Once I am in, it closes with a loud clang, and I panic looking for an escape and running into the sides of the box. I only stop because the box starts moving and I start feeling sick. There is a smell in the air that only comes with the humans, it is the stink from their metal contraptions. It makes me feel even worse on this never-ending nightmare. After what feels like an eternity, the box opens and I tentatively step out. I still feel sick, and am so scared, and they have placed me in an only slightly larger box. This one at least has a dirt ground and some food.

[28] Carey A. Williams, "The Basics of Equine Behavior," *Equine Science Center, Rutgers*, July 22, 2004, https://esc.rutgers.edu/fact_sheet/the-basics-of-equine-behavior/

[29] High Plains Journal, "Ranchers Use Donkeys to Protect Herds from Predators," *High Plains Journal*, August, 27, 2007, https://www.hpj.com/archives/ranchers-use-donkeys-to-protect-herds-from-predators/article_c4e984a0-fe8e-5816-a546-75b7020e3b51.html; Sydney Isenberg, "CPW Turns to Burros to Help Stop Wolf Depredation on Jackson County Rancher's Property," *Denver7 ABC*, March 18, 2022, https://www.thedenverchannel.com/news/local-news/cpw-turns-to-burros-to-help-stop-wolf-depredation-on-jackson-county-ranchers-property.

[30] Lisa Leiner and Markus Fendt, "Behavioural Fear and Heart Rate Responses of Horses After Exposure to Novel Objects: Effects of Habituation," *Applied Animal Behaviour Science* 131, no. 3–4 (May 2011): 108, https://doi.org/10.1016/j.applanim.2011.02.004.

I miss my family, though, and don't like all the humans so close and touching me. What will happen to me without a group, and confined to this small area?

C. What Would the Result be in the Animal Kingdom?

In the animal kingdom, the burros would happily share their lands with cattle. They are extremely social creatures, and bond easily to each other and other animals.[31] They are also more adapted to live in the wild than cattle are. This is why there are programs to include burros in livestock management in order to protect them from predators.[32] That said, the burros rely upon the land, while cattle can be supplemented with feed from ranchers. The ranchers are disinclined to spend money on feed because grazing rights are so much less expensive than taking on the feed costs themselves. So, due to the social nature of these animals, they would likely decide to share their land so long as enough forage was available.

D. How Does Viewing This Case from the Animal Perspective Inform Current Understandings?

Viewing this case from the animal perspective helps to highlight the family relationships among burros, which also exist among wild horses. It also highlights their tenuous grasp on the freedoms they have. Before the WFRHBA, these animals were labeled as "vermin" and were treated as creatures to be destroyed because they are in competition with us for resources.[33] So, the Act did help improve their station in the eyes of humans. However, they still suffer a lot at the hands of humans and for the benefit of ranching operations. Understanding how they live and their

[31] Leigh M.A. Murray, Katharine Byrne, and Richard B. D'Eath, "Pair-Bonding and Companion Recognition in Domestic Donkeys, *Equus Asinus*," *Applied Animal Behaviour Science* 143, no. 1 (January 2013): 72–73, https://doi.org/10.1016/j.applanim.2012.11.005.

[32] Jeffrey S. Green, "Donkeys for Predation Control," *Fourth Eastern Wildlife Damage Control Conference*, September, 25, 1989, https://static1.squarespace.com/static/52f6e70ae4b09d0c250122c6/t/54db383ee4b072be6d6ff3fa/1423652926948/guarddonkey2.pdf.

[33] We are indebted to an Animal Ethics Scholar for making this excellent point in a private communication: "Vermin is a very interesting and incredibly pejorative word. A word which often sanctifies and encourages extreme cruelty and murder. Used in warfare as well to denigrate and dehumanize the enemy. Vermin are such creatures as ferrets, stoats, sables, pigeons, and rodents. But, apparently, varmint are animals such as cayote, feral dogs, foxes, weasels, and sometimes wolves who 'come in from outside.' We make them vermin or varmint because they compete with us."

needs gives a better perspective for compassionate management and for future decision making for their benefit.

Gibbs v. Babbitt

A. Title of Case, Procedural History, How it is Typically Understood and Taught

In *Gibbs v. Babbitt* the court had to decide if it was constitutional for the federal government to enact a policy prohibiting the take of red wolves (listed as an endangered species at the time) on private land to preserve scarce natural resources.[34] In 1975 the red wolf was listed as endangered, placing limits and restrictions on "take" of the animals, mean accidental or purposeful killing or removal of the animals. When the red wolf was listed, the restrictions on taking of red wolves had exceptions under specific circumstances, which they laid out in § 17.84(c) of the statute.[35] In 1986 the U.S. Fish and Wildlife Service (FWS) started a reintroduction program of red wolves in Alligator River National Wildlife Refuge in North Carolina. In 1990 Richard Mann shot and killed a red wolf on his property, in violation of the taking exceptions outline under the Endangered Species listing. In response North Carolina passed a law titled "An Act to Allow the Trapping and Killing of Red Wolves by Owners of Private Land." This law allowed the take of red wolves on private property under special circumstances. The court ultimately decided the case in favor of the federal government's commerce clause authority, pointing out that red wolves are tied to interstate commerce in several ways, including scientific research, tourism, and the potential future in pelt trading. This case is commonly taught in Environmental Law courses, but it also is taught in Constitutional Law to illustrate the third prong of the *Lopez* test, where Congress may regulate activities that substantially affect interstate commerce.

B. First-Person Animal Perspective

My mate rises in the early light, trotting to weave around the edges of our den area, his tawny fur blending almost perfectly with the autumn-touched plants. He returns, unsettled, I sniff the wind and catch the scent that has disturbed him. An acrid smell, the small burnings that the two-legged

[34] Gibbs v. Babbitt, 214 F.3d 483, 487 (4th Cir. 2000).

[35] Special rules – Vertebrates, 50 C.F.R. § 17.84(c).

creatures carry in their hands.[36] I don't like one of them being so near but they also don't know where the den is, where our four pups roll about play fighting under the protection of our small pack. We all need to eat, and I need to stretch my legs, so I nudge my sister—the only other adult—to put the pups under her care and then head into the forest away from the smell. We hunt swiftly and silently, but there aren't many animals out. We end up circling around and heading back toward where the smell came from, I know there were prairie dogs in that area and my mate is headed toward their warren. Suddenly, he flushes out a rabbit and it takes off toward the edge of the woods. He follows swiftly, but not swiftly enough to kill it in the safety of the trees. He extends his stride to overtake the rabbit, but just as he is about to lunge, I hear an ear-splitting bang. At nearly the same moment my mate comes crashing to the ground. I smell his blood on the wind and see a two-legs approaching.[37] I want to go to him, but he isn't moving. I howl my grief, causing the two-legs to spin in fright, and then take off back toward the den.[38] I have nothing to offer my pups or my pack, no food, only grief and the knowledge that the winter will be a little bit harder.

C. What Would the Result be in the Animal Kingdom?

In the animal kingdom, the pack would decide how to address a killing of one of their own. They would also protect one of their own. There is also the added complication that the wolves would have wanted their territory rights acknowledged and respected. While animals don't own property in the same way that we do, they do have territory that overlays our property systems. If we looked at that territory, then the wolf that was on Mann's property might have been on his own property and had every right to be there. In which case Mann had murdered it. A solution for this could be that the federal government purchase a larger parcel to sustain a red wolf

[36] T. DeLene Beeland, *The Secret World of Red Wolves: The Fight to Save North America's Other Wolf* (Chapel Hill: University of North Carolina Press, 2013), 5; U.S. Fish and Wildlife Service "Red Wolf Recovery Program," U.S. Fish and Wildlife Service, last updated June, 2023, https://www.fws.gov/southeast/wildlife/mammals/red-wolf/#behavior-section.

[37] Suzanne W. Agan, Adrian Treves, and Lisabeth L. Willey, "Majority Positive Attitudes Cannot Protect Red Wolves (Canis Rufus) from a Minority Willing to Kill Illegally," *Biological Conservation* 262 (October 2021): 1, https://doi.org/10.1016/j.biocon.2021.109321.

[38] Regan Downey, "Grief Is Not Uniquely Human. Wolves Mourn Too," *Wolf Conservation Center*, September 26, 2016, https://nywolf.org/2016/09/grief-is-not-uniquely-human-wolves-mourn-too/.

population, and work to educate neighboring landowners on wildlife interactions and appropriate responses.

D. How Does Viewing this Case From the Animal Perspective Inform Current Understandings?

Viewing this case from the animal perspective offers the opportunity to see a wolf family dynamic attempting to live in their territory and trying to survive. The wolves' interests are not well represented in the current system because we don't take them into account beyond survival/preventing their extinction. This interest in survival doesn't give wolves inalienable rights that would see wolves as equals rather than a commodity. Viewing wolves as equals might offer them some sort of justice as well. Though the court case was decided in their favor, they didn't gain anything from it since in the long run the program failed and their cousins across the U.S. are still hunted. This is also an issue of education. Viewing things from the animal perspective can aid as an educational tool for understanding and respecting animals that are contentious, like predators.

Cetacean Community v. Bush* and *Winter v. Natural Resources Defense Council, Inc.

A. Title of Case, Procedural History, How it is Typically Understood and Taught

In the *Cetacean Community v. Bush* case the entirety of the cetacean community (an order of marine mammals, which includes whales, dolphins, and porpoises) sought an injunction against the United States Navy using the Surveillance Towed Array Sensor System Low Frequency Active Sonar (SURTASS LFAS) claiming that the use violated the Endangered Species Act, the Marine Mammal Protection Act, and the National Environmental Policy Act.[39] *Winter v. Natural Resources Defense Council, Inc.* was another effort to seek an injunction against "mid-frequency active" (MFA) sonar use by the U.S. Navy on the grounds that it violated the National Environmental Policy Act of 1969 (NEPA) and other laws.[40] These cases claimed that the sonar caused tissue damage and other harms and that it disrupted behaviors such as feeding, mating, and orientation within the

[39] Cetacean Cmty. v. Bush, 386 F.3d 1169, 1171–72 (9th Cir. 2004).
[40] Winter v. Nat. Res. Def. Council, Inc., 555 U.S. 7, 16–17 (2008).

water. One specific point the cases looked at was mass strandings of three different types of whales in the Bahamas in 2000.[41]

In *Cetacean Community*, the district court found that the Plaintiff lacked standing under those statutes and dismissed the case. The Plaintiff appealed and the Ninth Circuit Court of Appeals affirmed the lower court's decision. In *Winter*, the District Court granted a preliminary injunction against MFA sonar use. The Court of Appeals reversed and remanded the District Court—in its preliminary injunction—imposed a restriction on the use of sonar to when a marine mammal was not within 2200 yards from the Navy Vessel and decibel reduction requirement during "surface ducting" conditions.[42] Eventually the case went to the Supreme Court, where the Court granted certiorari and vacated the preliminary injunction, holding that the lower courts had applied a too lenient standard. The Supreme Court held that the harm of the inability of the Navy to deploy its fleets outweighed the possible harm to cetaceans. These cases may be taught in Environmental Law, and *Winter* is traditionally taught in Civil Procedure.

B. First-Person Animal Perspective

I move through the water effortlessly, slowly heading toward the surface through the increasing light. I've been holding my breath for two hours, my heart beating slow and steady to compensate for such a deep and long dive, and I am ready for a break.[43] My pod is around me at various states of ascension, we are a big family of a couple dozen beaked whales.[44] I hear the call of a cousin off in the distance. A couple of blue whales are feeding on krill, shooting at the patch and opening their mouths wide to take in as much as possible.[45] Then I notice a shape swimming below us. At first, I

[41] E. Christien Michael Parsons, "Impacts of Navy Sonar on Whales and Dolphins: Now beyond a Smoking Gun?" *Frontiers in Marine Science* 4 (2017).

[42] *Winter*, 555 U.S. at 7–8.

[43] S.A. Rommel, A.M. Costidis, A. Fernandez, P.D. Jepson, D.A. Pabst, W.A. McLellan, D.S. Houser, T.W. Cranford, A.L van Helden, D.M. Allen, and N.B. Barros, "Elements of Beaked Whale Anatomy and Diving Physiology and Some Hypothetical Causes of Sonar-Related Stranding," *Journal of Cretacean Research and Management* 7, no. 3 (April 2006): 205, https://repository.si.edu/bitstream/handle/10088/7475/da1.pdf?sequence=1&isAllowed=y.

[44] Rommel et al., "Elements of Beaked Whale Anatomy," 203.

[45] Jeremy A. Goldbogen, John Calambokidis, Robert E. Shadwick, Erin M. Oleson, Mark A. McDonald, and John A. Hildebrand, "Kinematics of Foraging Dives and Lunge-Feeding in Fin Whales," *Journal Experimental Biology* 209, no. 7 (April 2006): 1232, https://doi.org/10.1242/jeb.02135.

think it is another cousin, but it doesn't move right and it doesn't respond to the clicks I send out as a greeting. It is making a weird noise too; it is a meaningless hum that becomes piercing as it gets closer. My family is confused, and we are all trying to escape the noise, some of us diving again without renewed air. I don't know what to do and head toward the blue whales, but their food has disbursed, and they too are not sure what to do or how to escape the sound. I try to communicate with my family, but I can't hear them and I can't see them any longer. I rush to the surface to get to the air and to try to get as far from the weird creature as I can, but it hurts all through my body.[46] I think I'm going too fast, but I can't stop now. Once the creature has been gone for a while, I am on the surface trying to recover from the pain and the damage, and I can hear the blue whales crying in distress about the loss of food.[47]

C. What Would the Result be in the Animal Kingdom?

Where cetaceans' experiences and moral standing are recognized and validated, this suit would have been found in favor of the cetacean community. They wouldn't lack standing in their own groups as they were found to with our laws. They are also community animals that support one another. Particularly when looking at dolphins and porpoises, they will stand up against an attacker as a group for the best outcome for the community.

D. How Does Viewing This Case from the Animal Perspective Inform Current Understandings?

Looking at the animal perspective in this case outlines how much harm these animals suffer for military training. The Supreme Court decided what public favor would be but didn't take into account what public favor would be with a narrative from the animal perspective. Additionally, as with the example above regarding the wild horses and burros, education can go a long way toward swaying public opinion.

[46] Rommel et al., "Elements of Beaked Whale Anatomy," 189.

[47] The distance at which sonar can affect marine mammals is much greater than initially thought, and that is portrayed in this vignette. For the purpose of brevity we do not address the working theories and research here. Parson's article (reference xli) delves into this topic and discusses the subsequent impacts of sonar on marine mammals (such as groundings and physiological reactions).

Naruto v. Slater

A. Title of Case, Procedural History, How it is Typically Understood and Taught

The *Naruto v. Slater* case is about Naruto and his intellectual property rights.[48] Naruto is a crested black macaque (*Macaca nigra*)—a monkey that is typically under two feet tall, affectionate, and has a long face and a striking thatch of hair sticking up from the top of their head—who in 2018 lived in the Tangkoko Nature Reserve on the island of Sulawesi, Indonesia. Naruto ended up being named Plaintiff in a U.S. Court case because in 2018 photographer David Slater was taking photos and Naruto took the camera and took several photos of himself ("selfies"). Slater then used these photos in a book.

The People for the Ethical Treatment of Animals (PETA) and Dr. Antje Englehardt (a primatologist) brought a claim of copyright infringement on behalf of Naruto to the U.S. District Court. Initially they used the "next friend" doctrine to recover fair compensation for Naruto. The District Court dismissed the claim stating that Naruto lacked statutory standing; however, the Ninth Circuit Court of Appeals—while it too found that the Copyright Act does not grant standing to non-human plaintiffs—found that Naruto may have constitutional standing under Article III based on the earlier *Cetacean Community v. Bush* decision. After this was determined, Slater and PETA reached a settlement whereby Slater would donate twenty-five percent of future revenues to groups that protect crested macaques in Indonesia.[49] This case is relatively new, and is not frequently taught, but may be incorporated in more curriculums in the future.

B. First-Person Animal Perspective

I hear branches breaking as something heavy walks nearby. With other males from my troop, I was foraging for fruit for the entire troop. When we heard the snaps and tromping, we dropped our forage and fled to the safety of the canopy.[50] We watched silently as a giant hairless creature with

[48] Naruto v. Slater, 888 F.3d 418, 420 (9th Cir. 2018).

[49] Jason Slotkin, "'Monkey Selfie' Lawsuit Ends with Settlement Between PETA, Photographer," NPR, September 12, 2017, https://www.npr.org/sections/thetwo-way/2017/09/12/550417823/-animal-rights-advocates-photographer-compromise-over-ownership-of-monkey-selfie.

[50] Tyrrell et al., "Avoidant Social Style Among Wild Crested Macaque Males (Macaca Nigra) in Tangkoko Nature Reserve, Sulawesi, Indonesia," *Behaviour* 157, no. 5 (May 2020): 451–491, https://doi.org/10.1163/1568539X-bja10009.

a shifting hide came into the clearing. We've seen these creatures before, and are wary of them, but they haven't tried to eat us. There have been more of them recently at the forest edges, which keep getting closer to our home.[51] He stops and starts taking items out of a bag, and then attaches a couple of them together to make a black box with a tube. He holds this to his face while sitting there, sometimes pointing it at us. I watch him pressing parts and moving others while it makes a clicking sound. I see another black box behind him that looks the same, and I am so curious—and it looks like fun—I can't help but slowly approaching to check it out behind his back.[52,53] I picked up the box and checked out the tube coming out of it. To my great surprise I saw another monkey looking back at me! To show that I wanted to be peaceful, I silently bared my teeth at him and he did the same.[54] At the same time I pushed some buttons and the box made clicking noises. The creature shifted and so I set down the box and ran back up a tree. These creatures are always interrupting us when they are around and leaving things that smell funny, like them. I have other things to think about at home, I had to find a new social group a while after playing with the black box because I was too old to stay with my troop. It was a little scary, but also exciting.[55] Our social groups are everything to us, and I had to find a new group and figure out where I ranked among the other males. This is made more stressful by the fact that the places we can go are getting smaller.[56] There seems to be less forest lately.

[51] Jatna Supriatna et al., "Deforestation on the Indonesian Island of Sulawesi and the Loss of Primate Habitat," *Global Ecology and Conservation* 24 (December 2020): 1–10, https://doi.org/10.1016/j.gecco.2020.e01205.

[52] James R. Anderson, "Development of Tool-Use to Obtain Food in a Captive Group of Macaca Tonkeana," *Journal of Human Evolution* 14, no. 7 (November 1985): 637, https://doi.org/10.1016/S0047-2484(85)80072-5.

[53] This source describes a different macaque species, but this species comes from the same area of Indonesia and is closely related to the crested macaque.

[54] Bridget M. Waller, Jaime Whitehouse, and Jérôme Micheletta, "Macaques Can Predict Social Outcomes from Facial Expressions," *Animal Cognition* 19 (May 2016): 1031, https://doi.org/10.1007/s10071-016-0992-3.

[55] Pascal Marty et al., "Is Social Dispersal Stressful? A Study in Male Crested Macaques (Macaca Nigra)," *Hormones and Behavior* 87 (January 2017): 62, https://doi.org/10.1016/j.yhbeh.2016.10.018.

[56] Slotkin, "'Monkey Selfie.'"

C. What Would the Result be in the Animal Kingdom?

In the animal kingdom Naruto's rights would be dictated by his social standing. More dominant males make the decisions for the subordinate males in the group. Though even when there is a disagreement, the social connection in the group keeps interactions civil. They also mediate disputes through this connection to one another.[57] In the troop, if intellectual property was being disputed it would be a mediated disagreement with the likely outcome of whatever the alpha males deem is best for the group. Additionally, if the photo resulted in a boon for the troop, then that labor would make Naruto look good to the group.

D. How Does Viewing this Case from the Animal Perspective Inform Current Understandings?

While viewing this case from Naruto's perspective might not change the outcome, it might change how people see the rights of animals in the future. Particularly in cases where the animal is suffering a loss at the hands of humans. While the courts found in Slater's favor, he chose to donate some of the proceeds of the photos to the protection of black crested macaques. Seeing as how macaques are suffering habitat loss at the hands of humans, giving some compensation for profiting off their work is perhaps a fair trade.[58] Additionally, future lawmakers might consider the animals' perspective when looking at rights not conventionally associated with animals, like intellectual property.

Conclusion: Revisiting Legal Cases from a Critical Animal Studies Perspective

This chapter links critical animal studies and legal studies to re-consider famous legal cases from the perspective of animals they involve. Reconstructing the perspective and result of cases as they might be perceived from animal's perspectives invites readers to reimage cases they might already know in a new light—from a more-than-human perspective.

[57] Veronica B. Cowl, Susan L. Walker, and Susanne Shultz, "Reproductive Males Are Effective at Managing Conflict in Captive Sulawesi Crested Macaques (Macaca Nigra)," *American Journal of Primatology* 83, no. 7 (July 2021): 7, e23266, https://doi.org/10.1002/ajp.23266.

[58] Rebecca Mead, "Should Leopards Be Paid for Their Spots?" *New Yorker* (March 2022): https://www.newyorker.com/magazine/2022/03/28/should-leopards-be-paid-for-their-spots.

This is a tiny step in a much larger project of examining our implicitly anthropocentric legal institutions, forming new visions of a model that might incorporate notions of interspecies equity, and wielding the tools of our legal craft to lay pathways between the status quo and new realities.

Bibliography

Adams, Richard. *Watership Down*. London: Rex Collings Ltd, 1972.

Agan, Suzanne W., Treves, Adrian, and Lisabeth L. Willey. "Majority Positive Attitudes Cannot Protect Red Wolves (Canis Rufus) from a Minority Willing to Kill Illegally." *Biological Conservation* 262 (2021): 1-11. https://doi.org/10.1016/j.biocon.2021.109321.

Anderson, James R. "Development of Tool-Use to Obtain Food in a Captive Group of Macaca Tonkeana." *Journal of Human Evolution* 14, no. 7 (1985): 637-645. https://doi.org/10.1016/S0047-2484(85)80072-5.

Arizona State University. "ASU Charter." Arizona State University. Accessed September 9, 2022. https://president.asu.edu/asu-mission-and-goals#:~:text=ASU%20Charter,of%20the%20communities%20it%20serves.

Beeland, T. DeLene. *The Secret World of Red Wolves: The Fight to Save North America's Other Wolf*. Chapel Hill: University of North Carolina Press, 2013.

Bradshaw, Karen. *Wildlife as Property Owners: A New Conception of Animal Rights*. Chicago: University of Chicago Press, 2020.

Bradshaw, Karen. "Identifying Contemporary Rights of Nature in the United States." *Southern California Law Review* 96 (2022): 1437–1466.

Bradshaw, Karen. "Radically Reimagining the Human Relationship with Nature: The Current Status of Rights of Nature in the United States." *USC Law Review* (forthcoming 2023).

Broglio, Ron. *Animal Revolution*. Minneapolis: University of Minnesota Press, 2022.

Cetacean Cmty. v. Bush. 386 F.3d 1169, 1171–72 (9th Cir. 2004).

Compassion in World Farming. "Dairy Cow-Calf Separation and Natural Weaning." *Compassion in World Farming*, February 2013. https://www.compassioninfoodbusiness.com/media/5822658/cow-calf-separation-and-natural-weaning.pdf.

Cowl, Veronica B., Walker, Susan L., and Susanne Shultz, "Reproductive Males are Effective at Managing Conflict in Captive Sulawesi Crested Macaques (Macaca Nigra)." *American Journal of Primatology* 83, no. 7 (2021): 1–12. https://doi.org/10.1002/ajp.23266.

Deckha, Maneesha. "Animalizing Law, Humanizing Animals: The Diverse and Feminist-Informed Debates in Animal Law about How to Resist Legal Anthropocentrism." *Dalhousie Law Journal* 46, no. 1 (2023, forthcoming).

Downey, Regan. "Grief Is Not Uniquely Human. Wolves Mourn Too." Wolf Conservation Center. September 26, 2016. https://nywolf.org/2016/09/grief-is-not-uniquely-human-wolves-mourn-too/.

Fernandez, Angela. *Pierson v. Post, The Hunt for the Fox: Law and Professionalization in American Legal Culture.* Cambridge: Cambridge University Press, 2018.

Gibbs v. Babbitt. 214 F.3d 483, 487 (4th Cir. 2000).

Goldbogen, Jeremy A., Calambokidis, John, Shadwick, Robert E., Oleson, Erin M., McDonald, Mark A., and John A. Hildebrand. "Kinematics of Foraging Dives and Lunge-Feeding in Fin Whales." *Journal Experimental Biology* 209, no. 7 (2006): 1231–1244. https://doi.org/10.1242/jeb.02135.

Green, Jeffrey S. "Donkeys for Predation Control." Paper presented at Fourth Eastern Wildlife Damage Control Conference, September, 25, 1989. https://static1.squarespace.com/static/52f6e70ae4b09d0c250122c6/t/54db383ee4b072be6d6ff3fa/1423652926948/guarddonkey2.pdf.

Gubbins, Roberta M. "Mutual Mistake, the True Story of Rose of Aberlone—Otto Stockmeyer Reveals the Truth of Sherwood v Walker." *Detroit Legal News*, November 4, 2010. http://www.legalnews.com/ingham/771228.

High Plains Journal. "Ranchers Use Donkeys to Protect Herds from Predators." *High Plains Journal*, August, 27, 2007. https://www.hpj.com/archives/ranchers-use-donkeys-to-protect-herds-from-predators/article_c4e984a0-fe8e-5816-a546-75b7020e3b51.html.

Isenberg, Sydney. "CPW Turns to Burros to Help Stop Wolf Depredation on Jackson County Rancher's Property." *Denver7 ABC*, March 18, 2022. https://www.thedenverchannel.com/news/local-news/cpw-turns-to-burros-to-help-stop-wolf-depredation-on-jackson-county-ranchers-property.

Kleppe v. New Mexico. 426 U.S. 529, 531–35 (1976).

Landsberg, Gary M., and Sagi Denenberg. "Social Behavior of Cattle." In *Merck Veterinary Manual.* Last modified October 2022. https://www.merckvetmanual.com/behavior/normal-social-behavior-and-behavioral-problems-of-domestic-animals/social-behavior-of-cattle.

Leiner, Lisa and Markus Fendt. "Behavioural Fear and Heart Rate Responses of Horses After Exposure to Novel Objects: Effects of Habituation." *Applied Animal Behaviour Science* 131, no. 3–4 (2011): 104–109. https://doi.org/10.1016/j.applanim.2011.02.004.

Marty, Pascal R., Hodges, Keith, Heistermann, Michael, Agil, Muhammad, and Antje Engelhardt. "Is Social Dispersal Stressful? A Study in Male Crested Macaques (Macaca Nigra)." *Hormones and Behavior* 87 (2017): 62–68. https://doi.org/10.1016/j.yhbeh.2016.10.018.

Mead, Rebecca. "Should Leopards Be Paid for Their Spots?" *New Yorker*, March, 2022. https://www.newyorker.com/magazine/2022/03/28/should-leopards-be-paid-for-their-spots.

Milman, Oliver. "Suicides Indicate Wave of 'Doomerism' Over Escalating Climate Crisis." *Guardian*, May 19, 2022. https://www.theguardian.com/environment/2022/may/19/climate-suicides-despair-global-heating.

Mitchell, Les. "Farming: Animals or Machines?" *Southern African Linguistics and Applied Language Studies* 31, no. 3 (2013): 299–309.

Mitchell, Les. "Nonhumans and the Ideology of Purpose." *Anthrozoös* 25, no. 4 (2015): 491–502.

Murray, Leigh M.A., Byrne, Katharine, and Richard B. D'Eath. "Pair-Bonding and Companion Recognition in Domestic Donkeys, *Equus Asinus*." *Applied Animal Behaviour Science* 143, no. 1 (2013): 67–74. https://doi.org/10.1016/j.applanim.2012.11.005.

Murray, Les. "The Cows on Killing Day." in *Dog Fox Field: Poems*, 27–28. New York: Farrar Straus & Giroux, 1993.

Naruto v. Slater. 888 F.3d 418, 420 (9th Cir. 2018).

Nussbaum, Martha C. *Justice for Animals: Our Collective Responsibility*. New York: Simon & Schuster, 2022.

Parsons, E. Christien Michael. "Impacts of Navy Sonar on Whales and Dolphins: Now beyond a Smoking Gun?" *Frontiers in Marine Science* 4 (2017): 1–11. https://doi.org/10.3389/fmars.2017.00295.

Rommel, S.A., Costidis, A.M., Fernandez, A., Jepson, P.D., Pabst, D.A., McLellan, W.A., Houser, D.S., Cranford, T.W., van Helden, A.L., Allen, D.M., and N.B. Barros. "Elements of Beaked Whale Anatomy and Diving Physiology and Some Hypothetical Causes of Sonar-Related Stranding." *Journal of Cretacean Research and Management* 7, no. 3 (2006): 189–209. https://repository.si.edu/bitstream/handle/10088/7475/da1.pdf?sequence=1&isAllowed=y.

Rørvang, Maria Vilain, Nielsen Birte L., Herskin, Mette S., and Margit Bak Jensen. "Prepartum Maternal Behavior of Domesticated Cattle: A Comparison with Managed, Feral, and Wild Ungulates." *Frontiers in Veterinary Science* 5 (2018): 1–11. https://doi.org/10.3389/fvets.2018.00045.

Sherwood v. Walker. 33 N.W. 919 (Mich. 1887).

Simmons, Heather J. E. "By the Law Inspired." *Michigan Bar Journal* 92, no. 8 (2013): 58–59. http://www.michbar.org/file/journal/pdf/pdf4article2241.pdf.

Singer, Peter, and Jim Mason, *The Ethics of What We Eat: Why Our Food Choices Matter*. Emmaus: Rodale Books, 2007.

Skolimowski, Jerzy, dir. *EO*. 2022; Poland: Skopia Film.

Slotkin, Jason. "'Monkey Selfie' Lawsuit Ends with Settlement Between PETA, Photographer." NPR, September 12, 2017. https://www.npr.org/sections/thetwo-way/2017/09/12/550417823/-animal-rights-advocates-photographer-compromise-over-ownership-of-monkey-selfie.

Special rules—Vertebrates. 50 C.F.R. § 17.84(c).

Supriatna, Jatna, Shekelle, Myron, Fuad, Habiburrahman A.H., Winarni, Nurul L., Dwiyahreni, Asri A., Farid, Muhammad, Mariati, Sri, Margules, Chris, Prakoso, Bimo, and Zuliyanto Zakaria. "Deforestation on the Indonesian Island of Sulawesi and the Loss of Primate Habitat." *Global Ecology and Conservation* 24 (2020): 1–10. https://doi.org/10.1016/j.gecco.2020.e01205.

Tyrrell, Maura, Berman, Carol M., Duboscq, Julie, Agil, Muhammad, Sutrisno, Try, and Anje Engelhardt. "Avoidant Social Style Among Wild Crested Macaque Males (Macaca Nigra) in Tangkoko Nature Reserve, Sulawesi, Indonesia." *Behaviour* 157, no. 5 (2020): 451–491. https://doi.org/10.1163/1568539X-bja10009.

U.S. Fish and Wildlife Service. "Red Wolf Recovery Program." U.S. Fish and Wildlife Service. Last updated June, 2023. https://www.fws.gov/southeast/wildlife/mammals/red-wolf/#behavior-section.

von Keyserlingk, Marina A.G. and Daniel M. Weary. "Maternal Behavior in Cattle." *Hormones and Behavior* 52, no. 1 (2007): 106–113. https://doi.org/10.1016/j.yhbeh.2007.03.015.

Wagner, Kathrin, Seitner, Daniel, Barth, Kerstin, Palme, Rupert, Futschik, Andreas, and Susanne Waiblinger. "Effects of Mother Versus Artificial Rearing During the First 12 Weeks of Life on Challenge Responses of Dairy Cows." *Applied Animal Behaviour Science* 164 (2015): 1–11. DOI: https://doi.org/10.1016/j.applanim.2014.12.010.

Wild Free-Roaming Horses and Burros Act of 1971. Pub. L. No. 92–195, 85 Stat. 649 (1971).

Waller, Bridget M., Whitehouse, Jaime, and Jérôme Micheletta. "Macaques can Predict Social Outcomes from Facial Expressions." *Animal Cognition* 19 (2016): 1031–1036. https://doi.org/10.1007/s10071-016-0992-3.

Williams, Carey A. "The Basics of Equine Behavior." *Equine Science Center, Rutgers*, July 22, 2004. https://esc.rutgers.edu/fact_sheet/the-basics-of-equine-behavior/.

Winter v. Natural Resources Defense Council, Inc. 555 U.S. 7, 16–17 (2008).

CHAPTER 11

Stolen Children of the Endless Night: A Critical Account of the Lives of British Pit Ponies

Les Mitchell

Introduction

This chapter focuses on the lives of pit ponies in British coal mines in an attempt not only to recover what happened to those individuals but also to reconstruct something of their experiences as they were forced into years of relentless hard labour under abysmal conditions. It covers the period from circa 1700 to the mid-1960s. It is a critical account in that it makes a deliberate move to be situated with those who have least power, who are often treated as objects, as without sentience or agency and as discountable except for their usefulness to those with power.

L. Mitchell (✉)
International Studies Group, University of the Free State, Bloemfontein, South Africa
e-mail: les@peacecaravan.co.za

M. J. Glover, L. Mitchell (eds.), *Animals as Experiencing Entities*, The Palgrave Macmillan Animal Ethics Series,
https://doi.org/10.1007/978-3-031-46456-0_11

I examine the genesis of the large-scale use of small horses in British coal mines, their transportation, labour, conditions of work and confinement, and the likely pathological effects these had upon them as sentient beings.[1] The chapter closes with a description of the life of an archetypal pit pony. In the attempt to retrieve, however speculatively and fragmentarily, the ponies' lived experiences, I draw on a range of sources including official documents, journal articles, books, museum information, personal recollections from a variety of sources as well as scientific and veterinary information.

Dominant power lexicalises the world in its discourses, constructing a social reality to reflect its own ideology. The name *pit ponies* suggests that these individuals were at home in mines doing what they did because it was, in some way, natural for them. But in reality, they were coerced, abused and enslaved, and so I shall refer to them as *mine slave ponies* or *slave ponies* for short which is more congruent with the true state of affairs.[2] Some of these ponies were originally free living but the ancestors of others had been the captives of humans for generations. While ponies were increasingly used in coal mines during the industrial revolution, their coerced labour persisted deep into the twentieth century.

The Rise of Coal and the Huskar Tragedy

Around the year 1690, coal was being produced in Britain at the rate of three million tonnes per annum and output was increasing as coal became more extensively used and in a greater variety of ways. Coal was employed for domestic heating, in lime-burning and dyeing, by smiths and glassmakers in sugarhouses to fire brick and tile kilns and as a source of heat for breweries and distilleries.[3] In the eighteenth century coal was used to

[1] Ponies are small horses, about 146cm in height or less at the shoulder.

[2] Slavery is a social practice which has variations over time and varies between cultures. At its core is the assumption of the right of an individual or group to use the bodies of other sentient beings for their own purposes. It might also include the claim by that group of the right of ownership of those sentient beings and other ideological claims. This is a practice where the identities of the victims can differ widely but the essential practice remains constant and therefore identifiable. The social practice with respect to these animals and many others can rightfully be described as a form of slavery.

[3] John Butt, 'Coal Industry', in *The Oxford Companion to British History* (Oxford University Press, 2015), http://0.www.oxfordreference.com/view/10.1093/acref/9780199677832.001.0001/acref-9780199677832-e-1031.

make coke, a critical component in the smelting of iron, to manufacture gas for heating and illumination; it was also an important source of a range of chemicals and was used to fire the new steam engines. By 1800 annual coal output was over 11 million tonnes, and by 1854 output approached 65 million tonnes, an amount which had more than quadrupled by 1914 to reach 287.4 million tonnes, the highest output achieved by the industry.[4] Coal was a lucrative export and an important employer of men. And in the Cimmerian darkness, it was slave ponies who purchased that black gold with their lives.

We begin with an event in the middle of the nineteenth century that would change the lives of millions of small equines catastrophically and impact many women and children; it would inspire legislation and offer the British public a detailed and deeply disturbing insight into the conditions in their mines.

On the afternoon of Wednesday 4 July 1838, the heavy summer heat around the village of Silkstone Common, in what was then the West Riding of Yorkshire, was broken by an unusually violent thunderstorm. In the area were two coal mines, the Moor End and Huskar pits, both of which were owned by Mr Robert C. Clark.[5] Working underground that day were both adult miners and around forty children. So fierce was the storm that there was concern that the mine might begin to flood. But to complicate matters, the fire used for heating the steam engine, which operated the winder used for bringing coal and miners to the surface, had been extinguished by the rain and hail.[6]

The miners were uncertain about what was happening but, fearing there might be a risk of explosion from fire damp, they extinguished their lamps and made their way to the base of the shaft where they were asked

[4] Butt.

[5] '7 Jul 1838, 6—Sheffield and Rotherham Independent at Newspapers.Com', Newspapers.com, 6, accessed 2 June 2021, http://www.newspapers.com/image/?clipping_id=22840031&fcfToken=eyJhbGciOiJIUzI1NiIsInR5cCI6IkpXVCJ9.eyJmcmVlLXZpZXctaWQiOjQwMTE2Mjc5OSwiaWF0IjoxNjIyNjIyMDI5LCJleHAiOjE2MjI3MDg0Mjl9.ITHStsPQ2F6C-UaflgGiKERx8u1DcNIz-5vFbzy0Dzw.

[6] 'Huskar Pit Disaster', *The Penistone Archive* (blog), accessed 21 May 2021, https://penistonearchive.co.uk/huskar-pit-disaster/; 'The Coalmining History Resource Centre. 1820–183939.Pdf', accessed 2 June 2021, http://www.cmhrc.co.uk/cms/document/1820_39.pdf; '7 Jul 1838, 6—Sheffield and Rotherham Independent at Newspapers.Com'.

to assemble and wait.[7,8] They had already spent nine hours working in the pit.[9]

The children became particularly agitated and wanted to leave as soon as possible, and begged permission to do so, and for unclear reasons they eventually believed that this had been given to them.[10] Although they could not be hauled up the shaft, there was a side tunnel, a drift or day hole, into the mine from Nab's Wood. Sloping down into the pit, the drift was about four feet wide and up to six feet high, and was closed off from the mine by a trap door.[11] Compartmentalising sections of a mine was a common system for making sure the ventilation worked effectively and the trap door would only be used to let something or somebody through and then closed again.

The children found their way to the drift, went out through the trap door, closed it behind them and began to walk up the tunnel. As this was happening the rain and hail from the storm had continued to enter a local stream which was usually dry for most months of the year but on this day came into flood and burst its banks, pouring water into the drift. At the coroner's hearing, it was suggested that the flow of water was not so great in itself but that at one point in the drift it had pooled, making a small dam which eventually burst. With the water cascading down upon them, the children were trapped.[12] The door behind them had been closed and would be held ever more tightly by the flood as the lower section of the drift rapidly filled with water. A few older children managed to escape into 'slits' in the shaft but for the others there was no way out and twenty-six children lost their lives that day.[13]

Reports of the Huskar disaster shocked Britain, and Queen Victoria rapidly ordered an enquiry. Less than two years later, in 1840, a Royal Commission of Enquiry headed by Lord Ashley was tasked with investigating working conditions in mines and especially the employment of

[7] There is a suggestion that they might not have heard this instruction.

[8] 'Huskar Colliery 1838—Mining Accident Database', accessed 21 May 2021, http://mineaccidents.com.au/mine-accident/202/huskar-colliery-1838; '7 Jul 1838, 6—Sheffield and Rotherham Independent at Newspapers.Com'; 'Huskar Pit Disaster'.

[9] 'Huskar Colliery 1838—Mining Accident Database'.

[10] '7 Jul 1838, 6—Sheffield and Rotherham Independent at Newspapers.Com.' The reason why the children thought they had permission to leave is not clear.

[11] '7 Jul 1838, 6—Sheffield and Rotherham Independent at Newspapers.Com', 6.

[12] '7 Jul 1838, 6—Sheffield and Rotherham Independent at Newspapers.Com' column 3.

[13] '7 Jul 1838, 6—Sheffield and Rotherham Independent at Newspapers.Com'.

children. Four commissioners were appointed, as well as a number of assistant commissioners, who investigated mines for coal, iron, lead, tin and copper. Two years later in 1842, a report was published titled, 'The Condition and Treatment of the Children Employed in the Mines and Colliers of the United Kingdom. Carefully Compiled from the Appendix to the First Report of the Commissioners Appointed to Enquire into this Subject. With Copious Extracts from the Evidence, and Illustrative Engravings. [The preface signed: W.C.]'.[14] It is an account of the commission's investigations and findings and displays a strong commitment to detail, which affords a real sense of immediacy. The commissioners clearly took their work very seriously and literally risked life and limb going into the mines, getting wet from overhead water, crawling through low tunnels, slopping through standing water and being exposed to noxious gases and to the ever-present risks of cave-ins and explosions. They interviewed people who laboured down the mine and examined the social and physical conditions of the workers' lives. What the commissioners produced is a comprehensive document, displaying authority, authenticity and at times outrage. Although ponies, or 'horses', are mentioned frequently in the report,[15] how working in the mines affected them is not its focus but rather the impacts on children. However, since the ponies experienced many of the same conditions as the children, understanding what the children suffered offers critical insights into what the ponies also endured.[16]

What resulted from the report was the Mines and Collieries Bill of 1842, which forbade women, as well as boys and girls under the age of ten, from working under ground and it is worth noting that not all members of the British Parliament were entirely happy about this legislation.[17]

[14] Commissioners for Inquiring into the Employment and Condition of Children in Mines and Manufactories, *The Condition and Treatment of the Children Employed in the Mines and Colliers of the United Kingdom. Carefully Compiled from the Appendix to the First Report of the Commissioners ... With Copious Extracts from the Evidence, and Illustrative Engravings. [The Preface Signed: W.C.]*, 1842, http://access.bl.uk/item/viewer/ark:/81055/vdc_100054704592.0x000001.

[15] Pp 4, 5, 6, 11, 13, 15, 16, 20, 34, 38, 41, 42, 43, 44, 45, 51, 52, 53, 58, 62, 63, 70, 74, 82, 83, 85, 88, 89, 90.

[16] See especially pp 20, 41, 44, 52, 74, 89.

[17] 'UFR LCE—Département d'études Anglophones—The Mines Act, 1842', 21 July 2011, https://web.archive.org/web/20110721022926/http://anglais.u-paris10.fr/spip.php?article88; 'MINES AND COLLIERIES. (Hansard, 1 August 1842)', accessed 24 May 2021, https://api.parliament.uk/historic-hansard/lords/1842/aug/01/mines-and-collieries; 'MINES AND COLLIERIES. (Hansard, 14 July 1842)', accessed 21 May 2021, http://hansard.millbanksystems.com/lords/1842/jul/14/mines-and-collieries#s3v0065p0_18420714_hol_51.

The bill was also met with considerable dismay by some poor families, as it deprived them of their meagre but desperately needed incomes. From the perspective of small horses, it produced a dreadful effect, stimulating their significantly increased use underground as, to a great extent, they replaced those excluded women and children.

Underground Conditions and Types of Labour

There were many different types of coal mines at the time, some not much more than tunnels into the sides of hills, while others had deep shafts requiring lifting gear in order to bring the miners and coal to the surface. Coal mining operations were affected by such things as the thickness of the coal seams, the nature of the surrounding rock, the hydrology and local customary practices, such as who was allowed to work underground and at what age.[18]

Dr Mitchell, an assistant commissioner for the 1842 report, described his journey around one pit, which began with his descent after the skip of about 25 cwt (1270 kg) had been offloaded at the surface.[19] He was given a candle, which he held in a piece of clay, and he and his companions proceeded along a road or 'horse road' with rails laid in the centre to allow the horses to pull the wagons. The roof in this space was eight or nine feet high (about 2.6m) and would not be made any higher because rocks falling from a greater height would be even more dangerous. He saw the stables, observing that the horses looked fat and shiny and commented that this was thought to be due to the 'warm uniform temperatures' of the mine, and further recorded that a witness elsewhere stated that donkeys could not endure the heat of the coal mine.[20] Mr Symons, another

[18] Commissioners for Inquiring into the Employment and Condition of Children in Mines and Manufactories, *The Condition and Treatment of the Children Employed in the Mines and Colliers of the United Kingdom. Carefully Compiled from the Appendix to the First Report of the Commissioners … With Copious Extracts from the Evidence, and Illustrative Engravings. [The Preface Signed].*

[19] Commissioners for Inquiring into the Employment and Condition of Children in Mines and Manufactories, 5.

[20] Commissioners for Inquiring into the Employment and Condition of Children in Mines and Manufactories, 5.

assistant commissioner, noted that "it is difficult to describe the feeling of dark confinement and damp discomfort conveyed by a colliery, at first sight".[21]

One of the many problems faced in mines was the seepage of water, which could be particularly bad near the shafts themselves, although in some mines water continually seeped through the roofs of the workings, and often water had to be pumped out by hand or with powered machines. Miners sometimes had to work continually in wet conditions and the report stated that "[i]n a colliery at Mirfield the men were found actually working in water, and in many others the children's feet are never dry".[22] Of the conditions in the mine where she worked before the engine was "put up", Fanny Drake, aged 15, one of the witnesses of Mr Symons, said:

> I have had to hurry [pull the corf full of coal] up to my calves in water. It was as bad as this a fortnight at a time; and this was for half a year last winter. My feet were skinned and just as if they were scalded; for the water was bad—it had stood some time, and I was off my work owing to it and had a head-ache and bleeding at my nose.[23,24]

Margaret Hipps, 17 years old, spoke of her work dragging on her hands and feet, a bagie of coal of around 130 kg, to which she was hitched by a chain or rope, over a wet pavement in a tunnel about 28 inches (70cm) high and over a distance of between 200 and 400 yards (183 and 366m).[25] The Sub Commissioner commented:

> It is almost impossible that human beings can submit to such employment, crawling on hands and knees, harnessed like horses over soft slushy floors more difficult than dragging the same weight through our lowest-common

[21] Commissioners for Inquiring into the Employment and Condition of Children in Mines and Manufactories, 8.

[22] Commissioners for Inquiring into the Employment and Condition of Children in Mines and Manufactories, 9.

[23] Commissioners for Inquiring into the Employment and Condition of Children in Mines and Manufactories, 10.

[24] Water standing in contact with coal for some time might become acidic due to the presence of sulphur compounds as in acid mine drainage. This could explain Fanny Drake's skinned and 'scalded' feet—chemical burns.

[25] Commissioners for Inquiring into the Employment and Condition of Children in Mines and Manufactories, 52.

> sewers, and more difficult because of the inclination which is frequently one in three to one in six.[26]

Almost continually wet hooves and the accompanying discomfort were plausibly what the ponies endured too. It was also mentioned in the report, horses being covered in waxed cloths to offer some protection from water dripping from the roof.

Miners and horses were not the only occupants of mines. A witness, Henry Cannin, was reported as telling the commission:

> We have no mice in our pit but there are many in other pits. We have many rats, almost as big as rabbits, quite as big as half-grown rabbits. They rob our bait bags [personal food bags] and tear the candles sometimes. … They eat the horse's corn. We have thousands of gnats and thousands of spiders at the furthest part of the pit … and forty-legs and earwigs and black bats (beetles). Mushrooms will grow in the stables, fifty yards from the shafts.[27]

In another mine there were also 'chirping crickets'. The coal mines were multi-species places.

Another difficulty for mines was ensuring that there was efficient ventilation, without which there was a risk that parts of the workings would become very hot and humid and deficient in fresh air, but also that poisonous and/or explosive gases could accumulate. Many colliers worked half naked and some completely naked because of the heat and to spare their clothes from damage. A system of ventilation that was common involved two separate mine shafts, with one of those shafts containing a furnace which was continually kept burning. This fire created an up draft, drawing air out of the workings and caused air to be pulled down into the mine via the other shaft, thereby circulating the air through the different passages.

There was an ever-present risk of explosion and fire caused by the seepage of methane gas (fire damp) and the abundance of coal dust in the air. In some places carbon dioxide might pool at lower levels, meaning that below a certain level the 'air' would lack sufficient oxygen and this could cause suffocation. Carbon monoxide produced from fires or explosions was another hazard. Carbon monoxide is extremely poisonous, causing

[26] Commissioners for Inquiring into the Employment and Condition of Children in Mines and Manufactories, 52.

[27] Commissioners for Inquiring into the Employment and Condition of Children in Mines and Manufactories, 11.

death even at low concentrations and would have been undetectable. Any cocktail of these gases might have been in the air at any given time.

Light for individuals was provided by candles, often at the expense of the miners themselves and a number of candles would be required on a shift. Oil lamps might also have been used but both candles and oil lamps posed a significant risk of initiating an explosion; although later, the Davy Lamp reduced this threat. However, it is fair to say that illumination was extremely poor and with dust in the air the visibility must have been very low at times.

Rock falls were another significant hazard. Sometimes rocks falling directly down the mine shaft posed a threat to life and in other places there was the risk of cave-ins. In some mines wooden poles were used to try to stop the roof lowering over time and to prevent rock falls.

Since the ventilation system depended upon a particular pattern of air flow, and this required the use of doors (traps) in various places, which needed to be kept closed whenever possible, *trappers* were employed to control these doors. Once any traffic was allowed through, it was essential that the trap was closed again, or the ventilation system would be compromised. Children as young as five years of age did this critical job, sitting alone in the pitch dark for up to twelve hours a day, seeing only a few seconds of light as a corf approached and passed through their trap door.[28]

Miners, *holers* or *getters* dug out the coal and they were usually but not always male. They often worked in very confined spaces such as tunnels only 20 inches (50 cm) high—about knee height.[29] The work was hot, difficult and dangerous.

Coal was loaded into corfs or corves, rather like large baskets possibly with some kind of runners underneath. Later corfs would develop into something more like a wagon. *Hurriers*, *pullers* or *thrusters* would move these corves of coal through the narrow, low tunnels with the puller at the front, sometimes wearing a harness and with the thruster pushing from the back, often with their head.[30] This sometimes caused a bald patch to develop. The puller, if the harness system was used, was attached to the

[28] Commissioners for Inquiring into the Employment and Condition of Children in Mines and Manufactories, 38–42.

[29] Commissioners for Inquiring into the Employment and Condition of Children in Mines and Manufactories, 37, 38 and Preface, 'Illustrative Engravings'.

[30] Commissioners for Inquiring into the Employment and Condition of Children in Mines and Manufactories, 43–47.

corf via a chain, which went around their waist and between their legs.[31] Children, male and female, were employed for this work because the space where it was done was low and narrow.[32] It was hard work. Women, including women who were pregnant, also worked underground in many pits. Being shackled to the corf in order to pull it was especially difficult and painful for these women.

Margaret Harper, 13 years old, draws a telling parallel with slave ponies in her poignant testimony.

> We hurry the carts on the railroads by pushing behind; I frequently draw with ropes and chains as the horses do; it is dirty slavish work. And the water quite covers our ankles. I knock my head against the roofs, as they are not so high as I am, and they cause me to stoop, which makes my back ache.[33]

It appears that many miners did what they could to be kind to those children who worked with them. But getting as much coal out as quickly as possible in order to achieve the miner's quota was the objective; they were under huge pressure and in some cases the children bore the brunt.

William Dyson, a hurrier, gave this testimony about a girl who worked in his pit:

> [W]hen she is down she hurries in the same way that we do, without shoes or stockings; I have seen her thrashed many times when she does not please them; they rap her in the face and knock her down; I repeat, I have seen this many times. She does not like her work; she does not that; I have seen her cry many times; the men swear at her often; she says she will be killed before she leaves the coal-pit.[34]

The report documented, through first-hand testimony, the terrible violence that was often meted out by some men against children and women. In a violent and exploitive work environment the violence which could not be used against their exploiters was instead transferred down the levels of

[31] Commissioners for Inquiring into the Employment and Condition of Children in Mines and Manufactories, 47, 51.

[32] Commissioners for Inquiring into the Employment and Condition of Children in Mines and Manufactories, 26–30.

[33] Commissioners for Inquiring into the Employment and Condition of Children in Mines and Manufactories, 74.

[34] Commissioners for Inquiring into the Employment and Condition of Children in Mines and Manufactories, 30.

power to the most vulnerable. This was a situation where, it might reasonably be expected, animals, as the most vulnerable of all, would not fare well.[35]

Accidents were common, and it has been pointed out that children described not having been hurt *yet*, as if being hurt was to be expected at some time or other; the mines were extremely dangerous places for children.[36] Over the next 150 years, conditions in mines would improve in some respects but they were still extremely dangerous places. Even between 1900 and 1950 a recorded total of 84,331 UK miners lost their lives from mine work.[37]

The New Children: Procuring Victims, Abuse and Danger[38]

Like humans who worked the coal mines, horses are richly sentient beings. Horses detect high frequencies, beyond human hearing, and may suffer stress from noise; they have highly developed senses of smell and taste and are very sensitive to touch.[39] As mammals, and like birds, fish and reptiles, they have the sensory faculties and neurological features for felt sensory experiences.[40] Horses remember voices which have positive or negative associations for them, and different sides of their brains become activated

[35] The term *animal* is used here and elsewhere in the chapter to denote nonhuman animal.

[36] Commissioners for Inquiring into the Employment and Condition of Children in Mines and Manufactories, 74, 75.

[37] 'Coalmining Accidents and Deaths—Vanilla Circus—Maps, Poems and Searchable Databases for Mining in the UK', accessed 27 August 2021, http://www.cmhrc.co.uk/site/disasters/index.html.

[38] I will use the term abuse as defined by Agnew as any act that contributes to the pain or death of an animal or that otherwise threatens the welfare of an animal.

Such abuse may be physical (including sexual) or mental, may involve active maltreatment or passive neglect, may be direct or indirect, intentional or unintentional, socially approved or condemned and/or necessary or unnecessary (however defined). Robert Agnew, 'The Causes of Animal Abuse: A Social-Psychological Analysis', *Theoretical Criminology* 2, no. 2 (1998): 79.

[39] Maria Vilain Rørvang, Birte L. Nielsen, and Andrew Neil McLean, 'Sensory Abilities of Horses and Their Importance for Equitation Science', *Frontiers in Veterinary Science* 0 (2020), https://doi.org/10.3389/fvets.2020.00633.

[40] D. DeGrazia, 'Sentience and Consciousness as Bases for Attributing Interests and Moral Status: Considering the Evidence and Speculating Slightly Beyond', in L. Johnson, A. Fenton, and A. Shriver (eds.), *Neuroethics and Nonhuman Animals* (Cham: Springer, 2020), 17–31.

in each case, and their ears are held either forwards or backwards in response.[41] We know that horses exhibit rapid eye movement sleep (REM), as do many mammals, and this is the sleep humans enter most of the time when we are dreaming.[42] Horses can also exhibit vocalisations and movements while asleep. If horses do not get sufficient sleep, especially REM sleep, it can have a significant effect on them and they might even enter REM sleep while standing, leading to collapse and injury.[43] They are intelligent, sociable and have good memories.

Horses can see better in the dark than humans, once their eyes have adapted, but this does not happen quickly. In natural situations, light intensity changes relatively slowly and horses' eyes can adapt easily to such changes. However, when a horse goes from a bright situation into one which is relatively dark, perhaps from sunshine into a dark barn, the horse's sight becomes severely limited or it might not be able to see properly at all, causing significant symptoms of distress such as head shaking and even panic.[44]

When they are living free, horses need to avoid predators and have the largest eye of any land mammal, giving them a 350-degree field of vision. This wide field is almost totally monocular and not used for fine detail, but it is incredibly sensitive for detecting movement, which means horses are able to pick up even the slightest movement far in the distance.[45] However, horses do have a narrow strip of binocular vision when they look forward which means they can focus and see depth and detail, which is why a horse

[41] Serenella d'Ingeo et al., 'Horses Associate Individual Human Voices with the Valence of Past Interactions: A Behavioural and Electrophysiological Study', *Scientific Reports* 9, no. 1 (9 August 2019): 11568, https://doi.org/10.1038/s41598-019-47960-5.

[42] Mark S. Blumberg et al., 'What Is REM Sleep?', *Current Biology* 30, no. 1 (6 January 2020): R38–49, https://doi.org/10.1016/j.cub.2019.11.045; 'The Dangers of Sleep Deprivation in Horses | International Society for Equitation Science', accessed 12 August 2021, https://equitationscience.com/media/the-dangers-of-sleep-deprivation-in-horses; 'Do Horses Dream?', Of Horse, accessed 12 August 2021, https://www.ofhorse.com/view-post/Do-Horses-Dream.

[43] Discussion of this subject especially with reference to work by Christine Fuchs et al. in 'The Dangers of Sleep Deprivation in Horses | International Society for Equitation Science'.

[44] 'Point of View: Understanding How Horses See the World', Horse Sport, accessed 27 July 2021, https://horsesport.com/magazine/training/point-of-view-understanding-how-horses-see/; Christopher Murphy, 'Equine Vision. Equine Vision: Normal and Abnormal. In: Equine Ophthalmology (Ed. Gilger BC), Elsevier Saunders, St. Louis, 2010:396-434', 2010, 396–434.

[45] Patricia Evans, 'Equine Vision and Its Effect on Behavior', n.d., 3.

may move its head or even its body to look at something more clearly.[46] But if a horse's head is bent lower than the horizontal, for example when being forced to work in conditions where their neck is bent down, such as in a low tunnel, they are unable to see the path ahead and are forced to look at the area around their own front legs.[47]

A range of horses were used in mines and were selected for the type of work they would do and the conditions under which they would be forced to do it. A pony is a small horse, and types of small horses are found in various places, having evolved separately from their larger relatives. Shetland ponies, for example, are small and stocky and originally from the Isles of Shetland where they may have existed for some 2000 years.[48] During the nineteenth century thousands were imported into mainland Britain to be used as slave ponies but "often did not live very long".[49] Welsh ponies and possibly Fell ponies were also used in the coal mines, and as demand grew ponies were imported from abroad.[50]

> Both Shetland and Welsh ponies were common, as were Dale horses but breeds varied considerably throughout the coal fields. During times of high production and pony shortages very high prices could be paid for good animals meaning that ponies could be imported from as far away as the USA, Iceland and Russia.[51]

Their journeys to the coal mines would have involved abduction from their kin, forced treks and in some cases arduous and dangerous sea journeys. These sentient individuals were treated as commodities, objects of economic exchange who, at the terminal point, would morph into enslaved labourers. They were abducted, trafficked and then permanently enslaved. By these actions, perpetrated by men, they were brought to the end of their lives as they had known them from the day of their birth. They were

[46] Evans.

[47] Evans.

[48] Emily Donoho, 'The Shetland Pony: All You Need to Know about This Small Hardy Breed', Horse & Hound, 27 December 2018, https://www.horseandhound.co.uk/features/shetland-pony-facts-673878.

[49] Donoho.

[50] National Farmers Union, 'The Fell Pony', Countryside, 14 June 2018, https://www.countrysideonline.co.uk:443/hobbies-and-leisure/pets-and-animals/the-fell-pony43558/.

[51] Haig pit mining and colliery museum, 'Pit Ponies (Haig)', Haig Pit Mining And Colliery Museum., 7 May 2013, https://haigpit.wordpress.com/pit-ponies/.

forced to an interface through which many would pass and never return. They would cross from light into perpetual darkness.

Some horses would have entered relatively shallow mines by way of a side tunnel or drift, while others were sent into the mine by way of being tied, harnessed vertically, and lowered into the darkness of the shaft. Sometimes a cage was used. Dr Mitchell notes that "[i]t sometimes happens that when a horse is for the first time brought down into the coal-pit he falls in a fainting fit, which is supposed to arise from fear".[52] The description of horses being lowered into the pit and the panic and apparent loss of consciousness which ensued are a recurring theme in sources. The horses might fall into a dead faint or in some cases even break free from their restraints and fall hundreds of metres down the shaft.[53] Their fear and panic is easy to understand because the horses were roughly held as the restraints were applied; they were tipped up to a vertical position and fell into ever-increasing darkness, and experienced the attendant loss of vision due to their long adaption time. They would have experienced the noises of machinery, dust in the air, the sound of many human voices and activities, alien smells, moving pinpricks of light with leaping shadows and the sound of dripping water. They were now alone and powerless in an utterly alien world.

Slave ponies were used to do the jobs which previously had been carried out by children and women. They were forced to work in narrow, low tunnels. Although women and children had been forbidden to work underground because it was deemed to be completely unacceptable for them to do so, vulnerable nonhumans were forced to take their places. The anthropocentric, ethical and legal decisions here were determined by the species of the vulnerable labourer; for one species doing this work was unthinkable and illegal, and for the other it was acceptable and lawful.

Slave ponies would pull the corves of coal over the ground on sledges, or, in some mines, corves with wheels were used and pulled over boards which were fastened to the ground with iron pins, while types of rails were

[52] Commissioners for Inquiring into the Employment and Condition of Children in Mines and Manufactories, *The Condition and Treatment of the Children Employed in the Mines and Colliers of the United Kingdom. Carefully Compiled from the Appendix to the First Report of the Commissioners ... With Copious Extracts from the Evidence, and Illustrative Engravings. [The Preface Signed]*, 5.

[53] Steffan Rhys, 'In Darkness for Years, the Dangerous and Harsh Life of Pit Ponies', WalesOnline, 8 April 2008, https://www.walesonline.co.uk/news/wales-news/darkness-years-dangerous-harsh-life-2182018.

later developed.[54] The tunnels in which they did this were narrow and low, probably hot, possibly wet, the air dusty, likely containing noxious gases, extremely dark (coal has a very low albedo) and the walls, roof and floor uneven. As sentient, complex beings, the relentless work in such an environment would have taken a heavy toll on their bodies and minds.

In the 1870s, there were an estimated 200,000 slave ponies in Great Britain.[55] Life was very hard for them, many working up to 16 hours a day, often without food or water and there were widespread reports of abuse, injuries and sickness.[56] In a typical working day of eight to twelve hours, a pit pony would regularly pull 30 tonnes of coal.[57]

Bradley writes that following the 1911 Coal Mines Act an ostler was required to look after the ponies and could make decisions about their wellbeing and make sure they did not work more than seven consecutive shifts.[58] But he observes, "[o]f course in the 19th Century those rules did not exist and the poor creatures were worked until they dropped".[59] A reflection of the cramped working conditions for ponies is the requirements given for underground stables.

> The ponies had to be able to raise their heads in the stables, the height of the roof was to be seven feet when a five foot horse was in use and the horse should be able to relax its muscles because it had to work all day carrying its head low.[60]

[54] Jacqueline Hernigle, 'An Archipelago of Coal Pits: Predicting Archeological Features in the Richmond, Virginia Coalfield' (College of William and Mary—Arts and Sciences, 1991), 60, http://scholarworks.wm.edu/etd/1539625657 drawing on work by Ashton and Sykes 1967.

[55] Estimates vary and this is an upper estimate.

[56] Margaret Evans, 'Ghosts of the Coal Mines | Horse Journals', *Horse Journals*, 5 July 2019, https://www.horsejournals.com/popular/history-heritage/ghosts-coal-mines.

[57] Evans James Watt, a Scottish engineer who by 1776 had developed an improved steam engine, wanted to compare this engine to the work of horses. He studied small horses in mines and on farms and horses in factories and mills. Eventually, he and his business partner calculated that a horse could produce 33,000 foot pounds of work per minute. This they called one horsepower and the unit was used to help sell the engine. Later a new unit for power would be devised, one where work of 1 newton metre every second = 1 watt of power.

[58] 'Ponies and Ostlers', accessed 25 June 2021, http://www.healeyhero.co.uk/rescue/individual/Bob_Bradley/Ponies.html#top2.

[59] Robert Bradley, 'Ponies and Ostlers', Bob's History of Mining, nd, http://www.healeyhero.co.uk/rescue/individual/Bob_Bradley/Ponies.html.

[60] Haig pit mining and colliery museum, 'Pit Ponies (Haig)', Haig Pit Mining and Colliery Museum., 7 May 2013, https://haigpit.wordpress.com/pit-ponies/.

Brown (1989) records the following about mines in Shropshire:

> In recent times horses went down at four years of age and remained until their death. … In 1913 when there were 282 horses underground in Shropshire the Inspector of Mines recorded that 12 died or were destroyed following an accident, 5 died due to a disease, 82 cases of injury were reported together with 2 cases of ill treatment.[61]

A special mask was kept in the pit in case a pony had to be 'destroyed'. With the mask on the animal's head there was a specific hole where a sharp chisel blow could be made to the temple, which killed the pony.[62] Later a similar method might be used but with an explosive charge.

The horses faced injury to their feet and legs from slipping and from getting trapped under rails, as well as from the relentless pressure on their bones and joints. Eventually their 'legs would go' and they would be killed. Arthritis is relatively common in horses, especially as they age and, as in humans, two of the main causes are stress on joints and damage to the joints.[63]

We do not have detailed health information on slave ponies but Broster and colleagues investigated 227 'working horses' in India and Pakistan, all of whom pulled carts, some transporting goods and some people. They found that

> [a]ll horses examined were lame. Overall, 98% showed a gait abnormality in all 4 limbs and 87% had at least one limb scoring 3 or 4 on the lameness scale. Multiple pathological abnormalities within each limb were associated with lameness, with similar results in both countries. Chronic foot pathology was seen in every horse; 94% horses showed signs of chronic joint disease; 83% had digital flexor tendonitis in at least one limb. Lameness and

[61] Ivor J Brown, 'Horses in Shropshire Mines', *British Mining (Northern Mine Research Society)*, no. 39 (1989): 85.

[62] Brown, 85.

[63] 'Arthritis in Large Animals—Musculoskeletal System', Veterinary Manual, accessed 11 May 2021, https://www.msdvetmanual.com/musculoskeletal-system/arthropathies-in-large-animals/arthritis-in-large-animals; 'Pathophysiology of Osteoarthritis—VetFolio', accessed 17 August 2021, https://www.vetfolio.com/learn/article/pathophysiology-of-osteoarthritis.

pathological abnormalities were associated with specific pain responses in the feet, limbs and spine.[64]

Similar findings by other researchers using differing geographic sampling were noted in the same paper.[65]

It is difficult to imagine that slave ponies relentlessly hauling tonnes of coal day after day on uneven and often slippery surfaces would not have been subject to similar afflictions. Horses in wet conditions are also prone to getting fungal infections, especially those affecting the legs and hooves.

An excavation was made in the Central Bohemian region of the Czech Republic of skeletons of horses made to work in an iron mine hauling heavy carts between 1806 and 1914. The skeletons were found to be of thirteen individuals, most belonging to smaller breed types and estimated to be between four and seven years of age.[66] The normal life expectancy of a pony is around 25–30+ years. The skeletal pathologies recorded by the authors are numerous, including an irregular tooth pattern probably caused due to the use of a snaffle bit:

> [K]issing spine syndrome,[67] articular surface lipping,[68] synovial intervertebral inflamed line/osteoarthritis/intertransverse processes/ankylose osteophytosis as result of permanent heavy wagon pulling. Arthritis is present in several horses' with synostoses including lipping and nekroses of the tarsals. Very common are synostoses[69] between metacarpi II–IV and between metatarsi II–IV, which seem to result of mining wagon pulling stress, but also of the permanent hard bandage.[70]

Essentially the work forced on these individuals resulted in widespread and chronic damage to their skeletons, which would have caused extreme

[64] C. E. Broster et al., 'The Range and Prevalence of Pathological Abnormalities Associated with Lameness in Working Horses from Developing Countries', *Equine Veterinary Journal* 41, no. 5 (May 2009): 474, https://doi.org/10.2746/042516409X373907.

[65] Broster et al., 474, 475.

[66] Cajus G Diedrich, 'Pathologic Historic Mining Horses from Central Europe'. 1, no. 1 (2017): 26.

[67] Kissing spine syndrome—vertebrae overlap due to bony projections.

[68] Articular surface lipping—curled edge at the bearing surface of a joint and related to osteoarthritis.

[69] Synostoses—the union or fusion of adjacent bones by the growth of bony substance.

[70] Diedrich, 8.

pain, deformity, marked disability and almost certainly contributed to early deaths.

A Question of Vision

The eyesight of horses in coal mines is a source of a great deal of dispute. It was a widespread belief among miners and others that many of the horses used underground were blind or at least had deficient vision and that being underground led to this problem. However, a range of commentators, including the National Coal Board Museum, emphatically maintain that this was not true: "The popular belief that pit ponies go blind underground is untrue, and the use of blind ponies was prohibited".[71] While this is correct about the prohibition in the legislation promulgated in 1911, it raises the question of why legislation should specifically mention blind horses if this was not a known problem at the time. It might have been a response to allegations by activists who believed that blind horses were being used or it might be that blind horses were actually used.

There seem to be no allegations that blind horses were deliberately selected to be used underground, and there is little dispute that horses did have their eyes, heads and faces damaged due to accidents such as those from colliding with rocks protruding into tunnels from roofs and walls and from other obstructions. Indeed, this was so common that in some mines, horses had a type of hood put on them which was designed to protect the head and particularly the eyes from damage. Moving around in extremely low light conditions with multiple possible obstacles would never be a safe activity but the story turns out to be more complicated than this.

Some coal miners appear to have had their eyesight affected by working underground. In British coal mines there was a condition described as *miner's nystagmus* and there is much debate about its aetiology and even whether such a condition actually existed, although there were thousands of recorded cases and it is well documented. Llewelyn (1912) in his paper titled 'The Causes and Prevention of Miners' Nystagmus' writes,

> The chief symptom and physical sign is a rotatory osculation of the eyeballs. … The man first notices that he is unable to perform the more skilled

[71] 'WWMM—Pit Ponies 14/20', accessed 17 June 2021, http://www.wwmm.org/storie/storia.asp?id_storia=192&pagina=14&project=0.

> part of his work; he cannot notch timber well and fails to drive his wedge. Or strike with his pick the exact piece of coal aimed at. He next complains that the lamps dazzle his eyes, that he is unable to see anything at night time, and, finally that the lamps and all surrounding objects are going round and round. Headache varying from slight pain around the temples to attacks of extreme severity, giddiness on exertion and stooping, night blindness, dread of light ... all are found in a marked case of nystagmus.[72]

It was uncertain as to whether this was a disorder of the ocular mechanisms, a lesion of the brain stem caused by poor illumination or had some other cause. Fishman offers a fascinating and wide-ranging discussion on the subject and concludes that whatever the malady was, it seems certain that something was happening and it was very unlikely to be malingering.[73] Interestingly, Llewelyn notes that the disorder might be a cause of accidents, especially those resulting in injuries to the head and the eyes, and he also found a small number of his patients had *nyctalopia* or night blindness. Fishman cites Cogan's text *Neurology of the Ocular Muscles*, which makes this observation: "[t]he condition has been likened to that occurring in animals reared in the dark. The prognosis is favourable when the patient is removed to quarters having adequate illumination."[74] Llewellyn reports that he examined forty-six mine horses underground but did not find evidence of nystagmus, although no details are supplied of how his examination was carried out. The cause of the malady remains unknown, although he suggested that poor illumination was a crucial factor.

A possible cause of vision problems in slave ponies is vitamin A deficiency. The vitamin is essential for many processes in the body and the source of the vitamin for horses is found in plants containing beta carotene, a large molecule which is broken down in the body to produce vitamin A.[75] It is found in some fruits and in vegetables such as spinach,

[72] 'The Causes and Prevention of Miners' Nystagmus', n.d., 10.

[73] Ronald S. Fishman, 'Dark as a Dungeon: The Rise and Fall of Coal Miners' Nystagmus', *Archives of Ophthalmology* 124, no. 11 (1 November 2006): 1637, https://doi.org/10.1001/archopht.124.11.1637.

[74] Cogan 1956 p. 192 cited in Fishman, 1637.

[75] Tilman Grune et al., 'β-Carotene Is an Important Vitamin A Source for Humans123', *The Journal of Nutrition* 140, no. 12 (December 2010): 2268S–2285S, https://doi.org/10.3945/jn.109.119024.

broccoli, red peppers, squash, carrots and in grass. Deficiency of vitamin A causes significant health problems.

> Deficiency is characterized by nyctalopia [night blindness], lacrimation, keratinization of the cornea, susceptibility to pneumonia, abscesses of the sublingual gland, incoordination, impaired reproduction, capricious appetite, and progressive weakness in adult horses. Hooves may be deformed, with the horny layer unevenly laid down and unusually brittle.[76]

It can also cause diarrhoea. Nyctalopia or night blindness is an inability to see in low light conditions and has been known since antiquity, with its victims including mariners on long voyages, children in orphanages and soldiers in the American Civil War.[77] Despite a general link between malnutrition and eye disorders noted in earlier years, the chemical nature of vitamin A was only described and the vitamin crystallised, in the 1930s.[78]

In a passage particularly relevant to our enquiry, the MSD veterinary manual notes: "A vitamin A deficiency may develop if dried, poor-quality roughage is fed for a prolonged period (more than 6 months). If body stores of vitamin A are high, signs may not appear for several months."[79]

For those horses taken to the surface and allowed to graze, depending upon the condition of the grazing, their time out of the mine and the frequency with which they were brought above ground (perhaps once a year), they may or may not have been able to stave off vitamin A deficiency, but for those doing longer stints underground or living permanently in subterranean stables, it was likely a very different story. While food given to horses underground might have included some chopped hay, which would offer a variable amount of beta carotene depending on

[76] 'Nutritional Diseases of Horses and Other Equids—Management and Nutrition—Veterinary Manual', accessed 17 June 2021, https://www.msdvetmanual.com/management-and-nutrition/nutrition-horses/nutritional-diseases-of-horses-and-other-equids.

[77] R.D. Semba, *The Vitamin A Story: Lifting the Shadow of Death*, vol. 104, World Review of Nutrition and Dietetics (S. Karger AG, 2012), https://doi.org/10.1159/isbn.978-3-318-02189-9. For soldiers in the American Civil War the risk of vitamin A deficiency may have depended upon rank and to some extent race. The higher the rank the better the diet and the better the salary which could be used to supplement a soldier's diet.

[78] Richard D Semba, 'On the "discovery" of Vitamin A', *Annals of Nutrition & Metabolism* 61, no. 3 (2012): 192–98, https://doi.org/10.1159/000343124.

[79] 'Nutritional Diseases of Horses and Other Equids—Management and Nutrition—Veterinary Manual', accessed 17 June 2021, https://www.msdvetmanual.com/management-and-nutrition/nutrition-horses/nutritional-diseases-of-horses-and-other-equids.

pasture, curing, age, and the amount given, most diets were likely very low or completely deficient in beta carotene. In *A Text-book of Coal Mining* (1893) Hughes refers especially to a paper by Mr C. Hunting regarding the feeding of "colliery horses",

> in which the constituents of various foods are fully described, and the whole question gone into. It was long considered that oats alone were sufficient. Mr. Hunting points out that this is correct to a certain extent, as they contain more proportionate quantities of nutritious elements, but for very hard work, such as underground horses have to do, the consumption of muscle is far in excess of the waste of any other tissue, and food containing a heavy proportion of nitrogenous or flesh-forming material must be given. If the choice were limited to one article, oats are superior, but an equal weight of a proper mixture of beans and maize gives better results than oats alone; better in a double sense, because not only is its flesh-forming capacity greater, but it is considerably cheaper.[80]

There is discussion about the use of chopped hay, mixed straw and hard corn (the English meaning referring to various cereals such as wheat and rye and might even include beans) to appease hunger and the possibility of giving green feed for a while in summer.[81]

Another clue comes from a rather surprising source. In an article published in *Harper's Magazine* (1989) the author, Tom Wolf, refers to extensive research carried out by Emile Zola for his novel *Germinal* (1850), a political novel about coal miners in France. He describes how Zola visited a coal mine and beneath ground there was a huge horse, a Percheron, and Zola asked his guides how they get it in and out of the mine every day. At first, they thought he was joking but then realised he was serious and said:

> Mr. Zola, don't you understand? That horse comes down here once, when he's a colt, barely more than a foal, and still able to fit into the buckets that bring us down here. That horse grows up down here. He grows blind down here after a year or two, from the lack of light. He hauls coal down here until

[80] *A Text-Book of Coal-Mining* by Herbert W Hughes—1893 cited in 'Haulage by Horses', accessed 29 April 2021, http://www.healeyhero.co.uk/rescue/individual/Phil_Wyles/Ponies.html.

[81] 'HARD CORN | Definition of HARD CORN by *Oxford Dictionary* on Lexico.Com Also Meaning of HARD CORN', accessed 2 August 2021, https://www.lexico.com/definition/hard_corn; 'Haulage by Horses'.

he can't haul it anymore, and then he dies down here, and his bones are buried down here.[82]

Although this takes place in France and concerns a large horse, it echoes much of what would be the case for slave ponies in Britain. The phrase "He grows blind down here after a year or two" is particularly telling. It fits with the scenario where, for the first year or so, horses would draw on their stores of vitamin A and beta carotene and their night vision would be unaffected, but after these were exhausted, the night blindness as well as other symptoms would set in and they would not be able to see properly.

Drawing on the work of multiple researchers, Meyer in the journal *Pferdeheilkunde* (Equine Medicine) surveyed evidence of possible vitamin A deficiency as well as other medical problems in slave ponies.[83] He noted the frequent head and eye injuries sustained, the prevalence of eye diseases and the 'difficult' behaviour of many horses underground, which suggest that all of these might indicate vision problems in the poor light. He observes that many of the rations fed to horses who were used in mines were deficient in beta carotene, and also cites a number of sources which note that the fat of pit horses was noticeably white rather than having the normal yellowish tinge which is caused by the presence of beta carotene.[84] The absence of beta carotene suggests that it has all been metabolised to vitamin A and has not been replenished from dietary sources thus indicating a state of vitamin A deficiency.

Meyer suggests that not only would the horses be night blind but this lack of visual ability would in turn cause them to be much more susceptible to injuring their heads, faces and eyes, which would further compound their disability. The keratinisation of the cornea, mentioned earlier as another of the results of vitamin A deficiency, might also have been a factor, along with pain, ulceration and eventually permanent blindness. Meyer also says that there is some consensus that the 'working life" of slave ponies was around 5–7 years.[85]

In the book *Coal in Our Veins*, Thomas, recording a personal journey about her family history, visits *The Big Pit* near Merthyr Tydfil in South

[82] Tom Wolfe, 'Stalking the Billion-Footed Beast', *Harper's Magazine*, 1 November 1989, https://harpers.org/archive/1989/11/stalking-the-billion-footed-beast/.

[83] 'Meyer—Augenerkrankungenund Erblindungbei Grubenpferden-.Pdf', n.d.

[84] H Meyer, 'Augenerkrankungenund Erblindungbei Grubenpferden- eine retrospektive-Analyse', February 1999, 38–40.

[85] Meyer, 39.

Wales and learns from her guide, who is an ex miner, that the horses went down the pit at four years of age and were retired ten years later when they were brought out.[86] But, the miner tells her, if horses stayed down the mine most of their lives, when they came to the surface they were "blinded by the brightness".[87] She later learns the sad fate of such horses from Nathan, a resident of the area who says:

> "Let me tell you one thing. My father told me this. When the horses had worked ten years in the mine, they would retire them. They'd bring them up from the mine and release them into a field at night. When it was daylight they would be blinded, so they would walk the floor plan of the mine, wearing a path into the grass by walking it over and over."
>
> "Because they were blind", I repeated.
>
> "Yes", Nathan nodded. "Because it was all they knew."[88]

The horses were blinded and traumatised, walking the same path again and again, as they had done, relentlessly, day after day for ten years.

The Abuse Continues

It is tempting to think that the poor conditions for slave ponies may only have existed in the early part of the nineteenth century but this is not the case. In the late nineteenth and early twentieth centuries, organisations such as the Royal Society for the Prevention of Cruelty to Animals, the National Equine Defence League and the Scottish Society to Promote Kindness to Pit Ponies pressured the authorities about the treatment of slave ponies. This resulted in two pieces of legislation mentioning horses: the Coal Mines Regulation Act of 1887, which had just a small amount to say on the conditions of horses in mines, and later the Coal Mines Act of 1911, which is sometimes referred to as the pit pony's charter. However, the existence of legislation means little without its implementation and enforcement. This was reflected in some parliamentary discussions.

In the House of Lords, on 28 March 1928, Lord Danesford brought the attention of the House to the numbers of injuries and deaths of slave ponies and to the existing legislation, its shortcomings and lack of

[86] Erin Ann Thomas, *Coal in Our Veins: A Personal Journey* (Logan, Utah: Utah State University Press, 2012), 49.

[87] Thomas, 49.

[88] Thomas, 51.

implementation. Among other things he asked, "Will any of your Lordships say it is right to work a pony, or any horse, in a mine or anywhere else, for fourteen or sixteen hours a day?"[89]

He also drew attention to the allegation that

> the roofs are not kept up properly, but are allowed to sink, with the result that they are too low to allow the ponies to go under them; and the withers and backs of the ponies suffer very severely from what is known in the mines as "roofing", that is, rubbing against the roofs where they are too low.[90]

And referring to allegations which had been made, he said,

> [T]he readiest horse is always worked the most—or many of them, are worked ten, twelve, and even fourteen shifts per week.[91]

Lord Arnold also commented on "roofing" and posed the important question of,

> [H]ow far ponies are injured by the tubs while working on a fairly considerable gradient. In certain of the roadways there is a gradient, and the tub tends to attain a certain speed. I understand that a heavy tub will impinge upon the animal, and in some instances the animal will suffer broken limbs, or internal injuries, through the tub going down the gradient in that way and hurting the animal.[92]

The debate seemed to display genuine concern from members of the House about the welfare of ponies and exposed the ongoing legal and illegal abuse.

Twenty years later, on 27 October 1948, but this time in the House of Commons, Mr Peter Freeman (Newport) presented a number of grievances related to a particular pit pony (Ned) but also noted, "Members will be as anxious as I am to remedy these grievances, not only for this particular pit pony but for his 22,000 comrades who are also working down the

[89] 'PIT PONIES. (Hansard, 28 March 1928)', accessed 6 June 2021, https://api.parliament.uk/historic-hansard/lords/1928/mar/28/pit-ponies.

[90] 'PIT PONIES. (Hansard, 28 March 1928)'.

[91] 'PIT PONIES. (Hansard, 28 March 1928)'.

[92] 'PIT PONIES. (Hansard, 28 March 1928)'.

pits under very undesirable conditions".[93] Freeman refers to the case of a miner, Ronald Craddock, who reportedly at the end of a shift found a pony, "very exhausted, tired and in great pain" and that it

> was extremely difficult for this pony to get his breath, which was stertorous. He was often perspiring profusely, was very nervous, had many sears on his head, had much hair torn off, as because he was provided with no lantern, he did not see where he was going. He had bumped his head on many repeated occasions, had large lumps on his spine and had several sores.[94]

The miner, knowing this state of affairs had continued for some time, complained to the foreman saying that the pony should be looked at and that his workload was very heavy and carried out in difficult conditions. A week later, when nothing had happened in this regard, he repeated his complaint and, as a protest, made clear to his employers, The National Coal Board, he would refuse to work while such a situation existed. Later that day he was called to the manager's office and given two weeks' notice of dismissal. The reason for his dismissal was not stated at the time but later, when he applied for unemployment pay, this was refused because he had been dismissed for misconduct. Later his employer did offer him his job back if he would retract his allegations about the pony but he refused and was still out of work.

Freeman goes on to state that

> at the time under discussion another pony was treated underground for a smashed eye, another for a broken nose. Both were treated underground. Another pony, called "Wonder", apparently fell dead from exhaustion, or—I am unable to ascertain the facts—as a result of breaking his neck by accident.[95]

He also quotes Robert Smillie, the President of the National Miners Federation, who said that "the life of any pony down a mine was a life of hell itself" and an inspector of mines, Mr R. L Layfield, who said in 1930, "These little animals are suffering from heat, dust, dirt, rough conditions,

[93] 'Orders of the Day—Pit Ponies (Conditions)', TheyWorkForYou, accessed 29 April 2021, https://www.theyworkforyou.com/debates/?id=1948-10-27a.220.2.

[94] 'Orders of the Day—Pit Ponies (Conditions)'.

[95] 'Orders of the Day—Pit Ponies (Conditions)'.

dancing lights, and noise and the objectionable intolerable conditions at the best lead to intolerable suffering".[96]

Of his experience working in a mine in the early decades of the twentieth century, Coombs writes:

> I have seen a horse blind in one eye rubbing its nose against sharp stones so it could feel the way to turn and being whipped for not coming round fast enough. Every day horses are worked for sixteen and more hours straight off. What I have seen I know to be true-and I have seen it regularly. Very often they have no water at all during those long hours in the hot places of the mine. Within the last years I have seen horses drop from exhaustion several times during a shift.[97]

The system in the mines, a symptom of burgeoning capitalism, was fundamentally violent, exploitive and abusive, with slave ponies at the bottom of the hierarchy of power. The miners were also victims, often driven to work in the pit in order to survive and trying to look after their families. This would at times lead to some miners abusing animals as they were pressured to produce 'sufficient' coal.

> Often a man must choose between forcing an exhausted horse and being sent home on the dole. He must choose whether the horse shall suffer or his wife and children.[98]

Wendy Priest, Horse Keeper Supervisor at The National Coal Mining Museum for England, is cited by Evans as saying,

> Miners had to see their beloved ponies sometimes work for other miners who had no respect for the animals and often had no working knowledge of them. This frequently meant that the good, willing ponies were worked both shifts as men coming on shift would go for the easy animals first. Then there were those who were downright cruel. To these hardened miners, ponies were just a machine for moving coal.[99]

[96] 'Orders of the Day—Pit Ponies (Conditions)'.

[97] Bertie Coombes L, *These Poor Hands. The Autobiography of a Miner Working in South Wales* (Cardiff, UK: University of Wales Press, 2002).

[98] Coombes.

[99] Evans, 'Ghosts of the Coal Mines | Horse Journals'.

In one case, William Mostyn, the colliery deputy of Hawkesbury Colliery, ripped the tongue out of a pony's mouth while trying to get it to move.[100] The pony was euthanised and Mostyn was fined two pounds. He claimed to have done this so that the horse "did not hold up the movement of the coal".[101]

But there is little doubt that many miners cared deeply for the horses they looked after and did what they could for them, sometimes even giving them a treat of some kind and often referring to them as their mate.

Truner and Blackie cite this telling claim about miners made by Dr S. Scott Alison in 1842:

> The nervous system, including various parts of the brain, are comparatively little exercised, while that of the muscles is inordinately overworked … the collier becomes more a mining or working animal than a thinking being—more a machine than a rational creature.[102]

There was also the idea that miners were a special group, almost suggesting that they were made for mining.[103] It was thought by some that the work not only affected the miner's physical condition but their nervous system as well, "turning them into brutes".[104] Some commentators claimed miners had quite a healthy workplace with an "equitable climate", were generally healthy due to their "habitual cleanliness" and were able to endure fatigue.[105] G. Mallet, in the *Association Medical Journal* of 1885, argued that miners were able to recover from

> such an amount of injury as would in all human probability have proved fatal if it had occurred to individuals occupying a higher social position, or to those whose 'intellects [were] more highly developed'.[106]

[100] 'Pit Ponies—Stories from the Warwickshire Coalfield', *Our Warwickshire* (blog), accessed 25 October 2021, https://www.ourwarwickshire.org.uk/content/article/pit-ponies-stories-from-the-warwickshire-coalfield.

[101] 'Pit Ponies—Stories from the Warwickshire Coalfield'.

[102] S. Scott Alison cited in David M. Turner and Daniel Blackie, *Disability in the Industrial Revolution* (Manchester University Press, 2018), 58, https://doi.org/10.7765/9781526125774.

[103] Turner and Blackie, 57.

[104] Turner and Blackie, 58.

[105] Turner and Blackie, *Disability in the Industrial Revolution.*

[106] G Mallet cited in Turner and Blackie, 58.

Turner and Blackie comment, “The effect [of the work], according to this author, was not only that the majority of miners were ‘very low in the scale of intellectual culture’, but also that miners lacked nervous sensibility to pain”.[107] As Bourke points out, describing the ideology of a hierarchal world order, an idea prevalent since at least the time of Aristotle, “One aspect of this Chain of Being involved the perception of sensation. There was a parallel great Chain of Feeling, which placed male Europeans at one end and slaves and animals at the other.”[108] The elite in a hierarchy feel that they experience pain more acutely than those humans deemed to be below them and animals.[109] Another aspect of this mythical Chain of Being was that those lower down could rightfully be used by those above because this was their purpose for existing. Perhaps miners and ponies were held by the elite to be just a few links away from each other on this chain.

Conclusion: The Life of a Slave Pony

It is not possible to document a single history of what happened to all slave ponies in Britain, such was the diversity of the mines, the behaviour of individual miners and the changes which took place over more than 250 years of this abuse. But, drawing on the evidence, we can trace an archetypal life for a pony in a deep mine, who is one of the few to survive seven or eight years in the pit. We should not forget that while our pony is hypothetical, behind him stand perhaps a million of his comrades for whom this was essentially their lived existence as well.

It starts at around the age of four when he is abducted from his kin, sold and trafficked.[110] He is forced to walk for days, getting ever further from his home and family. He is lost and alone, the closeness of his kin has disappeared and he is in a new and fearful environment. The loading at the dockside is violent and he is terrified by the sea voyage.[111] Eventually he is forced off the ship and driven to a coal mine. This mine company is now, in human law, his rightful, lawful owner and he becomes a legal object, a commodity, a thing to be used for the mine’s benefit, a power source in the extractive process, a small part of the greater capitalist machine.

[107] Turner and Blackie, 58.

[108] Joanna Bourke, ‘The History of Pain Is Racist, Sexist, and Classist’, *The New Republic*, 20 June 2014, https://newrepublic.com/article/118283/cultural-history-pain.

[109] They might also be more stoic in bearing such pain.

[110] Almost all slave ponies were male.

[111] Sea voyages were extremely stressful and dangerous.

At this place, and in a very short time, there will be a calamitous event for this young, diminutive horse, because for him the life he has known, every strand of his old world, will exist no more and he will be plunged, literally, into a completely alien netherworld for the rest of his existence. Here is the portal between two worlds through which he will pass and never return; this is the place and time marking his final loss of home and family, the extinction of his identity and the snuffing out of all hope.

He is held by men, bound, strapped and hoisted into the air, head uppermost, vertical; he is swung out over the abyss and then lowered into it. He is petrified as he descends, the light ever diminishing until there is only impenetrable blackness. After an undefined time, the descent ends and he is swung aside, unhooked and eventually untied. This is his world now; his only world for the rest of his life and there is no escape. Because he is a horse his eyes have not yet become used to the low light conditions produced by the few candles or lamps and he cannot see properly and will not be able to do so for some time. There are flickering lights around him, movement, noises, shapes, dust and strange smells, water dripping on him and he stands with nowhere to go, alone of his kind and utterly powerless.

He will soon learn there is no grazing here, no smell of plants, no rhythm of the days or the seasons, day and night will no longer exist, no breeze or familiar smells, no warm sunshine or cool rain, no feeling the contact of kin or inhaling their smell, no grass beneath his feet. Only his own being remains, divested of all connections to that other world, a profoundly shocking, disorientating and cruel process. It is a method of cognitive and psychological breakdown which the operatives in one of today's black sites would be proud to call their own.

Soon his forced labour begins and it is a frightening and exhausting torment. His endless night, in which there are no days, consists of being hitched up to a cove or tub of coal, or a series of tubs, which he pulls with the young driver walking alongside or sitting on the tub forcing him on. Parts of the path are wet; some go up hill but other parts go steeply down and here the heavy load threatens to overwhelm his small body and he struggles desperately to hold it back. Some horses die this way. His head is down for much of the time in the low tunnels; it is unnatural and painful and finding his way is made even more difficult by this. He must bend his legs in some sections where the roof has sunk, so that he can avoid 'roofing'; his back and head have swellings and open sores from collisions with the protruding chunks of sharp rock above. The working shifts are long and arduous and often he does one after another, monotonously working hour after hour. His feeding and watering, when there is a rush on the work, are not priorities and he often goes for many hours without either.

The work goes on relentlessly, week in week out, year after year and in some way, he adapts in order to survive. He suffers fungal infections to his feet from walking continually on the wet ground, he injures his legs on the rails and his face on the projecting rocks. Infected wounds are a common problem. After two years the stores of vitamin A in his liver are getting exhausted, and there is little beta carotene in his food with which to replace it. His eyesight has been deteriorating for some time, and now he has great difficulty making out the shapes and distances in this dark and flickering world. His inability to see properly leads to more injuries to his head and eyes, and one eye becomes ulcerated due to the deficiency. Pulling the tubs, he feels the side of the tunnel with his face in order to guide him as he is whipped to move faster. The vitamin A deficiency also affects his lungs, already irritated by the dust; his coat is poor and he is now even more prone to infections and often suffers from diarrhoea.

The relentless hard physical labour begins to take its toll on his bones and joints. Although still young in years he has become an old horse. His joints, including those of his legs, are damaged and inflamed. The vertebrae in his back are calcifying, so that they touch or kiss, meaning his movement is affected and he is in severe pain. Arthritis is setting in.

In the stables, if he is fortunate, he will be able to stand with his head raised to a normal position. However, sleeping is difficult as his damaged skeleton means that lying down only brings pain. So, he stands and sleeps on his feet. This is dangerous because when he enters REM sleep, as he must, his body naturally becomes paralysed, meaning that he collapses, bringing him even greater pain and the possibility of severe damage to a limb.

If he continues to evade the many dangers of the mine, the time eventually comes, after some five to eight years down the pit, when he can no longer pull the coal. It is too painful; he cannot walk properly and he cannot see. He might then be sold for meat or killed.

And we can wonder if, in all of his suffering over those long years in darkness, he also has the added pain of remembering his past life, his kin and home. Given everything we know, it is very hard to believe he does not. Does he, in his subterranean stable, dream of these things? In that endless night does he endure a *hiraeth*, a Welsh word, difficult to translate into English but meaning a deep, painful longing for home, a longing which is touched with grief and sadness for something lost, a yearning for a home which no longer exists?

In Emile Zola's political novel about coal mining, *Germinal* (1885), one of the horses, Trompette, lies dying underground in the pit while the miners are on strike above and an old horse, Bataille, is with him. In this passage Zola poignantly captures something of the harrowing depth of the atrocities we committed.

"It was, in fact, Trompette. Since his descent he had never become acclimatized. He remained melancholy, with no taste for his task, as though tortured by regret for the light. In vain Bataille, the doyen of the mine, would rub him with his ribs in his friendly way, softly biting his neck to impart to him a little of the resignation gained in his ten years beneath the earth. These caresses increased his melancholy, his skin quivered beneath the confidences of the comrade who had grown old in darkness; and both of them, whenever they met and snorted together, seemed to be grieving, the old one that he could no longer remember, the young one that he could not forget. At the stable they were neighbours at the manger, and lived with lowered heads, breathing in each other's nostrils, exchanging a constant dream of daylight, visions of green grass, of white roads, of infinite yellow light."[112]

Even with such a brief survey as this, our crimes[113] against those small defenceless beings, the slave ponies of the British coal mines, are clear and, I propose, now deserve to be publicly acknowledged and owned.

Bibliography

Agnew, Robert. 'The Causes of Animal Abuse: A Social-Psychological Analysis'. *Theoretical Criminology* 2, no. 2 (1998): 177–209.

Blumberg, Mark S., John A. Lesku, Paul-Antoine Libourel, Markus H. Schmidt, and Niels C. Rattenborg. 'What Is REM Sleep?' *Current Biology* 30, no. 1 (6 January 2020): R38–49. https://doi.org/10.1016/j.cub.2019.11.045.

Bourke, Joanna. 'The History of Pain Is Racist, Sexist, and Classist'. *The New Republic*, 20 June 2014. https://newrepublic.com/article/118283/cultural-history-pain.

[112] Emile Zola, *Germinal* (London: J. M. DENT & SONS LTD., 1885), chap. V, https://www.gutenberg.org/files/56528/56528-h/56528-h.htm.

[113] *Crime* here as in the older meaning of the word, not tied to a legal prescript, as in an act of great wickedness.

Bradley, Robert. 'Ponies and Ostlers'. Bob's History of Mining, n.d. http://www.healeyhero.co.uk/rescue/individual/Bob_Bradley/Ponies.html.

Broster, C. E., C. C. Burn, A. R. S. Barr, and H. R. Whay. 'The Range and Prevalence of Pathological Abnormalities Associated with Lameness in Working Horses from Developing Countries'. *Equine Veterinary Journal* 41, no. 5 (May 2009): 474–81. https://doi.org/10.2746/042516409X373907.

Brown, Ivor J. 'Horses in Shropshire Mines'. *British Mining (Northern Mine Research Society)*, no. 39 (1989): 78–88.

Butt, John ButtJohn. 'Coal Industry'. In *The Oxford Companion to British History*. Oxford University Press, 2015. http://0.www.oxfordreference.com/view/10.1093/acref/9780199677832.001.0001/acref-9780199677832-e-1031.

'Coalmining Accidents and Deaths—Vanilla Circus—Maps, Poems and Searchable Databases for Mining in the UK'. Accessed 27 August 2021. http://www.cmhrc.co.uk/site/disasters/index.html.

Commissioners for Inquiring into the, Employment and Condition of Children in Mines and, and Manufactories. *The Condition and Treatment of the Children Employed in the Mines and Colliers of the United Kingdom. Carefully Compiled from the Appendix to the First Report of the Commissioners ... With Copious Extracts from the Evidence, and Illustrative Engravings. [The Preface Signed: W.C.]*, 1842. http://access.bl.uk/item/viewer/ark:/81055/vdc_100054704592.0x000001.

Coombes, Bertie L., *These Poor Hands. The Autobiography of a Miner Working in South Wales*. Cardiff, UK: University of Wales Press, 2002.

Diedrich, Cajus G. 'Pathologic Historic Mining Horses from Central Europe.' 1, no. 1 (2017): 26.

Donoho, Emily. 'The Shetland Pony: All You Need to Know about This Small Hardy Breed'. Horse & Hound, 27 December 2018. https://www.horseandhound.co.uk/features/shetland-pony-facts-673878.

Evans, Margaret. 'Ghosts of the Coal Mines | Horse Journals'. Horse Journals, 5 July 2019. https://www.horsejournals.com/popular/history-heritage/ghosts-coal-mines.

Evans, Patricia. 'Equine Vision and Its Effect on Behavior', n.d., 3.

Fishman, Ronald S. 'Dark as a Dungeon: The Rise and Fall of Coal Miners' Nystagmus'. *Archives of Ophthalmology* 124, no. 11 (1 November 2006): 1637. https://doi.org/10.1001/archopht.124.11.1637.

Grune, Tilman, Georg Lietz, Andreu Palou, A. Catharine Ross, Wilhelm Stahl, Guangweng Tang, David Thurnham, Shi-an Yin, and Hans K. Biesalski. 'β-Carotene Is an Important Vitamin A Source for Humans123'. *The Journal of Nutrition* 140, no. 12 (December 2010): 2268S-2285S. https://doi.org/10.3945/jn.109.119024.

Haig pit mining and colliery museum. 'Pit Ponies (Haig)'. Haig Pit Mining And Colliery Museum., 7 May 2013. https://haigpit.wordpress.com/pit-ponies/.

'HARD CORN | Definition of HARD CORN by Oxford Dictionary on Lexico. Com Also Meaning of HARD CORN'. Accessed 2 August 2021. https://www.lexico.com/definition/hard_corn.

'Haulage by Horses'. Accessed 29 April 2021. http://www.healeyhero.co.uk/rescue/individual/Phil_Wyles/Ponies.html.

Hernigle, Jacqueline. 'An Archipelago of Coal Pits: Predicting Archeological Features in the Richmond, Virginia Coalfield'. College of William and Mary—Arts and Sciences, 1991. http://scholarworks.wm.edu/etd/1539625657.

Horse Sport. 'Point of View: Understanding How Horses See the World'. Accessed 27 July 2021. https://horsesport.com/magazine/training/point-of-view-understanding-how-horses-see/.

'Huskar Colliery 1838—Mining Accident Database'. Accessed 21 May 2021. http://mineaccidents.com.au/mine-accident/202/huskar-colliery-1838.

Ingeo, Serenella d', Angelo Quaranta, Marcello Siniscalchi, Mathilde Stomp, Caroline Coste, Charlotte Bagnard, Martine Hausberger, and Hugo Cousillas. 'Horses Associate Individual Human Voices with the Valence of Past Interactions: A Behavioural and Electrophysiological Study'. *Scientific Reports* 9, no. 1 (9 August 2019): 11568. https://doi.org/10.1038/s41598-019-47960-5.

'Meyer—Augenerkrankungenund Erblindungbei Grubenpferden-.Pdf', n.d.

Meyer, H. 'Augenerkrankungenund Erblindungbei Grubenpferden- eine retrospektiveAnalyse', February 1999, 38–40.

'MINES AND COLLIERIES. (Hansard, 1 August 1842a)'. Accessed 24 May 2021. https://api.parliament.uk/historic-hansard/lords/1842/aug/01/mines-and-collieries.

'MINES AND COLLIERIES. (Hansard, 14 July 1842b)'. Accessed 21 May 2021. http://hansard.millbanksystems.com/lords/1842/jul/14/mines-and-collieries#s3v0065p0_18420714_hol_51.

Murphy, Christopher. Equine Vision. Equine Vision: Normal and Abnormal. In: Equine Ophthalmology (Ed. Gilger BC), Elsevier Saunders, St. Louis, 396–434, 2010.

Newspapers.com. '7 Jul 1838, 6—Sheffield and Rotherham Independent at Newspapers.Com'. Accessed 2 June 2021. http://www.newspapers.com/image/?clipping_id=22840031&fcfToken=eyJhbGciOiJIUzI1NiIsInR5cCI6IkpXVCJ9.eyJmcmVlLXZpZXctaWQiOjQwMTE2Mjc5OSwiaWF0IjoxNjI-yNjIyMDI5LCJleHAiOjE2MjI3MDg0Mjl9.ITHStsPQ2F6C-UaflgGiKERx8u1DcNIz-5vFbzy0Dzw.

'Nutritional Diseases of Horses and Other Equids—Management and Nutrition—Veterinary Manual'. Accessed 17 June 2021. https://www.msdvetmanual.com/management-and-nutrition/nutrition-horses/nutritional-diseases-of-horses-and-other-equids.

Of Horse. 'Do Horses Dream?' Accessed 12 August 2021. https://www.ofhorse.com/view-post/Do-Horses-Dream.
Our Warwickshire. 'Pit Ponies—Stories from the Warwickshire Coalfield'. Accessed 25 October 2021. https://www.ourwarwickshire.org.uk/content/article/pit-ponies-stories-from-the-warwickshire-coalfield.
'Pathophysiology of Osteoarthritis—VetFolio'. Accessed 17 August 2021. https://www.vetfolio.com/learn/article/pathophysiology-of-osteoarthritis.
'PIT PONIES. (Hansard, 28 March 1928)'. Accessed 6 June 2021. https://api.parliament.uk/historic-hansard/lords/1928/mar/28/pit-ponies.
'Ponies and Ostlers'. Accessed 25 June 2021. http://www.healeyhero.co.uk/rescue/individual/Bob_Bradley/Ponies.html#top2.
Rhys, Steffan. 'In Darkness for Years, the Dangerous and Harsh Life of Pit Ponies'. WalesOnline, 8 April 2008. https://www.walesonline.co.uk/news/wales-news/darkness-years-dangerous-harsh-life-2182018.
Rørvang, Maria Vilain, Birte L. Nielsen, and Andrew Neil McLean. 'Sensory Abilities of Horses and Their Importance for Equitation Science'. *Frontiers in Veterinary Science* 0 (2020). https://doi.org/10.3389/fvets.2020.00633.
Semba, R.D. *The Vitamin A Story: Lifting the Shadow of Death*. Vol. 104. World Review of Nutrition and Dietetics. S. Karger AG, 2012a. https://doi.org/10.1159/isbn.978-3-318-02189-9.
Semba, Richard D. 'On the "discovery" of Vitamin A'. *Annals of Nutrition & Metabolism* 61, no. 3 (2012b): 192–98. https://doi.org/10.1159/000343124.
'The Causes and Prevention of Miners' Nystagmus, n.d., 18.
'The Coalmining History Resource Centre. 1820–1839.Pdf'. Accessed 2 June 2021. http://www.cmhrc.co.uk/cms/document/1820_39.pdf.
'The Dangers of Sleep Deprivation in Horses | International Society for Equitation Science'. Accessed 12 August 2021. https://equitationscience.com/media/the-dangers-of-sleep-deprivation-in-horses.
The Penistone Archive. 'Huskar Pit Disaster'. Accessed 21 May 2021. https://penistonearchive.co.uk/huskar-pit-disaster/.
TheyWorkForYou. 'Orders of the Day—Pit Ponies (Conditions)'. Accessed 29 April 2021. https://www.theyworkforyou.com/debates/?id=1948-10-27a.220.2.
Thomas, Erin Ann. *Coal in Our Veins: A Personal Journey*. Logan, Utah: Utah State University Press, 2012.
Turner, David M., and Daniel Blackie. *Disability in the Industrial Revolution*. Manchester University Press, 2018. https://doi.org/10.7765/9781526125774.
'UFR LCE—Département d'études Anglophones—The Mines Act, 1842', 21 July 2011. https://web.archive.org/web/20110721022926/http://anglais.u-paris10.fr/spip.php?article88.

Union, National Farmers. 'The Fell Pony'. Countryside, 14 June 2018. https://www.countrysideonline.co.uk:443/hobbies-and-leisure/pets-and-animals/the-fell-pony43558/.

Veterinary Manual. 'Arthritis in Large Animals—Musculoskeletal System'. Accessed 11 May 2021. https://www.msdvetmanual.com/musculoskeletal-system/arthropathies-in-large-animals/arthritis-in-large-animals.

Wolfe, Tom. 'Stalking the Billion-Footed Beast'. *Harper's Magazine*, 1 November 1989. https://harpers.org/archive/1989/11/stalking-the-billion-footed-beast/.

'WWMM—Pit Ponies 14/20'. Accessed 17 June 2021. http://www.wwmm.org/storie/storia.asp?id_storia=192&pagina=14&project=0.

Zola, Emile. *Germinal*. London: J. M. DENT & SONS LTD., 1885. https://www.gutenberg.org/files/56528/56528-h/56528-h.htm.

Index[1]

[1] Note: Page numbers followed by 'n' refer to notes.

M. J. Glover, L. Mitchell (eds.), *Animals as Experiencing Entities*, The Palgrave Macmillan Animal Ethics Series,
https://doi.org/10.1007/978-3-031-46456-0

SPRINGER NATURE

GPSR Compliance

The European Union's (EU) General Product Safety Regulation (GPSR) is a set of rules that requires consumer products to be safe and our obligations to ensure this.

If you have any concerns about our products, you can contact us on ProductSafety@springernature.com

In case Publisher is established outside the EU, the EU authorized representative is:

Springer Nature Customer Service Center GmbH
Europaplatz 3
69115 Heidelberg, Germany

Kolibri 360 GmbH
Ferdinand-Jühlke-Straße 7
99095 Erfurt, Deutschland
produktsicherheit@kolibri360.de